T0235351

Springer Complexity

Springer Complexity is an interdisciplinary program publishing the best research and academic-level teaching on both fundamental and applied aspects of complex systems – cutting across all traditional disciplines of the natural and life sciences, engineering, economics, medicine, neuroscience, social and computer science.

Complex Systems are systems that comprise many interacting parts with the ability to generate a new quality of macroscopic collective behavior the manifestations of which are the spontaneous formation of distinctive temporal, spatial or functional structures. Models of such systems can be successfully mapped onto quite diverse "real-life" situations like the climate, the coherent emission of light from lasers, chemical reaction-diffusion systems, biological cellular networks, the dynamics of stock markets and of the internet, earthquake statistics and prediction, freeway traffic, the human brain, or the formation of opinions in social systems, to name just some of the popular applications.

Although their scope and methodologies overlap somewhat, one can distinguish the following main concepts and tools: self-organization, nonlinear dynamics, synergetics, turbulence, dynamical systems, catastrophes, instabilities, stochastic processes, chaos, graphs and networks, cellular automata, adaptive systems, genetic algorithms and computational intelligence.

The three major book publication platforms of the Springer Complexity program are the monograph series "Understanding Complex Systems" focusing on the various applications of complexity, the "Springer Series in Synergetics", which is devoted to the quantitative theoretical and methodological foundations, and the "SpringerBriefs in Complexity" which are concise and topical working reports, case-studies, surveys, essays and lecture notes of relevance to the field. In addition to the books in these two core series, the program also incorporates individual titles ranging from textbooks to major reference works.

Editorial and Programme Advisory Board

Henry Abarbanel, Institute for Nonlinear Science, University of California, San Diego, USA

Dan Braha, New England Complex Systems Institute and University of Massachusetts, Dartmouth, USA

Péter Érdi, Center for Complex Systems Studies, Kalamazoo College, USA and Hungarian Academy of Sciences, Budapest, Hungary

Karl Friston, Institute of Cognitive Neuroscience, University College London, London, UK

Hermann Haken, Center of Synergetics, University of Stuttgart, Stuttgart, Germany

Viktor Jirsa, Centre National de la Recherche Scientifique (CNRS), Université de la Méditerranée, Marseille, France

Janusz Kacprzyk, System Research, Polish Academy of Sciences, Warsaw, Poland

Kunihiko Kaneko, Research Center for Complex Systems Biology, The University of Tokyo, Tokyo, Japan

Scott Kelso, Center for Complex Systems and Brain Sciences, Florida Atlantic University, Boca Raton, USA

Markus Kirkilionis, Mathematics Institute and Centre for Complex Systems, University of Warwick, Coventry, UK

Jürgen Kurths, Nonlinear Dynamics Group, University of Potsdam, Potsdam, Germany

Andrzej Nowak, Department of Psychology, Warsaw University, Poland

Linda Reichl, Center for Complex Quantum Systems, University of Texas, Austin, USA

Peter Schuster, Theoretical Chemistry and Structural Biology, University of Vienna, Vienna, Austria

Frank Schweitzer, System Design, ETH Zurich, Zurich, Switzerland

Didier Sornette, Entrepreneurial Risk, ETH Zurich, Zurich, Switzerland

Stefan Thurner, Section for Science of Complex Systems, Medical University of Vienna, Vienna, Austria

Springer Series in Synergetics

Founding Editor: H. Haken

The Springer Series in Synergetics was founded by Herman Haken in 1977. Since then, the series has evolved into a substantial reference library for the quantitative, theoretical and methodological foundations of the science of complex systems.

Through many enduring classic texts, such as Haken's *Synergetics and Information and Self-Organization*, Gardiner's *Handbook of Stochastic Methods*, Risken's *The Fokker Planck-Equation* or Haake's *Quantum Signatures of Chaos*, the series has made, and continues to make, important contributions to shaping the foundations of the field.

The series publishes monographs and graduate-level textbooks of broad and general interest, with a pronounced emphasis on the physico-mathematical approach.

More information about this series at
http://www.springer.com/series/712

Lev N. Lupichev • Alexander V. Savin •
Vasiliy N. Kadantsev

Synergetics of Molecular Systems

Springer

Lev N. Lupichev
Department of Cybernetics
Moscow State Technical University
  of Radioengineering, Electronics
  and Automation
Moscow
Russia

Alexander V. Savin
Department of Polymer and Composite
  materials
Semenov Institute of Chemical Physics,
  RAS
Moscow
Russia

Vasiliy N. Kadantsev
Department of Cybernetics
Moscow State Technical University
  of Radioengineering, Electronics
  and Automation
Moscow
Russia

ISSN 0172-7389
ISBN 978-3-319-38186-2   ISBN 978-3-319-08195-3 (eBook)
DOI 10.1007/978-3-319-08195-3
Springer Cham Heidelberg New York Dordrecht London

© Springer International Publishing Switzerland 2015
Softcover reprint of the hardcover 1st edition 2015
This work is subject to copyright. All rights are reserved by the Publisher, whether the whole or part of the material is concerned, specifically the rights of translation, reprinting, reuse of illustrations, recitation, broadcasting, reproduction on microfilms or in any other physical way, and transmission or information storage and retrieval, electronic adaptation, computer software, or by similar or dissimilar methodology now known or hereafter developed. Exempted from this legal reservation are brief excerpts in connection with reviews or scholarly analysis or material supplied specifically for the purpose of being entered and executed on a computer system, for exclusive use by the purchaser of the work. Duplication of this publication or parts thereof is permitted only under the provisions of the Copyright Law of the Publisher's location, in its current version, and permission for use must always be obtained from Springer. Permissions for use may be obtained through RightsLink at the Copyright Clearance Center. Violations are liable to prosecution under the respective Copyright Law.
The use of general descriptive names, registered names, trademarks, service marks, etc. in this publication does not imply, even in the absence of a specific statement, that such names are exempt from the relevant protective laws and regulations and therefore free for general use.
While the advice and information in this book are believed to be true and accurate at the date of publication, neither the authors nor the editors nor the publisher can accept any legal responsibility for any errors or omissions that may be made. The publisher makes no warranty, express or implied, with respect to the material contained herein.

Printed on acid-free paper

Springer is part of Springer Science+Business Media (www.springer.com)

Contents

Chapter 1
Introduction

The methodology underlying the analysis of nonlinear dynamic systems has by now evolved into a new scientific paradigm called synergetics. This interdisciplinary science aims to reveal the general principles governing evolution and self-organization of complex systems in various scientific areas, and it is mainly based on the design and study of nonlinear dynamic models [1]. One of the key features of synergetics is the consideration of nonlinear collective interactions among the different components making up the model systems. It can be said that *synergetics is a multicomponent system + nonlinearity + cooperativity*. In physics, chemistry, biology, ecology, economics, and sociology, synergetics focuses on the traits of a system's spatial-temporal organization. It has been shown that cooperation of system components follows universal principles that are independent of the nature of the system itself.

In essence, synergetics is a relatively new paradigm in the study of dynamic systems possessing complex multicomponent structure, in which a powerful arsenal of nonlinear dynamics methods is used. The majority of models in synergetics originate from physics, where numerous physical phenomena cannot be described without taking into account the nonlinearity and cooperativity of the system. In this book we shall consider the basic models of nonlinear dynamics of molecular systems and discuss their applications in synergetics. Nowadays the nonlinear dynamics of molecular systems is evolving steadily and its progress is partly due to the fast development in computer technologies and computational methods of nonlinear dynamics. New models and ideas are emerging which have not yet been introduced in synergetics, but their future application to synergetics looks promising. One such model is the ratchet model or nonsymmetrical pendulum. It has been shown that the nonsymmetry of this system is a useful concept when studying mechanisms for deriving energy from colored noise. We shall consider this model in detail.

It is known that both integrable and non-integrable systems can have specific solutions that correspond to coherent formations or spatial-temporal structures,

© Springer International Publishing Switzerland 2015
L.N. Lupichev et al., *Synergetics of Molecular Systems*, Springer Series in Synergetics, DOI 10.1007/978-3-319-08195-3_1

such as solitons, kinks, breathers, and others. The emergence of such formations in molecular structures can provide a range of information on the properties of these structures and play a significant role in energy-related phenomena, energy transfer processes, topological defects, dislocations, and charges, as well as in structure transitions. This is why it is of considerable interest to study the conditions governing the emergence of spatial-temporal structures (self-organization) in various molecular systems and investigate the properties of these structures.

The equations which allow solutions in the form of solitons are well-known in mathematics as being the very same equations that possess the notable property of complete integrability. This type of equation is also known to be useful for modelling a broad spectrum of physical phenomena. However, their application to real systems in condensed matter physics, engineering, biophysics, etc., requires one to introduce perturbations into the equations to account for the complex features of any real structure. This means that one cannot confine attention to completely integrable systems. A systematic study of the different perturbations (boundary conditions, heterogeneity, and so on) is the next stage in the investigation of systems with soliton solutions, leading to an increased appreciation of the versatility of the synergetic (nonlinear) approach. It is worth noting that, in the general case, nonlinear structures such as kinks and others refer to very different types of solutions than linear or quasi-linear modes, and that these solutions should be considered equally fundamental.

The nonlinear equations (models) show significant generality, but at the same time a distinct equation can describe (in some approximation) a number of nonlinear phenomena as well as various nonlinear effects. For example, nonlinear models of molecular chains with a periodic substrate potential describe a wave spreading in the presence of the gripping potential.

Generally, an analytical description of system dynamics is barely possible in the case of one-dimensional systems which are specific in their properties and do not include many real physical conditions. For example, the need to introduce finite temperature into the model is a significant challenge in numerous tasks. Furthermore, even nonstable modes turn out to contribute significantly, and their impact depends on their half-life. The half-life has to be sufficiently high in order to obtain a noticeable response of the system to the external action. The longer the half-life, the more significant is the effect of the nonstable modes.

In real systems, there are solitary waves rather than solitons [2]. The Hamiltonians of real systems may be initially nonintegrable or become nonintegrable when physical perturbations are taken into account. One of the examples of integrable systems is the Toda chain [3]. This model is often used in applications, such as the theory of heat conduction, the analysis of shock wave propagation, and others.

The aim of synergetics is to reveal the general principles of evolution and self-organization in complex systems based on the development and analysis of nonlinear dynamics models in different scientific fields. Many effects of self-organization, typical to dissipative systems, are also inherent in conservative Hamiltonian systems where there is neither dissipation, nor energy gain, in contrast to active media. Thus, localised excitations (concentrated in a limited space region), like peculiar deformed particles, can exist and spread in the various media. A special feature of such spatial-temporal structures is that, in contrast to dissipative structures, dispersion plays the prevailing role in structure formation (*dispersion structures*), so dissipative effects are negligible. The concept of *the dispersion structures* was introduced by one of the authors, V.N. Kadantsev in [4, 5]. In general, the following three physical factors are significant in wave propagation: nonlinearity, dispersion, and dissipation. We restrict ourselves here to the case where there is no dissipation or negligibly small dissipation, i.e., in this book we will consider only conservative systems.

Recently, it was realised that the ideal method for transferring vibrational excitations, electrons, and protons in a medium is their transfer in the form of solitary waves, which retain their identity during collisions and are described by the solutions of some special equations. From now on we will refer to any localised excitations, such as particles, as solitons, when dealing with soliton-like excitations. The interest in solitons can be seen in various branches of science such as elementary particle physics, plasma physics, laser physics, hydrodynamics, biophysics, transmission line analysis, and so on. Furthermore, there are emerging applications of solitons in information storage and data transfer. Finally, a plethora of specific applications of solitary waves exists in science and technology: ranging from hydrodynamics to meteorology, computer technologies to shock wave dynamics, the theory of nonlinear filters to dislocation dynamics in crystals, and elementary particle physics to the theory of consciousness in neuroscience.

The formation of nonlinear waves in a continuous medium described by nonlinear equations is associated with spontaneous breaking of local symmetry in homogeneous systems, i.e., it is related to the autolocalization of excitation energy, the density of electrical charge, and other physical parameters.

Excitations in the form of solitary waves along with common extended waves are typical of many nonlinear dynamical systems. However, their analytical description has been developed in detail only in the case of one-dimensional systems.

Objectively evaluating the limited capabilities of both analytical and numerical approaches to the solution of nonlinear problems, Norman Zabusky concluded that there was a need for a united synthetic approach. According to him, the synergetic approach to nonlinear mathematical and physical problems may be defined as the joint application of common analysis and computational mathematics to solving reasonably formulated problems concerned with the mathematical and physical meanings of systems of equations [6].

One of the first computational experiments carried out on the first generation of computers – MANIAC I – aimed to test the hypothesis of a uniform distribution of energy over the degrees of freedom. The experiment, which focused on the numerical analogue of a system of cubic oscillators, led to unexpected results, raising the famous Fermi–Pasta–Ulam problem [7]: following the evolution of the energy distribution over the degrees of freedom for a sufficiently large cycle index, the authors observed no tendency for energy equipartition in the system.

The solution to the Fermi–Pasta–Ulam problem was first proposed by Zabusky and Kruskal in the early 1960s. They showed that the Fermi–Pasta–Ulam system represents a difference analogue of the Korteweg–de Vries equation, and the *soliton* (the term suggested by Zabusky), transferring the energy among different modes, prevents energy equipartition. Through heavy use of the computer, Stanislaw Ulam realised the importance and benefits of the synergy between the computer and its users, i.e., continuous human–computer interaction, as achieved in modern computer technology thanks to the design of a usable display.

Considering the complexity of the systems and states studied by Haken's synergetics, it is clear that Zabusky's synergetic approach, containing Ulam's synergy as a part, will occupy an important place among the other methods and tools of synergetics.

References

1. Haken, H.: Synergetics. An Introduction. Nonequilibrium Phase Transitions and Self-Organization in Physics, Chemistry and Biology. Springer, Berlin/Heidelberg/New York (1978)
2. Korteweg, D.J., de Vries, G.: On the change of form of long waves advancing in a rectangular canal, and on a new type of long stationary waves. Philos. Mag. **39**, 422 (1895)
3. Toda, M.: Theory of Nonlinear Lattices, 2nd edn. Springer, Berlin (1989)
4. Kadantsev, V.N., Lupichev, L.N., Savin, A.V.: Synergetics of molecular systems. I. Dynamics of an one-dimensional nonlinear lattice. In: Lupichev, L.N. (ed.) Proceedings of Nonlinear Phenomena in Open Systems, Moscow, 3 (2002)
5. Kadantsev, V.N.: Stability and Evolution of Dynamic Systems. Foundations of Synergetics, vols. 1, 2. IPR-Books, Saratov (2010)
6. Zabusky, N.J.: A synergetic approach to problems of nonlinear dispersive wave propagation and interaction. In: Ames, W.F. (ed.) Proceedings of the Symposium on Nonlinear Partial Differential Equations. Academic, New York 223 (1967)
7. Fermi, E., Pasta, J., Ulam, S.: Studies of nonlinear problems, I. Los Alamos Report LA 1940 (1955)

Chapter 2
Acoustic Solitons

The need to consider nonlinear terms in the different equations of solid state physics has long been realised. Ignoring lattice anharmonicity, it is impossible to explain the thermal expansion and heat conductivity of a solid body. Previously, in the framework of perturbation theory, only a small nonlinearity was taken into account, one that occasionally caused a loss of key features, determined by the nonlinearity of the problem. Anharmonicity of molecular lattice vibration reveals itself especially strongly in so-called quasi-one-dimensional crystals – systems composed of parallel chains the size of the molecular diameter (the nearest-neighbor distance inside one chain is substantially smaller than the distance between the atoms in different chains).

2.1 Fermi–Pasta–Ulam Problem (FPU)

A rigorous consideration of the nonlinear vibration of molecular chains began with the work of Fermi, Pasta, and Ulam (FPU) [1]. For the first time, a computational study of nonlinear dynamics was carried out. In a chain with harmonic interaction potential, the normal modes of vibration are mutually independent variables. The modes do not interact with each other (thermalisation of one mode does not lead to thermalisation of other modes). FPU considered that, if a nonlinearity were introduced into the interaction, an energy flux would occur, leading eventually to equipartition of energy, in accord with the principles of statistical mechanics. They thus set out to confirm this in a computational experiment, but it turned out that only a small part of the energy was redistributed. Such systems have been shown to return recurrently to their initial states.

Let us consider a one-dimensional chain of molecules, arranged along the x axis at interval a. All molecules in the chain are supposed to be of unitary mass M, and the interaction of molecules is described by a unified potential $V(r)$, where r

© Springer International Publishing Switzerland 2015
L.N. Lupichev et al., *Synergetics of Molecular Systems*, Springer Series
in Synergetics, DOI 10.1007/978-3-319-08195-3_2

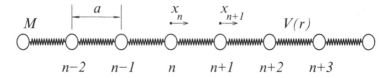

Fig. 2.1 Model of a one-dimensional molecular chain

is the displacement of a molecule with respect to its equilibrium position ($r = 0$ in equilibrium position). We also assume that only nearest-neighboring molecules interact with each other. A schematic view of this system is shown in Fig. 2.1. In the ground state, the nth molecule in the chain has coordinate $x = na$. Let $x_n(t)$ be the displacement of the nth molecule with respect to its equilibrium position at time t. Then the Hamiltonian of the system takes the form

$$\mathscr{H} = \sum_{n=-\infty}^{+\infty} \left[\frac{1}{2} M \dot{x}_n^2 + V(r_n) \right] , \qquad (2.1)$$

where a dot over x_n denotes differentiation with respect to time t, and $r_n = x_{n+1} - x_n$ is the elongation of the nth bond of the chain. The following equations of motion correspond to the Hamiltonian of the system (2.1):

$$M \ddot{x}_n = F(x_{n+1} - x_n) - F(x_n - x_{n-1}) , \quad n = 0, \pm 1, \pm 2, \pm 3, \dots , \qquad (2.2)$$

where the function $F(r) = dV/dr$.

If the force resulting from the bond deformation is proportional to the bond strain as in Hooke's law, i.e., $F(r) = Kr$, the string is said to be linear and the molecular interaction is described by the harmonic potential

$$V(r) = \frac{1}{2} K r^2 , \qquad (2.3)$$

where K is the rigidity of the intermolecular interaction. For an anharmonic potential, the rigidity is $K = V''(0)$. In the case of a harmonic interaction potential, the equations of motion (2.2) are linearised:

$$M \ddot{x}_n = K(x_{n+1} - 2x_n + x_{n-1}) , \quad n = 0, \pm 1, \pm 2, \pm 3, \dots . \qquad (2.4)$$

Any linear combination of solutions of equations (2.4) is also a solution of this system.

Consider a finite chain composed of $N + 2$ links ($n = 0, 1, 2, \dots, N, N + 1$) with fixed end particles ($x_0 \equiv 0$, $x_{N+1} \equiv 0$). Then Eqs. (2.4) have N linearly independent solutions (linear modes):

$$x_n^{(l)}(t) = A_1 \sin \frac{\pi l n}{N+1} \cos(\omega_1 t + \delta_1) ,$$

$$\omega_1 = 2\sqrt{K/M} \sin \frac{\pi l}{2(N+1)} , \quad l = 1, 2, \ldots, N . \tag{2.5}$$

The amplitudes A_1 and phases δ_1 of the modes do not depend on time and are defined by initial conditions. The modes do not interact with each other, so the linear chain is not ergodic.

In his early academic career, Fermi was engaged in a study of the ergodic problem, and when computers came on the scene, he returned to this theme because it was thought that one particular problem might be solved with the aid of a computer. He thought that, if a nonlinear term were to be introduced into the force acting between particles in a chain, the modes would exchange energy, causing the system to reach a statistical equilibrium state in which energy is uniformly distributed over the linear modes. It was this expectation that FPU believed would be confirmed by a computational simulation.

They modeled the chain dynamics using three potentials. The first potential involved a cubic anharmonic term (the FPU α-potential):

$$V(r) = K \left(\frac{1}{2}r^2 - \frac{1}{3}\alpha r^3 \right) , \tag{2.6}$$

the second, quartic anharmonic term (the FPU β-potential):

$$V(r) = K \left(\frac{1}{2}r^2 + \frac{1}{4}\beta r^4 \right) , \tag{2.7}$$

and the third, a piecewise continuous quadratic function:

$$V(r) = \begin{cases} \frac{1}{2}Kr^2 , & \text{if } |r| < r_0 , \\ \frac{1}{2}K'r^2 + \frac{1}{2}(K - K')r_0^2 , & \text{if } |r| \geq r_0 . \end{cases} \tag{2.8}$$

The results turned out to be qualitatively similar for all the potentials.

FPU considered a chain with fixed end points and a number of links N equal to 32 or 62. To model this system, one must put $n = 0, 1, \ldots, N + 1$, $x_0 = 0$, and $x_{N+1} = 0$ in the equations of motion (2.2). At the initial time, the lowest mode was excited, so the initial condition was

$$x_n(0) = A \sin \frac{\pi n}{N+1} , \quad x_n'(0) = 0 , \quad n = 0, 1, \ldots, N + 1. \tag{2.9}$$

The numerical integration of the equations of motion (2.2) with initial condition (2.9) showed that, after some time, almost all the energy was back in the initial mode. This is the so-called FPU recurrence phenomenon.

Computational simulation occasionally leads to utterly unexpected results, and the FPU recurrence phenomenon was one of these. Furthermore, this result was repeatedly confirmed. One may assert that, if the energy is low and the initial shape of the wave is sufficiently smooth, the recurrence phenomenon occurs. Norman Zabusky [2, 3] summarised the results in the empirical equation which determines the recurrence time at an initial excitation of the lowest mode:

$$t_r = \frac{0.44 N^{3/2} t_1}{\sqrt{|\alpha| A}} , \tag{2.10}$$

where $t_1 = 2N \sqrt{K/M}$ is the time taken by the wave of long wavelength to travel back and forth along the chain of N particles with fixed ends (or the period of the wave going round the closed chain of $2N$ particles). It was shown further that the recurrence phenomenon is related to the presence of localised solitary waves (solitons) in the chain, while (2.10) is associated with the characteristics of the soliton motion [4].

The FPU recurrence phenomenon is also observed in the case of a finite chain with periodic boundary conditions. Let us consider in detail this phenomenon occurring in the closed chain. For convenience of numerical modeling, we introduce the following dimensionless variables: displacement $u_n = x_n/a$, time $\tau = t\sqrt{K/M}$, and energy $H = \mathscr{H}/Ka^2$. Then the cyclic chain of N particles can be described by the dimensionless Hamiltonian

$$H = \sum_{n=0}^{N-1} \left[\frac{1}{2} u_n'^2 + U(\rho_n) \right] , \tag{2.11}$$

where the prime denotes differentiation with respect to the dimensionless time τ, while $\rho_n = u_{n+1} - u_n$ is the relative displacement, and the dimensionless interaction potential $U(\rho_n) = V(a\rho_n)/Ka^2$ is normalised according to the conditions

$$U(0) = 0 , \quad U_\rho(0) = 0 , \quad U_{\rho\rho}(0) = 1 . \tag{2.12}$$

The Hamiltonian (2.11) leads to the following finite system of equations of motion:

$$u_n'' = F(u_{n+1} - u_n) - F(u_n - u_{n-1}) , \quad n = 0, 1, \ldots, N-1 , \tag{2.13}$$

where the function $F(r)$ is defined here as $F(r) = dU/d\rho$, while $n + 1 = 0$ if $n = N - 1$ and $n - 1 = N - 1$ if $n = 0$.

The linear mode of the cyclic chain takes the form

$$u_n(\tau) = A \exp \left[i(qn - \omega\tau) \right] , \tag{2.14}$$

where A, $q \in [0, 2\pi]$, and $\omega = 2\sin(q/2)$ are the amplitude, wavenumber, and frequency of the mode, respectively. In the case of the cyclic chain of N molecules, the wavenumbers can only take N values, viz.,

$$q_k = 2\pi k / N , \quad k = 0, 1, \ldots, N - 1 .$$

The amplitude of the kth mode is defined by the equation

$$A_k = \frac{1}{\sqrt{N}} \sum_{n=0}^{N-1} u_n \exp(-iq_k n) . \tag{2.15}$$

It follows from this equation that $A_{N-k} = \bar{A}_k$. For the chain with harmonic interaction potential, the amplitude of the mode is constant, as opposed to the chain with anharmonic potential, for which the amplitude depends on time according to

$$A_k' = \frac{1}{\sqrt{N}} \sum_{n=0}^{N-1} u_n' \exp(-iq_k n) . \tag{2.16}$$

In the case of a harmonic chain, the energy of the kth mode is described by the equation $E_k = \omega_k^2 |A_k|^2$, where the mode frequency is specified by the relation $\omega_k = 2 \sin(q_k/2)$. Let us define the energy of the mode for an anharmonic chain as $E_k = (|A_k'|^2 + \omega_k^2 |A_k|^2)/2$ and integrate the system (2.13) with the initial condition corresponding to the linear mode of wavenumber q_1:

$$u_n(0) = A \sin \frac{2\pi n}{N} , \qquad u_n'(0) = 0 . \tag{2.17}$$

Note that, given the initial condition, the problem is equivalent to the dynamics of a chain of $N/2 - 1$ links with fixed ends, when the first linear mode is excited.

From now on, we write the FPU β-potential of intermolecular interaction in the form

$$U(\rho) = \frac{1}{2}\rho^2 + \frac{1}{4}\beta\rho^4 . \tag{2.18}$$

Furthermore, let us fix the amplitude of the mode $A = 0.1$, but allow the anharmonicity parameter β to change.

The equations of motion are linear when $\beta = 0$, so all the energy of the initial excitation remains in the given mode. The rest of the modes are not excited in this scenario. When $\beta > 0$, the modes can exchange energy. The numerical simulation showed that a portion of the energy drifts from the first mode, but periodically recurs to it (Fig. 2.2a left), whence the initial shape of the wave is recovered. This is the FPU recurrence phenomenon. The greater the anharmonicity of the interaction potential, the higher the amplitude of the change in the mode energy. For strong nonlinearity, the periodic energy exchange between the modes becomes unstable and recurrence does not proceed, resulting in the energy spreading

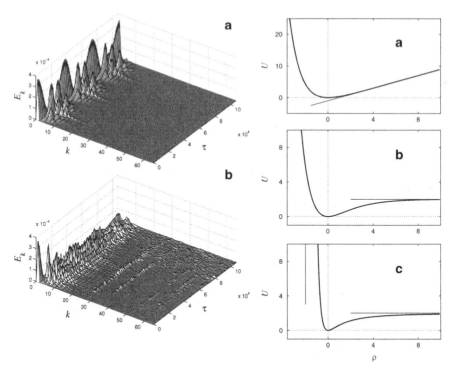

Fig. 2.2 *Left*: Energy exchange between the linear modes in the anharmonic cyclic chain ($N = 132$) for initial excitation of only the first mode $k = 1$ ($q = 2\pi/N$ and the amplitude of the mode is $A = 0.1$). The dependence of the energy distribution E_k over the linear modes of the chain on the time τ is also shown (k is the mode number). (**a**) Weak nonlinearity $\beta = 1,000$. Periodic recurrence of the energy to the initial mode occurs (the FPU recurrence phenomenon). (**b**) Strong nonlinearity $\beta = 2,000$. The energy is redistributed over the other modes. *Right*: Interstitial potentials $U(\rho)$. (**a**) Toda potential (2.34) with $b = 1$. (**b**) Morse potential (2.41) with $\epsilon = 2$, $\beta = 0.5$. (**c**) Lennard-Jones potential (4,2) (2.42) with $\epsilon = 2$, $a = 2$. *Lines* show the asymptotics of the potentials

among the various modes (Fig. 2.2b left). The same effect takes place if one fixes the nonlinearity parameter and changes the mode amplitude. The FPU recurrence phenomenon is observed only when the mode amplitude is less than an arbitrary threshold value.

For the recurrence phenomenon to exist, the initial wave shape must be smooth (the wavelength of the mode $\lambda = 2\pi/q \gg 1$). Excitation of a single short wavelength mode in the anharmonic chain leads to fast thermalisation of the rest of the modes. The recurrence phenomenon is associated with the existence of stable, elastically-interacting solitary waves in the chain, i.e., solitons. At an energy below the threshold value, the initial deformation of the chain transforms into solitons which are periodically assembled together, forming the initial shape of the wave.

2.2 Solitary Waves

Consider an anharmonic chain. The dimensionless Hamiltonian of the chain takes the form (2.11), where the sum runs over all integer indices ($n = 0, \pm 1, \pm 2, \ldots$). Taking into account the normalization conditions (2.12), the dimensionless interaction potential $U(\rho) = V(a\rho)/Ka^2$ can be expanded in a series:

$$U(\rho) = \frac{1}{2}\rho^2 - \frac{1}{3}\alpha\rho^3 + \frac{1}{4}\beta\rho^4 + \cdots , \qquad (2.19)$$

where $\alpha = -U^{(3)}(0)/2$ and $\beta = U^{(4)}(0)/6$ are nonlinearity parameters. The equations of motion take the form (2.13) with $n = 0, \pm 1, \pm 2, \ldots$ and

$$F(\rho) = \rho - \alpha\rho^2 + \beta\rho^3 + \cdots . \qquad (2.20)$$

In the case of small displacements, one can neglect all the anharmonic terms in the series (2.19). As a result, we obtain the harmonic potential $U(\rho) = \rho^2/2$ and the equations of motion become linear:

$$u_n'' = u_{n+1} - 2u_n + u_{n-1} , \quad n = 0, \pm 1, \pm 2, \ldots , \qquad (2.21)$$

The solution of (2.21) can be represented as a sum of linear waves

$$u_n(\tau) = A \exp i(qn - \omega\tau) , \qquad (2.22)$$

where A, $q \in [-\pi, \pi]$, and $\omega = 2\sin(q/2)$ are the wave amplitude, wavenumber, and wave frequency, respectively. The wavelength $\lambda = 2\pi/q$ tends to infinity as $q \to 0$ and the wave velocity is

$$s(q) = \frac{\omega(q)}{q} = \frac{\sin(q/2)}{q/2}$$

which tends to the dimensionless velocity of the long-wavelength phonon, $s_0 = 1$.

In the case of anharmonic potential, the linear wave solution (2.22) is no longer the explicit solution of the infinite-dimensional system of equations of motion (2.13). Let us search for the solution of this system as a solitary wave of constant shape, i.e.,

$$u_n(\tau) = u(n - s\tau) ,$$

where s is the dimensionless velocity of the wave, and the wave shape $u(n)$ depends smoothly on the discrete variable n. To use the continuum approximation, the following parameter must be small:

$$\mu = \max_n \left| \frac{du(n)}{dn} \right| .$$

This parameter describes the reciprocal width of the solitary wave (soliton). It follows that all the derivatives obey the relationship $d^m u/dn^m = O(\mu^n)$. Therefore in the continuum approximation, the following partial differential equation corresponds to the discrete equations of motion (2.13):

$$(1 - s^2)u_{zz} + \frac{1}{12}u_{zzzz} + \frac{1}{360}u_{zzzzzz} - \alpha\left(2u_z u_{zz} + \frac{1}{3}u_{zz}u_{zzz} + \frac{1}{6}u_z u_{zzzz}\right) 11$$

$$+\beta\left(3u_z^2 u_{zz} + u_z u_{zzz} + \frac{1}{4}u_{zz}^2 + \frac{1}{4}u_z^2 u_{zzzz}\right) + O(\mu^6) = 0,$$

where $z = n - s\tau$ is the continuous wave variable approximating the discrete variable n. Considering only the terms smaller than μ^5, this equation takes the form

$$(1 - s^2)u_{zz} + \frac{1}{12}u_{zzzz} - 2\alpha u_z u_{zz} + 3\beta u_z^2 u_{zz} = 0. \tag{2.23}$$

Let us change from the absolute displacement $u(z)$ to a relative displacement $\rho = u_z$. Then (2.23) takes the form

$$(1 - s^2)\rho_z + \frac{1}{12}\rho_{zzz} - 2\alpha\rho\rho_z + 3\beta\rho^2\rho_z = 0. \tag{2.24}$$

For a solitary wave, the state of the chain at infinity must be the ground state, and therefore

$$\rho, \rho_z, \rho_{zz} \longrightarrow 0 \quad \text{as } z \to \pm\infty. \tag{2.25}$$

Integrating (2.24) once and considering the boundary conditions (2.25) leads to the well-known Boussinesq equation:

$$(1 - s^2)\rho + \frac{1}{12}\rho_{zz} - \alpha\rho^2 + \beta\rho^3 = 0. \tag{2.26}$$

This has explicit analytical solutions only for the FPU α-potential (the cubic anharmonic potential, $\beta = 0$):

$$\rho(z) = -A/\cosh^2(\mu z), \quad A = 3(s^2 - 1)/2\alpha, \quad \mu = \sqrt{3(s^2 - 1)}, \tag{2.27}$$

and the FPU β-potential (the quartic anharmonic potential, $\alpha = 0$):

$$\rho(z) = A/\cosh(\mu z), \quad A = \pm\sqrt{2(s^2 - 1)/\beta}, \quad \mu = \sqrt{12(s^2 - 1)}. \tag{2.28}$$

The soliton solution of the Boussinesq equation (2.24) has been studied by Toda and Waddati [5]. In the chain with cubic anharmonicity ($\beta = 0$), the soliton solution (2.27) exists for any value of the anharmonicity parameter $\alpha \neq 0$ and velocity $s > 1$. The bell-shaped function (2.27) describes a region of localized compression (extension) in the chain for the anharmonicity parameter $\alpha > 0$ ($\alpha < 0$). This region moves along the chain, retaining its shape with velocity $s > s_0$, where $s_0 = 1$ is the dimensionless velocity of sound (the velocity of the long-wavelength phonon) in the chain. Such localized excitation of the chain is called a *supersonic acoustic soliton*.

In terms of the absolute displacement, the acoustic soliton is described by the following solution of the equations of motion for the chain:

$$u_n(\tau) = \frac{A}{\mu}\left\{1 - \tanh\left[\mu(n - s\tau)\right]\right\}. \tag{2.29}$$

Since the displacement $u_n \to -2A/\mu$ as $n \to -\infty$ and $u_n \to 0$ as $n \to +\infty$, the chain moves as a whole to the right through a distance $2A/\mu$ (total compression of the chain).

The energy of the acoustic soliton (2.29) in the chain with cubic anharmonicity is given by the relation

$$E = \sum_{n=-\infty}^{+\infty}\left[\frac{1}{2}u_n'^2 + U(\rho_n)\right] = \int_{-\infty}^{+\infty}\left[\frac{1}{2}(1 + s^2)\rho^2(z) - \frac{1}{3}\alpha\rho^3(z)\right]dz$$

$$= \frac{1}{\mu}\left[\frac{2}{3}(1 + s^2)A^2 + \frac{16}{45}\alpha A^3\right], \tag{2.30}$$

and the root-mean-square width of the soliton is defined by

$$L = 2\left(\frac{1}{R}\sum_{n=-\infty}^{+\infty}n^2\rho_n\right)^{1/2} = 2\left[\frac{1}{R}\int_{-\infty}^{+\infty}z^2\rho(z)dz\right]^{1/2} = \frac{\pi}{\sqrt{3}\mu}, \tag{2.31}$$

where the total compression of the chain is

$$R = \sum_{n=-\infty}^{+\infty}\rho_n = \int_{-\infty}^{+\infty}\rho(z)dz = \frac{2A}{\mu}.$$

It follows from (2.27), (2.30), and (2.31) that the energy of the acoustic soliton $E(s)$ and its amplitude $A(s)$ steadily increase when its velocity increases, while the soliton width $L(s)$ steadily decreases: as $s \to 1$, $E(s) \searrow 0$, $A(s) \searrow 0$, and $L(s) \nearrow \infty$, and as $s \to \infty$, $E(s) \nearrow \infty$, $A(s) \nearrow \infty$, and $L(s) \searrow 0$.

Equations (2.27) and (2.29)–(2.31) have been obtained in the continuum approximation, which can only be used if the soliton width $L(s) \gg 1$. The continuum approximation gives acceptable results for $L(s) = \pi/\mu(s)\sqrt{3} = \pi/3\sqrt{s^2 - 1} > 5$, that is, for $s < \sqrt{1 + (\pi/15)^2} = 1.022$. At high velocities, the continuum approximation cannot be applied, but this does not mean that the discrete equations of motion do not admit soliton solutions for $s > 1.022$. Indeed, soliton solutions exist for all values of the velocity $s > 1$. To obtain the explicit shape of the soliton with width comparable to the lattice spacing ($L \sim 1$), the numerical Eilbeck–Flesh pseudospectral method [6, 7] can be used.

In a chain with quartet anharmonicity ($\alpha = 0$), the acoustic soliton exists only for positive anharmonicity $\beta > 0$. According to (2.28), the soliton solution exists for each value of the velocity $s > 1$. By symmetry of the interaction potential, there are two similar solutions: soliton and antisoliton, with amplitudes $A > 0$ and $A < 0$, respectively. In a localization region of the acoustic soliton (antisoliton), compression (extension) of the chain takes place.

At a fixed velocity $s > 1$, the two types of supersonic soliton have the same energy

$$
E = \sum_{n=-\infty}^{+\infty} \left[\frac{1}{2}u_n'^2 + U(\rho_n) \right] = \int_{-\infty}^{+\infty} \left[\frac{1}{2}(1 + s^2)\rho^2(z) + \frac{1}{4}\beta\rho^4(z) \right] dz
$$

$$
= \frac{1}{\mu} \left[(1 + s^2)A^2 + \frac{1}{3}\beta A^4 \right] , \tag{2.32}
$$

the same absolute value of total chain compression (extension)

$$
R = \left| \sum_{n=-\infty}^{+\infty} \rho_n \right| = \left| \int_{-\infty}^{+\infty} \rho(z)dz \right| = \frac{A\pi}{\mu} ,
$$

and the same width

$$
L = 2 \left(\frac{1}{R} \sum_{n=-\infty}^{+\infty} n^2\rho_n \right)^{1/2} = 2 \left[\frac{1}{R} \int_{-\infty}^{+\infty} z^2\rho(z)dz \right]^{1/2} = \frac{\pi}{\mu} . \tag{2.33}
$$

It follows from (2.28), (2.32), and (2.33) that the energy of the acoustic soliton $E(s)$ and its amplitude $A(s)$ steadily increase when its velocity increases, while the soliton width $L(s)$ steadily decreases: as $s \to 1$, $A(s) \searrow 0$, $E(s) \searrow 0$, and $L(s) \nearrow \infty$, and as $s \to \infty$, $E(s) \nearrow \infty$, $A(s) \nearrow \infty$, and $L(s) \searrow 0$.

The continuum approximation used here for the chain with quartet anharmonicity is applicable when

$$
L(s) = \frac{\pi}{\mu(s)} = \frac{\pi}{\sqrt{12(s^2 - 1)}} > 5 ,
$$

that is, when $s < \sqrt{1 + \pi^2/300} = 1.016$. Yet, using the pseudospectral method [6, 7], it can be shown that the discrete equations of motion (2.13) have a soliton solution for all supersonic values of the velocity $s > 1$. However, the explicit soliton solution cannot be obtained analytically. It can only be obtained numerically, although with any desired precision, for velocities $s > 1$. An explicit analytical equation can only be obtained in the case of the chain with the Toda potential [8].

2.3 Solitons in the Toda Chain

A nonlinear chain with exponential interaction has been studied by Toda and Waddati [5] and Toda [9–13]. The results of the study are most comprehensibly represented in his book [8]. It has been shown that the equations describing the dynamics of such a lattice admit explicit N-soliton solutions. These equations also have an infinite set of integrals of motion, and constitute a completely integrable Hamiltonian system which can be solved by the inverse scattering method.

The dimensionless Toda potential, normalized by the conditions (2.12), has the form

$$U(\rho) = \frac{1}{b}\left\{\rho + \frac{1}{b}\left[\exp(-b\rho) - 1\right]\right\} , \qquad (2.34)$$

where $b > 0$ is the anharmonicity parameter. The potential is given in Fig. 2.2a (right). The potential increases exponentially as $b^{-2}\exp(-b\rho)$ when $\rho \to -\infty$ and linearly as $\rho/b - 1/b^2$ when $\rho \to +\infty$. The chain with the interaction potential (2.34) is called the Toda chain. The parameter b in the potential describes its anharmonicity. For small displacements, $b\rho \ll 1$, the interaction potential takes the form

$$U(\rho) = \frac{1}{2}\rho^2 - \frac{1}{6}b\rho^3 + \cdots .$$

The dimensionless equations of motion (2.13) are

$$u_n'' = \frac{1}{b}\left[\exp(-b\rho_{n-1}) - \exp(-b\rho_n)\right], \quad n = 0, \pm 1, \pm 2, \dots . \qquad (2.35)$$

Taking into account the relative displacements $\rho_n = u_{n+1} - u_n$, the equation of motion (2.35) can be rewritten in terms of the relative displacements:

$$\rho_n'' = -\frac{1}{b}\left[\exp(-b\rho_{n-1}) - 2\exp(-b\rho_n) + \exp(-b\rho_{n+1})\right], \quad n = 0, \pm 1, \pm 2, \dots . \qquad (2.36)$$

It can readily be shown that the partial solution of (2.36) which tends to zero exponentially as $n \to \pm\infty$ is the solitary wave solution

$$\exp(-b\rho_n) - 1 = \frac{\sinh^2 q}{\cosh^2\left[q(n - s\tau)\right]} , \qquad (2.37)$$

where $s > 1$ is the velocity of the wave and the parameter q is determined from the dispersion equation

$$s = \frac{\sinh q}{q} . \qquad (2.38)$$

The acoustic soliton (2.37) can have any supersonic velocity. It follows from (2.38) that the velocity $s \rightarrow 1 + 0$ as $q \rightarrow 0$ and s tends to infinity as $q \rightarrow \infty$. The parameter q determines the reciprocal width of the soliton.

It is easily shown that the displacement has the form [8]

$$u_n(\tau) = \frac{1}{b} \ln \left[\frac{1 + \exp(-2q) \exp 2q(n - s\tau)}{1 + \exp 2q(n - s\tau)} \right] + \text{const.} \qquad (2.39)$$

The relative displacement is

$$\rho_n = u_{n+1} - u_n = -\frac{1}{b} \ln \frac{1 + \sinh^2(q)}{\cosh^2 q(n - s\tau)} < 0 \,,$$

whence the chain is compressed in a region of soliton localization. The total chain compression is

$$u_{-\infty} - u_{+\infty} = \frac{2q}{b} \,.$$

As a result of chain compression, there is an excess of mass in the chain that allows us to attribute a mass to the soliton (soliton mass $m = 2q/b$ in dimensionless units). The energy of the soliton is

$$E = \sum_{n=-\infty}^{+\infty} \frac{1}{2} u_n'^2 + U(\rho_n) = \frac{2}{b^2} \left[\sinh(q) \cosh q - q \right] . \qquad (2.40)$$

The profile of the acoustic soliton in the Toda chain is shown in Fig. 2.3 (left). In terms of absolute u_n or relative ρ_n displacements, the acoustic soliton is described, respectively, by the step function (Fig. 2.3a left) or the bell-shaped function, corresponding to chain compression (Fig. 2.3b left). When the soliton velocity increases, its width decreases and the amplitude grows. At all values $s > 1$, the solitons interact with each other as elastic particles. When they collide, they repel each other (see Fig. 2.3 right). Furthermore, the solitons exchange momentum as perfect rigid particles, the only difference being a minor delay that occurs during their repulsion [8].

The Toda potential (2.34) appropriately describes molecular interaction during compression of intermolecular bonds, but it is not suitable for describing bond expansion. As can be seen in Fig. 2.2a (right), the potential rises exponentially for negative deformation ($\rho < 0$), and linearly for positive deformation ($\rho > 0$). Hence, it is not generally used to model molecular chain dynamics.

The deformation of valence bonds is more frequently described by the Morse potential

$$U(\rho) = \varepsilon \left[\exp(-\beta\rho) - 1 \right]^2 , \qquad (2.41)$$

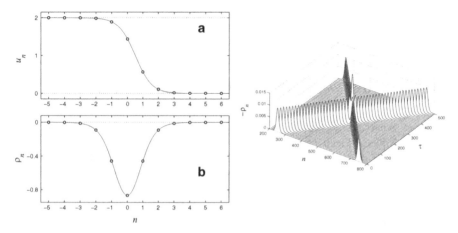

Fig. 2.3 *Left*: Acoustic soliton in the Toda chain with $b = 1$, $q = 1$, $(s = 1.1752)$. The distribution of the absolute u_n (**a**) and relative ρ_n (**b**) displacements of the links in the chain is shown. *Right*: Collision of the two acoustic solitons in the Toda chain with $b = 1$, $s = 1.017$ $(q = 0.1)$. The distribution of the relative chain displacements ρ_n is shown as a function of time τ

where the parameter ε corresponds to the bond energy and the parameter β describes the anharmonicity of the potential. The potential profile is shown in Fig. 2.2b (right). The potential grows exponentially as $\varepsilon \exp(-2\beta\rho)$ when $\rho \to -\infty$ and tends to ε when $\rho \to +\infty$.

The disadvantage of the Morse potential is that it allows the molecules to pass through each other since it is defined for all values of the displacements ρ. To prevent this, one must add a repulsive core to the potential. Given the equilibrium intermolecular bond length a, the energy of the potential must tend to infinity when the bond compression reaches the value a. The Lennard-Jones potential $(2n, n)$, being a potential with a repulsive core, is most frequently used:

$$U(\rho) = \varepsilon \left[\left(1 + \frac{\rho}{a} \right)^{-n} - 1 \right]^2 , \qquad (2.42)$$

where $n = 1, 2, 3, \ldots$. At $n = 6$, the potential (2.42) describes the weak nonvalence van der Waals interaction of molecules with reasonable accuracy. The potential profile is shown in Fig. 2.2c (right). It is defined only for the relative displacement $\rho > -a$. The potential tends to $U(\rho) \to +\infty$ as $\rho \to -a$ and $U(\rho) \to \varepsilon$ as $\rho \to +\infty$.

In contrast to the Toda chain, a chain with the Morse interaction potential (2.41) or the Lennard-Jones potential (2.42) is not a completely integrable system. Nevertheless, there exist acoustic solitons in these systems at all supersonic values of the velocity, although some peculiarities manifest themselves in the dynamics. As the Lennard-Jones potential differs most significantly from the Toda potential, the peculiarities must be more pronounced in this case. The profile of the soliton (solitary wave) in the Morse and Lennard-Jones chains can only be obtained numerically.

2.4 Numerical Methods for Finding Soliton Solution

A solitary wave can be found in a discrete chain with a high degree of accuracy using the Eilbeck–Flesh pseudospectral method [6, 7]. In terms of the absolute displacement u_n, the chain dynamics is described by infinite-dimensional systems of discrete equations (2.13). In terms of the relative displacement $\rho_n = u_{n+1} - u_n$, the equations of motion have the form

$$\rho_n'' = F(\rho_{n+1}) - 2F(\rho_n) + F(\rho_{n-1}) , \quad n = 0, \pm 1, \pm 2, \ldots . \quad (2.43)$$

Let us search for a solution of this system as a solitary wave of unchanged profile which moves with constant velocity s: $\rho_n(s) = \rho(n - s\tau) = \rho(z)$, where $z = n - s\tau$ is the continuum wave variable $z = n - s\tau$. The wave profile $\rho(z) \to 0$ and its derivative $\rho'(z) \to 0$ as $z \to \infty$.

Replacing the discrete variable n by a continuous one z, (2.43) takes the form

$$s^2 \frac{d^2\rho}{dz^2}\bigg|_{z=n} = F\big(\rho(n + 1)\big) - 2F\big(\rho(n)\big) + F\big(\rho(n - 1)\big) , \quad n = 0, \pm 1, \pm 2, \ldots . \quad (2.44)$$

The main idea of this method is to approximate the explicit soliton solution $\rho(z)$ by the finite Fourier series on a finite interval $-L/2 \le z \le L/2$:

$$\rho(z) \approx R(z) = \sum_{k=0}^{K} a_k c_k(z) , \quad (2.45)$$

where $c_k(z) = \cos(2\pi kz/L)$, $k = 0, 1, 2, \ldots, K$. Substituting (2.45) into (2.44) leads to the continuous equation

$$\mathscr{F}(z) = s^2 \sum_{k=0}^{K} a_k \left(\frac{2\pi k}{L}\right)^2 c_k(z) + G(z + 1) - 2G(z) + G(z - 1) = 0 , \quad (2.46)$$

where

$$G(z) = F\left(\sum_{k=0}^{K} a_k c_k(z)\right) .$$

The Fourier coefficients $\{a_k\}_{k=0}^{K}$ can be found numerically as the roots of the system of K nonlinear equations

$$\rho(L/2) = \sum_{k=0}^{K} a_k c_k(L/2) = 0 \,,$$

$$\mathcal{F}(z_i) = 0 \,, \quad i = 0, 1, \ldots, K - 1 \,,$$

(2.47)

where $z_i = iL/2K$ and the function $\mathcal{F}(z)$ is given by (2.46).

This method can unambiguously answer the question about the existence of a soliton for any value of the velocity s. The absence of a soliton solution of (2.47) implies the impossibility of soliton motion for a given value of s. When solving (2.47) numerically, it suffices to put $K = 100$ and $L = 10D$, where D is the diameter of the soliton solution given by

$$D = 2 \left[\frac{\int_0^{L/2} z^2 \rho(z) dz}{\int_0^{L/2} \rho(z) dz} \right]^{1/2} .$$

The value $A = -\rho(0)$ describes the amplitude of the soliton (the maximum relative displacement of a chain link in a region of soliton localization). The soliton energy is

$$E = \sum_{n=-N_L}^{N_L} \frac{1}{2} v_n^2 + V\left(r(n)\right) \,,$$

where N_L is an integer part of $L/2$ and the velocity is

$$v_{-N_L-1} = 0, \quad v_{n+1} = v_n + s \sum_{k=1}^{K} \frac{2\pi k a_k}{L} \sin \frac{2\pi k n}{L} \,, \quad n = -N_L, -N_L + 1, \ldots, N_L.$$

2.5   Solitons in the Lennard-Jones Chain

Here we consider a chain with the Lennard-Jones potential (12,6) and parameters $a = 1$ and $\varepsilon = 1/72$. Note that, at these parameter values, the potential rigidity is $U''(0) = 72\varepsilon/a^2 = 1$. Numerical solution of (2.47) has shown that acoustic solitons exist in the chain at all supersonic values of the velocity $s > 1$. The dependencies of the soliton amplitude A, diameter D, and energy E, on the velocity s, are shown in Fig. 2.4 (left). As can be seen from Fig. 2.4a and b (left), the amplitude A tends

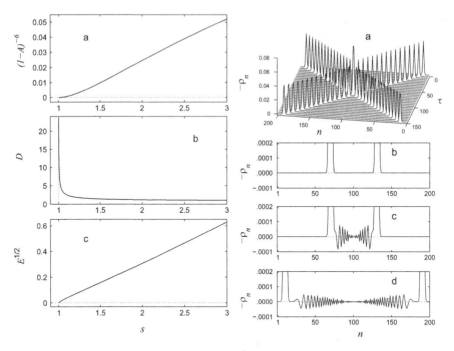

Fig. 2.4 *Left*: Dependence of the function $(1-A)^{-6}$ of the amplitude A (**a**), the width D (**b**), and the square root of the energy $\sqrt{E}$ (**c**) on the velocity of the acoustic soliton s in the Lennard-Jones chain. *Right*: Collision of acoustic solitons in the Lennard-Jones chain at velocity $s = 1.2$ (**a**). The soliton profile before the collision at time $\tau = 50$ (**b**) and after collision at $\tau = 100$ (**c**) and $\tau = 150$ (**d**)

steadily to 1 as $1 - O(s^{-1/6})$, while the soliton diameter D decreases monotonically. The soliton width becomes less than 10 at $s = 1.0055$, 5 at $s = 1.0226$, 2 at $s = 1.175$, and finally less than 1 at $s = 3.7$, at which point it is less than the chain spacing. The soliton energy increases steadily as s^2 (see Fig. 2.4b left).

As the Lennard-Jones chain is not a completely integrable system, the soliton interaction is no longer elastic. The collision of two solitons is accompanied by the emission of low-amplitude waves (phonons). The non-elasticity of the soliton interaction can be characterized by the energy loss $p = \left[(E_1 - E_2)/E_1\right]100\,\%$, where E_1 and E_2 are the soliton energies before and after collision, respectively. The dependence of the energy loss p on the velocity s is given in Table 2.1. The maximum energy loss $p = 0.00083\,\%$ is observed for $s = 1.23$. The soliton collision is shown in Fig. 2.4 (right). The phonon emission can be observed only at a high magnification. Therefore, the weak 'non-elasticity' of the soliton interaction is related to the proximity of the Lennard-Jones chain to a completely integrable system. The energy loss at the collision tends to zero as $s \to 1 + 0$ (in this limit, the soliton dynamics is described by the continuous integrable Boussinesq equation)

Table 2.1 Dependence of the energy loss (%) in a soliton collision in the Lennard-Jones chain on their velocity s

s	1.03	1.06	1.1	1.2	1.23
p (%)	1.4×10^{-5}	1.1×10^{-4}	3.7×10^{-4}	8.1×10^{-4}	8.3×10^{-4}
s	1.3	2.0	2.5	3.0	3.5
p (%)	7.6×10^{-4}	5.6×10^{-5}	7.9×10^{-6}	9.8×10^{-7}	4.6×10^{-8}

Fig. 2.5 Formation of the acoustic soliton in the chain with the Morse potential ($\varepsilon = 0.5$, $\beta = 1$) as a result of the compression of the first bond in the molecular chain of $N = 500$ molecules. Dependence of the distribution of the local displacement ρ_n in the chain on the time τ

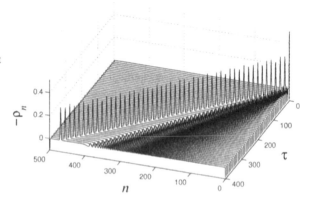

and as $s \to \infty$ (where the dynamics is described by an integrable system of rigid spheres).

Thus, the acoustic solitons in the Lennard-Jones chain exist at all supersonic values of the velocity $s > 1$. The solitons interact virtually as rigid particles. The maximum soliton energy loss observed at collision for $s = 1.23$ is only 0.00083 % of their energy. The solitons in the chain with the Morse interaction potential can be considered to be interacting elastically.

The acoustic solitons are formed as a result of the compression of molecular bonds. To model the formation of the soliton, it is best to consider a finite chain of N molecules, subjected to the compression of its end link by a value $\rho_0 > 0$ at the initial moment of time. The chain dynamics will be described by the equations

$$u''_n = F(u_{n+1} - u_n) - F(u_n - u_{n-1}), \quad n = 2, 3, \ldots, N - 1, \quad u_1 \equiv \rho_0, \quad u_N \equiv 0,$$
$$(2.48)$$

with initial conditions $u_n(0) = 0$ and $u_n'(0) = 0$. Let us take here the Morse potential (2.41) with parameters $\varepsilon = 0.5$, $\beta = 1$. For these parameters, the potential rigidity $U''(0) = 1$. Numerical simulation of the soliton dynamics shows that the compression of a single finite link leads to the formation of an acoustic soliton in the chain, along with a spreading wave packet moving with the velocity of sound (see Fig. 2.5). The bigger the initial compression of the link, the bigger fraction of the energy accumulated by the soliton. The fraction of the soliton energy relative to the total energy is $p = 0.504$ for $\rho_0 = 0.1$, $p = 0.718$ for $\rho_0 = 0.2$, and $p = 0.922$ for

$\rho_0 = 0.5$. Thus, in the anharmonic chain, the energy of the dynamic compression of the chain end is effectively transferred along the chain by the acoustic soliton.

2.6   Solitons in the Diatomic Chain

The diatomic chain with nonlinear intermolecular interaction has become a subject of close attention in connection with modeling thermal conductivity in nonmetallic crystals [14–16]. Anomalies in the thermal conductivity of nonlinear systems were first observed in the notable work of Fermi, Pasta, and Ulam [1]. It has been realized by now that the impact of nonlinearity, in the context of the classical theory of thermal conductivity, does not reduce to the inelastic phonon–phonon interaction. Experimentally observed thermal solitons in the quasi-one-dimensional system [17] can significantly modify the character of thermal conductivity. Moreover, no one has yet succeeded in deriving the thermal conductivity equation from first principles.

The diatomic Toda lattice has the property of complete integrability in the case of equal masses [8] and exhibits stochastic behavior for a specific mass ratio [15]. To understand the features of the dynamical behavior of this system and the mechanism underlying its thermal conductivity, one must determine the dynamical properties of the acoustic solitons. We thus investigate numerically the soliton dynamics in the diatomic Toda lattice. We will show that the soliton motion in a given system is always accompanied by a phonon emission which is insignificantly small in the range of the sound velocity and steadily increases when the soliton velocity increases.

2.6.1   Model of the Diatomic Chain

Consider a chain consisting of particles of masses m_1 and m_2, which are located at a fixed interval a from each other. The model is shown schematically in Fig. 2.6. We describe the interaction of the neighboring particles by the Toda potential

$$V(\rho) = Kb^{-1}\left\{\rho + b^{-1}\left[\exp(-b\rho) - 1\right]\right\},$$

where ρ is the relative change in the intermolecular distance, K is the rigidity coefficient, and b is the potential anharmonicity parameter.

The Hamiltonian of the system can be represented in the form

$$H = \sum_n \left[\frac{1}{2}m_1\dot{x}_{2n}^2 + \frac{1}{2}m_2\dot{x}_{2n+1}^2 + V(\rho_{2n}) + V(\rho_{2n+1})\right], \qquad (2.49)$$

where the dot denotes differentiation with respect to time t, x_n is the displacement of the nth site from its equilibrium position, and $\rho_n = x_{n+1} - x_n$ is the relative

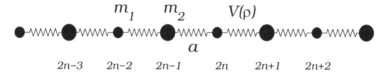

Fig. 2.6 Schematic view of the diatomic molecular chain

displacement of the nth site. The following equations of motion correspond to the Hamiltonian (2.49):

$$m_1 \ddot{x}_{2n} = F(\rho_{2n}) - F(\rho_{2n-1}), \quad m_2 \ddot{x}_{2n+1} = F(\rho_{2n+1}) - F(\rho_{2n}), \quad n = 0, \pm 1, \pm 2, \ldots, \tag{2.50}$$

where $F(\rho) = dV/d\rho$. For small-amplitude waves, the dispersion equation is readily found to be

$$\omega^4 m_1 m_2 - 2K(m_1 + m_2)\omega^2 + 4K^2 \sin^2(\lambda a) = 0 ,$$

where ω is the wave frequency and $0 \leq \lambda \leq \pi/a$ is the wavenumber. This gives the velocity of sound in the form (the velocity of long-wavelength, small-amplitude waves)

$$v_0 = \lim_{\lambda \to 0} \frac{\omega(\lambda)}{\lambda} = a\sqrt{2K/M} ,$$

where the mass $M = m_1 + m_2$.

For the convenience of calculation, we introduce the dimensionless displacement $u_n = x_n/a$, time $\tau = t\sqrt{2K/M}$, and energy $\mathcal{H} = H/Ka^2$. Then the Hamiltonian of the chain (2.49) takes the form

$$\mathcal{H} = \sum_n \left[\frac{1}{2}\mu_1 u'^2_{2n} + \frac{1}{2}\mu_2 u'^2_{2n+1} + U(r_{2n}) + U(r_{2n+1}) \right] ,$$

where the prime denotes differentiation with respect to time τ, $\mu_i = 2m_i/M$, $i = 1, 2$, is the dimensionless mass, $r_n = u_{n+1} - u_n$ is the relative displacement, and the potential is

$$U(r) = \beta^{-1}\left\{ r + \beta^{-1}[\exp(-\beta r) - 1] \right\} , \tag{2.51}$$

with $\beta = ab$. In terms of the dimensionless variables, the equations of motion (2.50) take the form

$$\mu_1 u''_{2n} = G(r_{2n}) - G(r_{2n-1}), \quad \mu_2 u''_{2n+1} = G(r_{2n+1}) - G(r_{2n}), \quad n = 0, \pm 1, \pm 2, \ldots,$$
$$(2.52)$$

where $G(r) = dU/dr = \beta^{-1}[1 - \exp(-\beta r)]$.

2.6.2 Continuum Approximation

Hereafter, for simplicity, we shall put $\beta = 1$. Then the dimensionless Toda potential (2.51) for the small-amplitude displacements has the form

$$U(r) \approx \frac{1}{2} r^2 - \frac{1}{6} r^3 .$$

Let us use the continuum approximation $u_n(\tau) = u(x, \tau)|_{x=n}$. Then the equations of motion (2.52) lead to the well-known Boussinesq equation [18, 19]:

$$u_{\tau\tau} = u_{xx} - u_x u_{xx} + \frac{1}{12} c u_{xxxx} . \qquad (2.53)$$

We will search for its solution in the form of a wave of constant shape $u_n(\tau) = u(\xi)$, where $\xi = n - s\tau$ is the wave variable and $s > 1$ is the wave velocity.

As a result of the series of elementary approximations, (2.53) leads to the following equation in terms of the relative displacement $r = u_\xi$:

$$(1 - s^2) r^2 - \frac{1}{3} r^3 + \frac{1}{3} c r_\xi^2 = 0 ,$$

where the coefficient $c = 1 - 3\mu_1 \mu_2 / 4$. The solution of this equation has the form

$$r(\xi) = -\frac{A}{\cosh^2(\alpha\xi)} , \qquad (2.54)$$

where $A = 3(s^2 - 1)$ and $\alpha = \sqrt{3(s^2 - 1)/4c}$ are the amplitude and the reciprocal width of the soliton, respectively.

2.6.3 Numerical Simulation of Soliton Dynamics

Let us consider the dynamics of an acoustic soliton in a chain with $\mu_1 = \mu$ and $\mu_2 = 2 - \mu$, where $0 < \mu \le 1$. We numerically integrated the equations of motion (2.52) with $n = 1, \ldots, N$, $N = 300$ and boundary conditions ($u'_1 \equiv 0$, $u_N \equiv 0$) at the fixed ends. The soliton solution (2.54) was taken as the initial condition.

Let the center of the soliton be located at the site $n = N/2$ at the initial time. To model the soliton dynamics in an infinite chain, we shift the soliton back just as it passes 100 links of the chain, i.e., we carry out the substitution

$$u_n(\tau) = u_{n+100}, \quad n = 1, \ldots, N - 100, \quad u_n(\tau) = 0, \quad n = N - 99, \ldots, N$$
$$u'_n(\tau) = u'_{n+100}, \quad n = 1, \ldots, N - 100, \quad u'_n(\tau) = 0, \quad n = N - 99, \ldots, N.$$

At each such point in time, the soliton is described by the current velocity $s = 100/\tau_1$, where τ_1 is the time of the soliton passage over 100 links and the energy is

$$E = \sum_{n=1}^{N/2-1} \left[\frac{1}{2}\mu_1 u'^2_{2n} + \frac{1}{2}\mu_2 u'^2_{2n+1} + U(r_{2n}) + U(r_{2n+1}) \right].$$

The numerical integration showed that the soliton dynamics depends significantly on the ratio of particle masses $\kappa = \mu_1/\mu_2$. For $\kappa = 1$ ($\mu_1 = \mu_2 = 1$) and $\kappa = 0$ ($\mu_1 = 0$, $\mu_2 = 2$), the equations of motion become completely integrable. Therefore, the soliton motion occurs without phonon emission. The soliton moves with a constant velocity. However, the intermediate value of κ is more interesting. In the intermediate region, the soliton motion is always accompanied by phonon emission (see Fig. 2.7 left), leading to the slowing down of the soliton. This phenomenon is conveniently described by the fraction of its energy which is lost

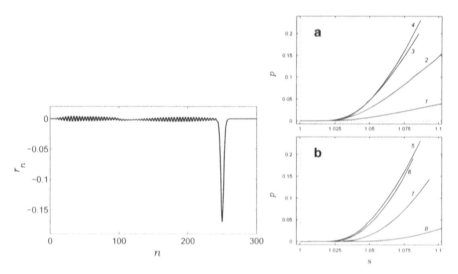

Fig. 2.7 *Left*: Phonon emission of the acoustic soliton in the diatomic chain. Soliton velocity $s = 1.03$ and mass ratio $\kappa = 0.65$. *Right*: Dependence of the soliton energy loss p on the velocity s at (**a**) $\kappa = 0.9, 0.8, 0.7$, and 0.65 (lines 1, 2, 3, and 4, respectively) and (**b**) $\kappa = 0.65, 0.6, 0.5$, and 0.4 (lines 5, 6, 7, and 8, respectively)

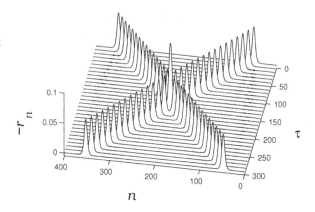

Fig. 2.8 Elastic collision of acoustic solitons in the diatomic chain ($\kappa = 0.65$) at the velocity $s = 1.01$

when it passes 100 links of the chain:

$$p(\tau) = \frac{E(\tau) - E(\tau + \tau_1)}{E(\tau)} .$$

The dependence of the soliton energy loss p on its velocity s is shown in Fig. 2.7 (right) for different values of κ. As can be seen, when the velocity decreases, the energy loss decreases proportionally to $(1 - s)^2$. The value of the proportionality coefficient depends on κ. With κ decreasing from 1 to 0.65, the energy loss grows steadily. The maximal loss is observed at $\kappa = 0.65$. A further decrease in κ no longer leads to a monotonic decrease in the energy loss (see Fig. 2.7 right).

For velocities with $1 < s < 1.015$, phonon emission by the soliton becomes negligibly small. In this case the solitons actually move with constant velocities and interact as rigid particles (see Fig. 2.8).

The modeling performed here has shown that, in the diatomic chain, acoustic soliton motion is always accompanied by phonon emission. The emission is negligibly small in the range of the sound speed ($1 < s < 1.015$), but as the soliton velocity increases, the fraction of the energy emitted by the phonons grows proportionally to $(1-s)^2$. These results allow us to conclude that the equipartition of energy in diatomic chains happens as a result of the intensive emission of phonons by the acoustic solitons. Maximal emission is reached at the ratio of particle masses of the chain $\kappa = m_1/m_2 = 0.65$. It is for this value of κ that the effect of chaos in the system dynamics is expected to be the most pronounced.

2.7 Acoustic Solitons in a Helix Chain

The development of modern nonlinear physics has led to the discovery of new elementary mechanisms which determine the behavior of many physical processes in crystals and other ordered molecular systems at the molecular level. Today, the role of acoustic solitons, ensuring the most efficient mechanisms of energy transfer in such processes as heat conductivity, fracture of solids [20–23], and signal propagation in biological macromolecules, is quite clear [24].

One pioneering theoretical study of the nonlinear dynamics of macromolecular chains [1, 4, 12, 25] considered the one-dimensional (spatial-linear) models with positive anharmonicity, in which only the longitudinal displacements of atoms (molecules) in the chain were taken into account. In this case, when neighboring sites of a chain approach each other, the repulsive force between them increases faster than in the harmonic approximation. One of the consequences of this is the existence of dynamically stable solitary waves of compression which are referred to as supersonic acoustic solitons.

Essentially, acoustic solitons do not interact with longitudinal acoustic phonons, so they transfer energy in a loss-free manner over long distances. The process changes dramatically if the transverse and longitudinal displacements are taken into account. In this case, the soliton will have a finite path length and its motion will be accompanied by emission of transverse and orientational phonons of the chain.

The effect of the transverse molecular oscillations in a chain on the soliton dynamics was considered for the first time in [26]. Solitons turned out to be highly sensitive to longitudinal perturbations. This problem was investigated comprehensively in [27–32]. The soliton interaction with orientational molecular oscillations was analysed in [33].

For a series of biomolecular chains, it is hard to understand the way they function without considering the transverse motion of the chain links. Thus, in the DNA molecule, the stretching of base pairs in the transverse direction makes denaturation possible. The Peyrard–Bishop model of DNA melting [34–36] considered only the transverse motion of complimentary base pairs. Although the DNA molecule (having both longitudinal and transverse degrees of freedom) is considered as an isolated object, this model actually describes the one-dimensional dynamics of the molecular chain in an effective substrate potential. A comprehensive review of models of DNA nonlinear dynamics is given in [37].

The geometric structure of biomolecular systems requires use of two- and three-dimensional models. This is the only way to take into account the system's anharmonicity, which is determined by its molecular geometry. For example, in the framework of the simplest cluster model of the α-chymotrypsin enzyme, it was shown that geometric anharmonicity in the two-dimensional system makes energy transfer between degrees of freedom possible, even for small amplitudes [38, 39].

Applying current computational power to the analysis of nonlinear molecular systems dynamics, one can move from simple one-dimensional models to more complex two- and three-dimensional models, which take into account the geometrical structure of the system in a realistic way. The simplest and most convenient objects from this point of view are the zigzag molecular chains for which the nonlinear dynamics is considered in detail in [40–44]. To understand the mechanisms at work in the majority of biological systems, a two-dimensional model is inappropriate. The simplest example of such a system is a protein α-helix macromolecule.

Consider the three-dimensional dynamics of a free molecular chain. Clearly, the chain, in the absence of a substrate, will have a ground state with a regular stable structure only if the interaction between the remote neighbours is taken into account in addition to the short-range interaction. The inclusion of the long-range interaction results in the appearance of a *secondary structure*, of the chain, which is often encountered in many macromolecules (DNA, proteins, and the like). Geometrically, the secondary structure is realized in the form a helix.

The three-dimensional dynamics of a helix chain was analysed in [45], where the existence of the three-dimensional acoustic soliton of compression was shown. The existence of this soliton is associated with the physical anharmonicity (anharmonicity of intermolecular interaction).

Here we consider in detail the soliton dynamics in the helix chain. We will show that, along with a soliton of longitudinal compression, there also exists another type of soliton in the helix, namely the soliton of torsion. In this case, the existence of this soliton is associated with the geometrical anharmonicity of the helix. This anharmonicity is determined by three-dimensional structure of the helix and manifests itself even if all the intermolecular interaction forces are harmonic. The geometrical anharmonicity was first studied in [46], where it was shown that this anharmonicity can ensure the existence of a breather-like excitation in a linear molecular chain.

2.7.1 Model of a Helix Chain

Consider a molecular helix chain shown in Fig. 2.9 (left), in which each molecule (peptide group) of the chain interacts with its six nearest neighbors. The interaction between first neighbors is the strongest. It is mainly due to the deformation of the rigid valence bonds. The interaction between second neighbors results from the deformation of the softer valence angles. The interaction between third neighbors is the weakest. It is described by the non-valence molecular interaction. These three types of interaction stabilize the three-dimensional helix chain, provided that each molecule of the chain can move in three (x, y, z) directions.

The geometry of the helix chain is uniquely given by a set of three parameters: the radius R_0, the angular spacing ϕ, and the longitudinal spacing Δz of the helix. In the equilibrium position, the radius vector of the nth site of the helix is

$$\mathbf{R}_n = R_0\big(\cos(n\phi), \sin(n\phi), nh\big) , \quad n = 0, \pm 1, \dots ,$$

where $h = \Delta z / R_0$ is the dimensionless longitudinal spacing of the helix. The angular spacing of the helix satisfies $|\phi| \le \pi$. The helix is right-handed for $\phi > 0$ and left-handed for $\phi < 0$. When $\phi = \pm\pi$, the three-dimensional helix degenerates into a planar zigzag structure.

The helix can also be uniquely specified by the distances between three neighboring molecules, i.e., the distances D_1, D_2, and D_3 between the first, second, and

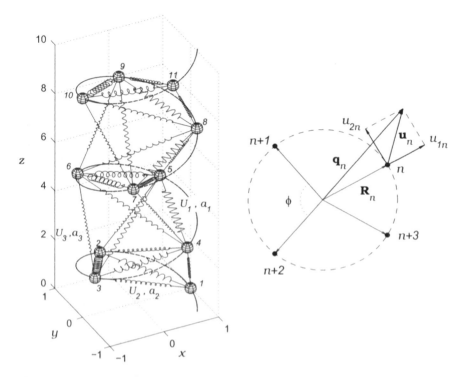

Fig. 2.9 *Left*: Fragment of the helix chain consisting of 11 molecules. The geometry of the chain corresponds to an α-helix ($\phi = 100°$). The intermolecular potentials U_j, $j = 1, 2, 3$ are shown schematically by the springs of different diameters (spring thickness corresponds to interaction rigidity). *Right*: Local coordinate system in the xy plane

third neighbors. The dimensionless nearest-neighbor intermolecular distance is

$$a_j = \frac{D_j}{R_0} = \sqrt{2[1 - \cos(j\phi)] + j^2 h^2}, \quad j = 1, 2, 3 . \tag{2.55}$$

It follows from (2.55) that the angular spacing ϕ of the helix obeys

$$\frac{3 - 4\cos\phi + \cos(2\phi)}{8 - 9\cos\phi + \cos(3\phi)} = \frac{4a_1^2 - a_2^2}{9a_1^2 - a_3^2} . \tag{2.56}$$

After solving (2.56), we readily obtain

$$R_0 = \frac{\sqrt{4D_1^2 - D_2^2}}{4\sin^2(\phi/2)}, \qquad \Delta z = \frac{\sqrt{D_2^2/4 - D_1^2\cos^2(\phi/2)}}{\sin(\phi/2)} .$$

The Hamiltonian of the helix chain has the form

$$H = \sum_n \left[\frac{1}{2} M \left(\dot{x}_n^2 + \dot{y}_n^2 + \dot{z}_n^2 \right) + K R_0^2 \sum_{j=1,2,3} U_j(r_{jn}) \right], \tag{2.57}$$

where M is the mass of a single link of the chain, and x_n, y_n, and z_n are the displacements of the nth chain link from its equilibrium position. The dot denotes differentiation with respect to time t. The constant K defines the rigidity of intermolecular interaction. The dimensionless potential $U_j(r_{jn})$ describes the interaction between the nth and the $(n + j)$th molecules, and $r_{jn} = R_{jn}/R_0$ is the dimensionless distance between them. The interaction potentials are normalized by the conditions $U_j(a_j) = 0$, $U_j'(a_j) = 0$, $j = 1, 2, 3$.

We describe the molecular interaction by the Morse potentials

$$U_j(r) = \frac{1}{2} \frac{\kappa_j}{\gamma_j^2} \left\{ 1 - \exp\left[-\gamma_j(r - a_j) \right] \right\}^2$$

$$= \frac{1}{2} \kappa_j (r - a_j)^2 \left[1 - \gamma_j(r - a_j) + \cdots \right], \quad j = 1, 2, 3, \tag{2.58}$$

where $\kappa_j = K_j/K = U_j''(a_j)$ is the dimensionless rigidity of the interaction and γ_j is the anharmonicity parameter. In the limit $\gamma_j \to 0$, the potential (2.58) turns into the harmonic potential

$$U_j(r) = \frac{1}{2} \kappa_j (r - a_j)^2, \quad j = 1, 2, 3.$$

For further calculation, it is convenient to introduce the dimensionless time

$$\tau = \omega_0 t, \qquad \omega_0 = \sqrt{K/M},$$

and dimensionless displacement vectors

$$\mathbf{q}_n = (q_{1n}, q_{2n}, q_{3n}) = \frac{\mathbf{R}_n}{R_0} + \mathbf{v}_n, \quad \mathbf{v}_n = (v_{1n}, v_{2n}, v_{3n}) = \frac{1}{R_0}(x_n, y_n, z_n). \tag{2.59}$$

Then the dimensionless distances $r_{jn} = |\mathbf{q}_{n+j} - \mathbf{q}_n|$ and the Hamiltonian of the chain (2.57) can be rewritten in the dimensionless form

$$\mathscr{H} = \frac{H}{K R_0^2} = \sum_n \left[\frac{1}{2} \left(\frac{d\mathbf{q}_n}{d\tau} \right)^2 + \sum_{j=1,2,3} U_j \left(|\mathbf{q}_{n+j} - \mathbf{q}_n| \right) \right]. \tag{2.60}$$

The equations of motion corresponding to the Hamiltonian (2.60) take the form

$$\frac{d^2 \mathbf{q}_n}{d\tau^2} = \sum_{j=1,2,3} \left[W_j(r_{jn})(\mathbf{q}_{n+j} - \mathbf{q}_n) - W_j(r_{j,n-j})(\mathbf{q}_n - \mathbf{q}_{n-j}) \right],$$

$$n = 0, \pm 1, \pm 2, \dots, \tag{2.61}$$

where $W_j(r_{jn}) = U'_j(r_{jn})/r_{jn}$.

2.7.2 Dispersion Equation

It is more reasonable to consider the relative molecular displacement with respect to the equilibrium position locally for each molecule. For the equilibrium position of the nth molecule, we consider the coordinate system formed by the normal and the tangent to the circle $z/R_0 = nh$, $|\mathbf{q}| = 1$ in the xy plane, as shown in Fig. 2.9 (right). We denote the displacement vector $\mathbf{v}_n$ in this coordinate system by $\mathbf{u}_n = (u_{1n}, u_{2n}, u_{3n})$, where u_{1n} and u_{2n} are the normal and tangential projections of the displacement vector, respectively, and $u_{3n} = v_{3n}$ is the longitudinal coordinate. The new local coordinate system can be obtained by a rotational transformation, specified by

$$\mathbf{T}_n \mathbf{v}_n = \mathbf{u}_n , \quad \mathbf{T}_n = \begin{pmatrix} \cos(n\phi) & \sin(n\phi) & 0 \\ -\sin(n\phi) & \cos(n\phi) & 0 \\ 0 & 0 & 1 \end{pmatrix} . \tag{2.62}$$

The orthogonal operators $\mathbf{T}_n$ form a group: $\mathbf{T}_m \mathbf{T}_n = \mathbf{T}_{m+n}$, where $\mathbf{T}_0 = \mathbf{I}$ is the identity operator.

Substituting the expression $\mathbf{q}_n = \mathbf{R}_n / R_0 + \mathbf{T}_n^{-1} \mathbf{u}_n$ (see (2.59) and (2.62)) into the equations of motion (2.61) yields

$$\frac{d^2 \mathbf{u}_n}{d\tau^2} = \sum_{j=1,2,3} \left[\mathbf{T}_j^{-1} \mathbf{F}_j(\mathbf{u}_n, \mathbf{u}_{n+j}) - \mathbf{F}_j(\mathbf{u}_{n-j}, \mathbf{u}_n) \right], \quad n = 0, \pm 1, \pm 2, \dots,$$

where the intermolecular forces are defined by

$$\mathbf{F}_j(\mathbf{u}_n, \mathbf{u}_{n+j}) = W_j(r_{jn})(\mathbf{c}_j + \mathbf{u}_{n+j} - \mathbf{T}_j \mathbf{u}_n) . \tag{2.63}$$

Here, the distance r_{jn} between the nth and the $(n+j)$th molecules is

$$r_{jn} = \left| \mathbf{a}_{jn} + \mathbf{T}_{n+j}^{-1} \mathbf{u}_{n+j} - \mathbf{T}_n^{-1} \mathbf{u}_n \right|, \quad \mathbf{a}_{jn} = \frac{\mathbf{R}_{n+j} - \mathbf{R}_n}{R_0}, \tag{2.64}$$

and the constant vectors are defined by

$$\mathbf{c}_j = \left(1 - \cos(j\phi), \sin(j\phi), jh\right) . \tag{2.65}$$

As can be seen from (2.63)–(2.65), the right-hand side of the equations of motion (2.63) is not a function of the difference between the vectors $\mathbf{u}_n$ and $\mathbf{u}_{n+j}$.

In the harmonic approximation ($\gamma_j \to 0$, $j = 1, 2, 3$), for all the intermolecular forces, we have

$$\mathbf{F}_j\left(\mathbf{u}_n, \mathbf{u}_{n+j}\right) = \alpha_j\langle(\mathbf{u}_{n+j} - \mathbf{T}_j\mathbf{u}_n), \mathbf{c}_j\rangle\mathbf{c}_j + \cdots ,$$

where $\alpha_j = \kappa_j/a_j^2$ and $\langle\cdot, \cdot\rangle$ denotes the inner product. Substituting this expansion into (2.63) gives the linearized equations of motion:

$$\frac{d^2\mathbf{u}_n}{d\tau^2} = \sum_{j=1,2,3} \alpha_j \left[\langle\mathbf{u}_{n+j} - \mathbf{T}_j\mathbf{u}_n, \mathbf{c}_j\rangle\mathbf{T}_j^{-1}\mathbf{c}_j - \langle\mathbf{u}_n - \mathbf{T}_j\mathbf{u}_{n-j}, \mathbf{c}_j\rangle\mathbf{c}_j\right],$$

$$n = 0, \pm 1, \pm 2, \dots . \tag{2.66}$$

Substituting the plane wave

$$\mathbf{u}_n = \mathbf{A}_n \exp\left[i(kn - \Omega\tau)\right]$$

into the linear equations (2.66), we obtain the following dispersion law:

$$\begin{vmatrix} \Omega^2 - c_{11} & -ic_{12} & -ic_{13} \\ ic_{12} & \Omega^2 - c_{22} & -c_{23} \\ ic_{13} & -c_{23} & \Omega^2 - c_{33} \end{vmatrix} = 0 , \tag{2.67}$$

where the coefficients are

$$c_{11} = 2\sum_j \alpha_j[1 - \cos(j\phi)]^2[1 + \cos(jk)] ,$$

$$c_{12} = 2\sum_j \alpha_j[1 - \cos(j\phi)]\sin(j\phi)\sin(jk) ,$$

$$c_{13} = 2\sum_j \alpha_j jh[1 - \cos(j\phi)]\sin(jk) ,$$

$$c_{22} = 2\sum_j \alpha_j \sin^2(j\phi)[1 - \cos(jk)] , \tag{2.68}$$

$$c_{23} = 2\sum_j \alpha_j jh \sin(j\phi)[1 - \cos(jk)] ,$$

$$c_{33} = 2\sum_j \alpha_j j^2h^2[1 - \cos(jk)] .$$

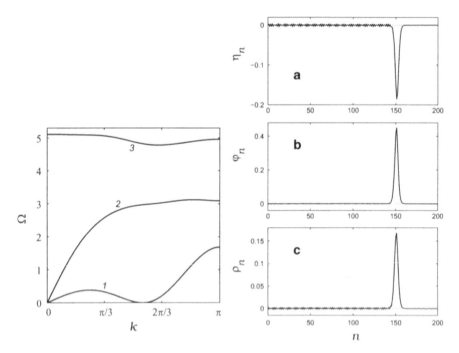

Fig. 2.10 *Left*: Dependence of the frequencies Ω_t (*line 1*), Ω_l (*line 2*), and Ω_{op} (*line 3*) on the wavenumber k, $0 \leq k \leq \pi$. *Right*: The three-component profile of the torsion soliton at the velocity $s = 1.5s_t$ ($\gamma_3 = 0$). Phonon emission by the soliton is notable

More explicitly, the dispersion equation (2.67) has the form

$$\Omega^6 - (c_{11} + c_{22} + c_{33})\Omega^4 + (c_{11}c_{22} + c_{11}c_{33} + c_{22}c_{33} - c_{12}^2 - c_{13}^2 - c_{23}^2)\Omega^2$$
$$+ (c_{11}c_{23}^2 + c_{22}c_{13}^2 + c_{33}c_{12}^2 - c_{11}c_{22}c_{33} - 2c_{12}c_{13}c_{23}) = 0 . \qquad (2.69)$$

Using the explicit form of (2.68) and (2.69), it can be shown that there are three non-degenerate, non-negative roots of the cubic equation (2.69) with respect to Ω^2 for $0 < k \leq \pi$. In the long-wavelength limit, $k \to 0$, the free term and the coefficient of Ω^2 in (2.69) tend to zero. Therefore, two of the three solutions of this equation correspond to the acoustic branches of the dispersion curve. These two roots, $\Omega_l(k)$ and $\Omega_t(k)$ (see Fig. 2.10 left), are related to the longitudinal and torsional molecular oscillations in the helix chain, respectively. The third root gives an optical branch $\Omega_{op}(k)$ corresponding to the transverse oscillations of the molecules. At $k = 0$, we have

$$\Omega_{op} = \sqrt{c_{11} + c_{22} + c_{33}} = 2\sqrt{\sum_j \alpha_j [1 - \cos(j\phi)]^2} .$$

The presence of the two acoustic branches leads to the existence of two speeds of sound: the longitudinal speed v_l and the torsional speed v_t. In the dimensionless form, they can be defined as the limits

$$s_{l,t} = v_{l,t}/v_0 = h \lim_{k \to 0} \frac{\Omega_{l,t}(k)}{k} ,$$

where $v_0 = (K/M)^{1/2} R_0$ is the characteristic velocity of the small-amplitude waves in the helix macromolecule.

For the numerical helix chain dynamics simulation, we take the following parameter values:

$$\phi = 100° , \quad h = 1 , \quad \kappa_1 = 10 , \quad \kappa_2 = 5 , \quad \kappa_3 = 1 . \tag{2.70}$$

The value of the angle ϕ corresponds to the angular spacing of the α-helix protein molecule. The rigidity constants k_1, k_2, and k_3, approximately follow the ratios of the valence bond, angle, and hydrogen bond.

The form of the dispersion curves for the parameter set (2.70) is shown in Fig. 2.10 (left). For $k = 0$, the frequencies are $\Omega_l = \Omega_t = 0$ and $\Omega_{op} = 5.1098$. It follows from (2.68) that there is a value of the wave number $k = k_0 = 1.74795$ for which the free term in the dispersion equation (2.69) becomes zero. For this value, a soft torsional mode emerges $\Omega_t(k_0) = 0$. As can be seen from Fig. 2.10 (left), the frequency spectrum of the helix chain includes a separate optical zone and two acoustic zones, with the frequency spectrum of the torsional oscillations lying inside the frequency spectrum of the longitudinal oscillations. Moreover, the longitudinal velocity of sound $s_l = 3.39475$ significantly exceeds the torsional velocity of sound $s_t = 0.750411$.

2.7.3 Numerical Methods for Finding the Soliton Solution

Here we consider a numerical scheme for finding solitary waves with a stationary profile, where the wave profile is found as the steady state of a certain discrete functional [42]. A necessary condition for the application of this scheme is the smooth dependence of the wave profile on the number of chain links. The main problem in applying such a scheme is finding a discrete functional that is optimal with respect to its numerical realization. To find the narrow soliton solutions, one should use the more complex pseudospectral method, suggested for one-dimensional chains by Eilbeck and Flesch [6], and used for analysis of soliton dynamics in a series of publications [7,47,48]. Once found, the soliton solutions are then used as the initial conditions for the numerical helix chain dynamics simulation. However, we shall demonstrate the absence of narrow solitons in the chain, so there would be no point applying the pseudospectral method, with the associated complications in its numerical realization.

The soliton solution of the equation of motion (2.61) is conveniently analysed in the cylindrical coordinate system in which

$$q_{1n} = (1 + \eta_n) \cos(n\phi + \theta_n) \,,$$
$$q_{2n} = (1 + \eta_n) \sin(n\phi + \theta_n) \,,$$
$$q_{3n} = nh + \beta_n \,,$$

where the variable η_n describes the radial displacement of the nth molecule from the cylinder surface, which spans all the sites of the helix chain at their equilibrium positions. The displacement is positive if a molecule is moving outside the helix and negative if it is moving inside. The second generalized coordinate θ_n describes the azimuthal displacement of the nth molecule with respect to its equilibrium position. The third coordinate β_n is the z coordinate of the displacement.

In terms of the new coordinate system, the Lagrangian of the helix chain has the form

$$\mathcal{L} = \mathcal{L} \left\{ \frac{d\eta_n}{d\tau}, \eta_n; \frac{d\theta_n}{d\tau}, \theta_n; \frac{d\beta_n}{d\tau}, \beta_n \right\}$$

$$= \sum_n \left\{ \frac{1}{2} \left[\left(\frac{d\eta_n}{d\tau} \right)^2 + (1 + \eta_n)^2 \left(\frac{d\theta_n}{d\tau} \right)^2 + \left(\frac{d\beta_n}{d\tau} \right)^2 \right] - \sum_{j=1,2,3} U_j(r_{jn}) \right\} \,, \tag{2.71}$$

where the distance is

$$r_{jn} = |\mathbf{q}_{n+j} - \mathbf{q}_n|$$
$$= \left[(1 + \eta_n)^2 + (1 + \eta_{n+j})^2 - 2(1 + \eta_n)(1 + \eta_{n+j}) \cos(j\phi + \theta_{n+j} - \theta_n) \right.$$
$$\left. + (jh + \beta_{n+j} - \beta_n)^2 \right]^{1/2} \,.$$

The corresponding equations of motion take the form

$$\frac{d^2 \eta_n}{d\tau^2} = (1 + \eta_n) \left(\frac{d\theta_n}{d\tau} \right)^2$$

$$- \sum_{j=1,2,3} \left\{ W_j(r_{j,n-j}) \left[1 + \eta_n - (1 + \eta_{n-j}) \cos(j\phi + \theta_n - \theta_{n-j}) \right] \right.$$

$$\left. + W_j(r_{jn}) \left[1 + \eta_n - (1 + \eta_{n+j}) \cos(j\phi + \theta_{n+j} - \theta_n) \right] \right\} \,.$$

$$\tag{2.72}$$

$$\frac{d^2\theta_n}{d\tau^2} = \frac{1}{1+\eta_n}\left\{ -2\frac{d\eta_n}{d\tau}\frac{d\theta_n}{d\tau} \right.$$

$$+ \sum_{j=1,2,3}\left[W_j(r_{jn})(1+\eta_{n+j})\sin(j\phi+\theta_{n+j}-\theta_n)\right.$$

$$\left.\left. - W_j(r_{j,n-j})(1+\eta_{n-j})\sin(j\phi+\theta_n-\theta_{n-j})\right]\right\} ,$$

$$(2.73)$$

$$\frac{d^2\beta_n}{d\tau^2} = \sum_j\left[W_j(r_{jn})(jh+\beta_{n+j}-\beta_n) - W_j(r_{j,n-j})(jh+\beta_n-\beta_{n-j})\right]. \quad (2.74)$$

We assume that there exists a solution in the form of a wave with a stationary profile: $\eta_n = \eta(nh-s\tau)$, $\theta_n = \theta(nh-s\tau)$, and $\beta_n = \beta(nh-s\tau)$, where $s = v/v_0$ is the dimensionless velocity. As shown in Fig. 2.10 (left), there are three types of linear modes: one optical and two acoustic modes. Therefore, there is no need to take into account the dispersion of the optical mode, and we can approximate the first and second time derivatives of the variable η_n by the simplest finite differences as follows:

$$\frac{d\eta_n}{d\tau} = -s\eta'(n-s\tau) \simeq -s\frac{\eta_{n+1}-\eta_{n-1}}{2h} ,$$

$$\frac{d^2\eta_n}{d\tau^2} = s^2\eta''(n-s\tau) \simeq s^2\frac{\eta_{n+1}-2\eta_n+\eta_{n-1}}{h^2} . \quad (2.75)$$

However, for the longitudinal and torsional displacements, we need to take into account the dispersion arising from the discreteness of the chain. The time derivatives of the displacements θ_n and β_n require the use of a more precise finite-difference approximation. Introducing the relative displacements $\varphi_n = \theta_{n+1}-\theta_n$ and $\rho_n = \beta_{n+1}-\beta_n$, we can rewrite

$$\frac{d\theta_n}{d\tau} = -s\theta'(n-s\tau)$$

$$\simeq -s\left(\frac{\theta_{n+1}-\theta_{n-1}}{2h} - \frac{\theta_{n+2}-3\theta_{n+1}+3\theta_n-\theta_{n-1}}{6h}\right)$$

$$= s(\theta_{n+2}-6\theta_{n+1}+3\theta_n+2\theta_{n-1})/6h$$

$$= s(\varphi_{n+1}-5\varphi_n-2\varphi_{n-1})/6h , \quad (2.76)$$

$$\frac{d^2\theta_n}{d\tau^2} = s^2\theta''(n-s\tau)$$

$$\simeq s^2 \left(\frac{\theta_{n+1} - 2\theta_n + \theta_{n-1}}{h^2} - \frac{\theta_{n+2} - 4\theta_{n+1} + 6\theta_n - 4\theta_{n-1} + \theta_{n-2}}{12h^2} \right)$$

$$= -s^2(\varphi_{n+1} - 15\varphi_n + 15\varphi_{n-1} - \varphi_{n-2})/12h^2 , \tag{2.77}$$

$$\frac{d\beta_n}{d\tau} \simeq s(\beta_{n+2} - 6\beta_{n+1} + 3\beta_n + 2\beta_{n-1})/6h$$

$$= s(\rho_{n+1} - 5\rho_n - 2\rho_{n-1})/6h , \tag{2.78}$$

$$\frac{d^2\beta_n}{d\tau^2} = s^2\beta''(n - s\tau)$$

$$\simeq -s^2(\rho_{n+1} - 15\rho_n + 15\rho_{n-1} + \rho_{n-2})/12h^2 . \tag{2.79}$$

Using the finite-difference approximations (2.75)–(2.79), we rewrite the equations of motion (2.72)–(2.74) as discrete equations for the relative displacements η_n, φ_n, and ρ_n :

$$\mathscr{F}_{n,1} = \frac{s^2}{h^2} \left[\eta_{n+1} - 2\eta_n + \eta_{n-1} - (1 + \eta_n)(\varphi_{n+1} - 5\varphi_n - 2\varphi_{n-1})^2/36 \right]$$

$$+ \sum_{j=1,2,3} \left\{ W_j(r_{j,n-j}) \left[1 + \eta_n - (1 + \eta_{n-j}) \cos \left(j\phi + \sum_{i=1}^{j} \varphi_{n-j+i-1} \right) \right] \right.$$

$$\left. + W_j(r_{jn}) \left[1 + \eta_n - (1 + \eta_{n+j}) \cos \left(j\phi + \sum_{i=1}^{j} \varphi_{n+i-1} \right) \right] \right\} = 0 ,$$

$$\tag{2.80}$$

$$\mathscr{F}_{n,2} = \frac{s^2}{12h^2} \left[(1 + \eta_n)(\varphi_{n+1} - 15\varphi_n + 15\varphi_{n-1} - \varphi_{n-2}) \right.$$

$$\left. + 2(\eta_{n+1} - \eta_{n-1})(\varphi_{n+1} - 5\varphi_n - 2\varphi_{n-1}) \right]$$

$$+ \sum_{j=1,2,3} \left[W_j(r_{jn})(1 + \eta_{n+j}) \sin \left(j\phi + \sum_{i=1}^{j} \varphi_{n+i-1} \right) \right.$$

$$\left. - W_j(r_{j,n-j})(1 + \eta_{n-j}) \sin \left(j\phi + \sum_{i=1}^{j} \varphi_{n-j+i-1} \right) \right] = 0 ,$$

$$\tag{2.81}$$

$$\mathscr{F}_{n,3} - \mathscr{F}_{n-1,3} = \frac{s^2}{12h^2} (\rho_{n+1} - 15\rho_n + 15\rho_{n-1} - \rho_{n-2})$$

$$+ \sum_{j=1,2,3} \left[W_j(r_{jn}) \left(jh + \sum_{i=1}^{j} \rho_{n+i-1} \right) \right.$$

$$\left. - W_j(r_{j,n-j}) \left(jh + \sum_{i=1}^{j} \rho_{n-j+i-1} \right) \right] = 0 .$$

$$(2.82)$$

Equation (2.82) can be integrated once with the result

$$\mathscr{F}_{n,3} = \frac{s^2}{12h^2} (\rho_{n+1} - 14\rho_n + \rho_{n-1})$$

$$+ \sum_{j=1,2,3} \sum_{l=1}^{j} W_j(r_{j,n-j+l}) \left(jh + \sum_{i=1}^{j} \rho_{n-j+l+i-1} \right) = 0 . \quad (2.83)$$

The system of the discrete equations (2.80), (2.81), and (2.83) was solved numerically. Our aim was to find the soliton solutions of this system, i.e., the solutions $\{\eta_n, \varphi_n, \rho_n\}_{n=1}^{N}$ which depend smoothly on the number of chain sites n, and have asymptotic behavior at the chain ends.

The first approximation to the soliton solution can be conveniently found as the minimum of the functional

$$F = \frac{1}{2} \sum_{n=4}^{N-3} \left(\mathscr{F}_{n,1}^2 + \mathscr{F}_{n,2}^2 + \mathscr{F}_{n,3}^2 \right) , \quad (2.84)$$

where N is the number of chain sites. The problem for the conditional minimum

$$\mathscr{F} \rightarrow \min : \quad \eta_n = \varphi_n = \rho_n = 0 , \quad n = 1, 2, 3, N-2, N-1, N , \quad (2.85)$$

was solved numerically using the variable metric method. The initial point was taken in the form of the bell-shaped pulse

$$\eta_n = A_\eta / \cosh^2 \left[\mu(n - N/2) \right] ,$$

$$\varphi_n = A_\varphi / \cosh^2 \left[\mu(n - N/2) \right] ,$$

$$\rho_n = A_\rho / \cosh^2 \left[\mu(n - N/2) \right] ,$$

where the parameter μ describes the reciprocal width and A_η, A_φ, and A_ρ are the amplitudes of the initial approximation to the soliton solution. The number of sites N must chosen to be approximately ten times larger than the soliton width. In this case, the soliton shape will not depend on the boundary conditions. We took $N = 400$, which is appropriate for finding the broad soliton solutions.

Because of surface roughness, corresponding to the functional

$$\mathscr{F} = \mathscr{F}(\eta_4, \ldots, \eta_{N-3}; \varphi_4, \ldots, \varphi_{N-3}; \rho_4, \ldots, \rho_{N-3}),$$

the search for the soliton solution as a minimum of the functional (2.84) leads to a slow convergence of the numerical minimization procedure. Therefore, the final form of the solution was found as a numerical solution of the system of $3(N-6)$ nonlinear equations (2.80), (2.81), and (2.83) with respect to the variables $\{\eta_n, \varphi_n, \rho_n\}_{n=4}^{N-3}$, where $\eta_n, \varphi_n, \rho_n \equiv 0$ for $n = 1, 2, 3, \ldots, N-2, N-1, N$. The system of nonlinear equations was solved numerically using the modified hybrid method (with the standard program from the package MINPACK). The point obtained by solving the constrained minimum problem (2.85) was used as the initial point for this method.

In addition to the velocity s, the soliton solution $\{\eta_n, \varphi_n, \rho_n\}_{n=1}^{N}$ is also characterised by its energy

$$E = \sum_{n=2}^{N-1} \left\{ \frac{s^2}{8h^2} \left[(\eta_{n+1} - \eta_{n-1})^2 + \frac{1}{9}(1 + \eta_n)^2 (\varphi_{n+1} - 5\varphi_n - 2\varphi_{n-1})^2 \right. \right.$$

$$\left. \left. + \frac{1}{9}(\rho_{n+1} - 5\rho_n - 2\rho_{n-1})^2 \right] + \sum_{j=1}^{3} U_j(r_{jn}) \right\},$$

which follows from (2.71), (2.75), (2.76), and (2.78), and the amplitudes are

$$A_\eta = \eta_{n'}, \quad \text{where} \quad |\eta_{n'}| = \max_{1 \leq n \leq N} |\eta_n|,$$

$$A_\varphi = \varphi_{n'}, \quad \text{where} \quad |\varphi_{n'}| = \max_{1 \leq n \leq N} |\varphi_n|,$$

$$A_\rho = \rho_{n'}, \quad \text{where} \quad |\rho_{n'}| = \max_{1 \leq n \leq N} |\rho_n|.$$

The root-mean-square width is

$$L = 2 \left[\sum_{n=1}^{N} (n - n_c)^2 \rho_n / R \right]^{1/2},$$

where

$$R = \sum_{n=1}^{N} \rho_n$$

is the total compression of the helix chain and

$$n_c = \frac{1}{2} + \sum_{n=1}^{N} \frac{n\rho_n}{R}$$

is the position of the soliton center.

2.7.4   Results of Numerical Analysis

Let us find the soliton solutions for the chain with parameters (2.70). The nonlinearity of the dynamics observed in α-chain protein macromolecules is caused mainly by their three-dimensional geometry and the nonlinearity of the soft hydrogen bonds. Therefore, we take into account the nonlinearity of the interaction only between distant neighbors, i.e., we set $\gamma_1 = 0$, $\gamma_2 = 0$, and $\gamma_3 \geq 0$. To distinguish between the effects of the geometrical and physical anharmonicity on the chain nonlinear dynamics, we consider the soliton solutions for the four values of the anharmonicity $\gamma_3 = 0$, 0.1, 1, and 10. For $\gamma_3 = 0$, physical anharmonicity is absent and the nonlinear dynamics is defined by the geometrical anharmonicity of the chain. As the value of γ_3 increases, the geometrical anharmonicity effect decreases, while the physical anharmonicity effect increases.

The numerical analysis performed for the discrete system of (2.80), (2.81), and (2.83) showed that two types of acoustic solitons can exist in the helix chain: solitons of torsion and compression. The torsional soliton is a localized nonlinear packet of torsional phonons, moving with the supersonic velocity $s > s_t$. The compression soliton is a localized nonlinear packet of longitudinal phonons moving with the supersonic velocity $s > s_l$.

The form of the torsional soliton is shown in Fig. 2.10 (right). The soliton has a bell-shaped profile in all three coordinates η_n, φ_n, and ρ_n. In a region of soliton localization, untwisting of the helix chain occurs ($\varphi_n > 0$). The radius of the helix chain decreases ($\eta_n < 0$) and its length increases ($\rho_n > 0$). The existence of the soliton is associated with the geometrical anharmonicity. The physical anharmonicity has the opposite sign and so opposes the formation of the torsion soliton. An increase in the physical anharmonicity parameter, i.e., $\gamma_3 \geq 0$, leads to narrowing of the soliton velocity spectrum. With weak physical anharmonicity ($\gamma_3 = 0, 0.1$), the spectrum of the soliton velocity is $1 < s/s_t < 1.59$. With medium physical anharmonicity ($\gamma_3 = 1$), the spectrum is narrower by a factor of two, viz., $1 < s/s_t < 1.32$, and with strong physical anharmonicity ($\gamma_3 = 10$), the torsional soliton no longer exists in the helix chain (the physical anharmonicity is stronger than the geometrical anharmonicity).

The numerical simulation showed that the soliton's motion is always accompanied by phonon emission (see Fig. 2.10 right). This emission is considerably more intensive at the maximal velocity. As a result of the emission, the energy and the velocity of the soliton steadily decrease (see Fig. 2.11 left). The energy E and the velocity s of the soliton depend on the time τ. To estimate emission intensity, it is

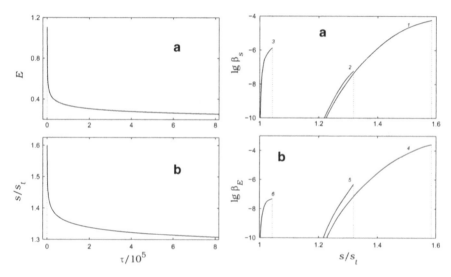

Fig. 2.11 *Left*: Dependence of the energy E and velocity s of the torsion soliton on the dimensionless time τ. The anharmonicity parameter is $\gamma_3 = 0$. *Right*: Dependence of the relaxation coefficients of the velocity β_s and energy β_E of the torsion soliton on its velocity s for $\gamma_3 = 0$ (*lines 1 and 4*) and $\gamma_3 = 1$ (*lines 2 and 5*). Dependence of the relaxation coefficients for the localized breather-like state at $\gamma_3 = 10$ (*lines 3 and 6*)

convenient to define the relaxation coefficients $\beta_s = -s'/s_t$ and $\beta_E = -E'/E$, where the prime denotes differentiation with respect to τ. As the soliton shape is uniquely determined by its velocity s, the relaxation coefficients β_s and β_E also uniquely depend on s. When the soliton velocity s decreases, the intensity of phonon emission tends exponentially to zero and becomes negligibly small at $s = 1.22s_t$ (see Fig. 2.11 right). At velocities $1 < s/s_t < 1.22$, the phonon emission vanishes completely and the soliton moves with a constant velocity.

With an increase in soliton velocity, its energy E and the absolute values of its three amplitudes A_η, A_φ, and A_ρ increase steadily, while its width L decreases monotonically (see Table 2.2). At $s/s_t < 1.2$, its width significantly exceeds the chain spacing that allows the use of the continuum approximation. The discreteness effects of the helix chain become apparent through the noticeable phonon emission at higher velocities. These effects manifest themselves more strongly as the soliton width decreases.

With strong physical anharmonicity ($\gamma_3 = 10$), the torsion soliton does not exist. The numerical simulation showed that, instead of the torsion soliton, there is a localized breather-like excitation (see Fig. 2.12 left), which has a narrow velocity spectrum $1 < s/s_t < 1.043$. The helix chain twisting occurs in the excitation localization region ($\varphi_n < 0$), decreasing its length ($\rho_n < 0$) and increasing its radius ($\eta_n > 0$). The motion of the excitation is also accompanied by phonon emission. With decreasing excitation velocity, the emission intensity tends exponentially to zero (see Fig. 2.11 right).

Table 2.2 Dependence of the energy E, width L, amplitudes A_η, A_φ, and A_ρ, and transmission coefficient p of the torsion soliton in the helix chain on its velocity s for $\gamma_3 = 1$

s/s_t	E	L	A_η	A_φ	A_ρ	p
1.04	0.01016	14.17	−0.0194	0.0363	0.0185	0.643±0.020
1.08	0.03117	10.54	−0.0385	0.0735	0.0364	0.463±0.020
1.12	0.06247	9.25	−0.0576	0.1121	0.0543	0.347±0.018
1.16	0.10585	9.11	−0.0770	0.1531	0.0727	0.269±0.016
1.20	0.16499	9.06	−0.0976	0.1966	0.0916	0.216±0.014

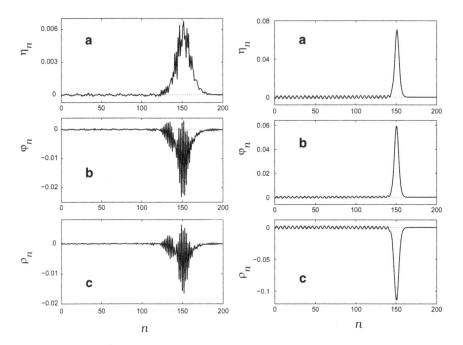

Fig. 2.12 *Left*: The three-component profile of the breather-like excitation at velocity $s = 1.042s_t$ ($\gamma_3 = 10$). *Right*: The three-component profile of the compression soliton at velocity $s = 1.146s_l$ ($\gamma_3 = 1$). Phonon emission by the soliton is clearly visible

The profile of the compression soliton is shown in Fig. 2.12 (right). The soliton has a bell-shaped profile in all three coordinates η_n, φ_n, and ρ_n. In the soliton localization region, compression ($\rho_n < 0$) and slight untwisting of the helix chain ($\varphi_n > 0$) occur, these being accompanied by an increase in the helix chain radius ($\eta_n > 0$). The existence of the compression soliton is associated with the physical anharmonicity. In the absence of physical anharmonicity ($\gamma_3 = 0$), the soliton does not exist. The spectrum of soliton velocities broadens steadily when the anharmonicity parameter is increased. Thus, for $\gamma_3 = 0.1$, the soliton velocity spectrum is $1 < s/s_l < 1.065$, while for $\gamma_3 = 1$, it is twice as broad $1 < s/s_l < 1.143$. A further increase in the anharmonicity parameter does not lead

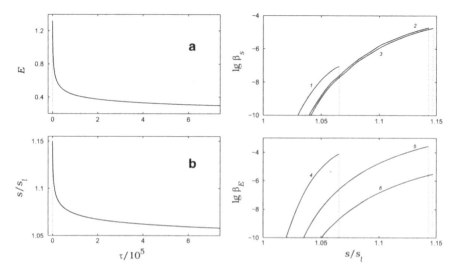

Fig. 2.13 *Left:* Dependence of the energy E and the velocity s on the dimensionless time τ. The anharmonicity parameter is $\gamma_3 = 1$. *Right:* Dependence of the relaxation coefficients of the velocity β_s and the energy β_E for the compression soliton on its velocity s for $\gamma_3 = 0.1$ (*lines 1* and *4*), $\gamma_3 = 1$ (*lines 2* and *5*), and $\gamma_3 = 10$ (*lines 3* and *6*)

to a significant change in the velocity spectrum: the spectrum is $1 < s/s_1 < 1.147$ for $\gamma_3 = 10$.

The numerical simulation of the soliton dynamics showed that the soliton motion is always accompanied by phonon emission (see Fig. 2.12 right). This emission is more intensive at its maximal velocity. As a result of the emission, the energy and the velocity of the soliton both decrease monotonically (see Fig. 2.13 left). We define the velocity relaxation coefficient as $\beta_s = -s'/s_1$. When the soliton velocity s decrease, the phonon emission intensity tends exponentially to zero (see Fig. 2.13 right). As can be seen, phonon emission is completely absent for $s/s_1 < 1.02$ when $\gamma_3 = 0.1$, $s/s_1 < 1.035$ when $\gamma_3 = 1$, and $s/s_1 < 1.05$ when $\gamma_3 = 10$.

When the soliton velocity s decreases, its energy E and the absolute values of all three coordinates A_η, A_φ, and A_ρ increase steadily, while its width L decreases monotonically (see Table 2.3). As can be seen, for $1 < s/s_1 < 1.05$, the soliton width significantly exceeds the chain spacing. Comparing Tables 2.2 and 2.3, we may conclude that the helix chain deformation in the torsion soliton localization region is caused mainly by its untwisting, while in the compression soliton localization region, it is due to its longitudinal compression.

Table 2.3 Dependence of the energy E, width L, amplitudes A_η, A_φ, and A_ρ, and transmission coefficient p of the compression soliton on its velocity s for $\gamma_3 = 1$

s/s_1	E	L	A_η	A_φ	A_ρ	p
1.01	0.02022	20.78	0.0054	0.0049	−0.0093	0.814±0.020
1.02	0.05763	15.18	0.0107	0.0096	−0.0183	0.708±0.020
1.03	0.10670	12.55	0.0158	0.0143	−0.0271	0.627±0.018
1.04	0.16557	11.03	0.0208	0.0188	−0.0358	0.560±0.016
1.05	0.23329	10.03	0.0258	0.0232	−0.0441	0.506±0.014

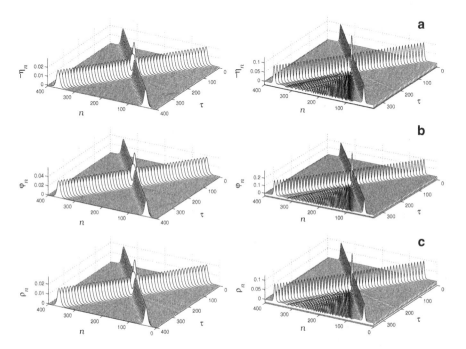

Fig. 2.14 *Left*: Elastic collision of two torsion solitons. The velocities of the first and second solitons are $s_1 = 1.04s_t$ and $s_2 = -1.04s_t$, respectively ($\gamma_3 = 1$). *Right*: Inelastic collision of two torsion solitons. The velocities of the first and second solitons are $s_1 = 1.2s_t$ and $s_2 = -1.2s_t$, respectively ($\gamma_3 = 1$)

2.7.5 Soliton Interaction

At velocities close to the sound velocity s_t, torsion solitons interact with each other as elastic particles. Their collision leads to elastic repulsion without phonon emission and a change of shape (see Fig. 2.14 left). At higher velocities, the soliton interaction becomes inelastic, and the collision is accompanied by phonon emission (see Fig. 2.14 right). Compression solitons moving with a velocity close to the speed of sound s_1 also interact as elastic particles. For other velocity values, the soliton

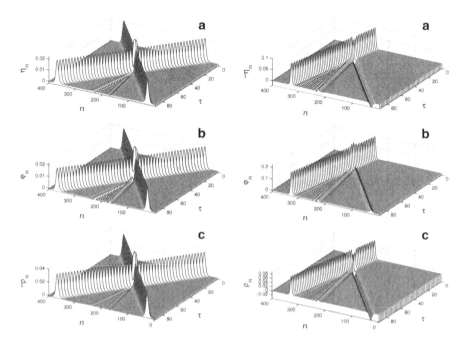

Fig. 2.15 *Left*: Inelastic collision of two compression solitons. The velocities of the first and second solitons are $s_1 = 1.04s_1$ and $s_2 = -1.04s_1$, respectively ($\gamma_3 = 1$). *Right*: Inelastic collision of a torsion soliton (velocity $s_1 = 1.2s_t$) with a compression soliton (velocity $s_2 = -1.04s_1$). The anharmonicity parameter is $\gamma_3 = 1$

repulsion is accompanied by torsional phonon emission (see Fig. 2.15 left). Solitons of different types interact with each other as inelastic particles. Even at velocities close to s_t and s_1, their collision leads to inelastic repulsion followed by phonon emission. The greater the velocity, the more intensive the emission (see Fig. 2.15 right). This way, both types of acoustic solitons interact with each other as elastic particles in the helix molecule. A soliton collision leads to their repulsion, followed by nonessential phonon emission.

2.7.6 Modeling Acoustic Soliton Formation

In finite helix chains, acoustic solitons can be formed by deformation of chain end-links. Here we model this process. Consider the chain dynamics under the torsion deformation of three chain end-links. We integrate the equations of motion (2.72)–(2.74) with fixed end boundary conditions $\eta'_n \equiv 0$, $\theta'_n \equiv 0$, and $\beta'_n \equiv 0$ for $n = 1, \ldots, N$ and the following initial conditions:

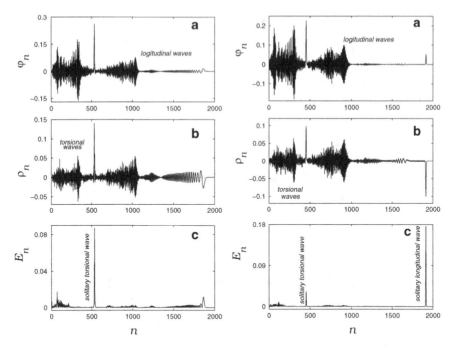

Fig. 2.16 *Left*: Formation of the torsion soliton and two wave packets in the helix chain under untwisting of three chain end-links at the initial time ($\tau = 0$). The amplitude of the initial deformation is $A_\theta = \pi/2$ and the anharmonicity parameter is $\gamma_3 = 0$. The distributions of the relative angular deformation φ_n, relative displacement ρ_n, and energy E_n in the helix chain at time $\tau = 550$ are shown. *Right*: Formation of a torsion soliton, compression soliton, and two wave packets in the helix chain under untwisting of three chain end-links at the initial time ($\tau = 0$). The amplitude of the initial deformation is $A_\theta = \pi/2$ and the anharmonicity parameter is $\gamma_3 = 0$. The distributions of the relative angular deformation φ_n, relative displacement ρ_n, and energy E_n in the helix chain at time $\tau = 550$ are shown

$$\eta_n(0) = 0 , \quad \eta'_n(0) = 0 , \quad \beta_n(0) = 0 , \quad \beta'_n(0) = 0 , \quad n = 1, 2, \ldots, N ,$$
$$\theta_1(0) = \theta_2(0) = \theta_3(0) = -A_\theta , \quad \theta_n(0) = 0 , \quad \theta'_n(0) = 0 , \quad n = 4, \ldots, N ,$$

where A_θ is the amplitude of the torsional deformation and $N = 2{,}000$ is the number of chain sites.

Numerical simulation of the dynamics has shown that initial deformation of the chain end-links leads to the formation of two oscillating wave packets and a torsion soliton in the chain for $\gamma_3 = 0$ (see Fig. 2.16 left). For the deformation amplitude $A_\theta = -\pi/2$, the torsion soliton accumulates more than 20 % of the initial deformation energy. For $\gamma_3 = 1$, the chain end-link deformation leads to the formation of two acoustic solitons (see Fig. 2.16 right). The torsion and compression solitons accumulate 9 and 48 % of initial deformation energy, respectively. The rest of the energy is spent on the formation of a wave packet of torsion solitons.

Table 2.4 Dimensional M_k (given in proton mass) and dimensionless μ_k masses of 20 amino acid residues of an α-helix protein molecule

Notation	Gly	Ala	Val	Leu	Ile	Phe	Pro
M	57	71	99	113	113	147	125
μ	0.474	0.591	0.824	0.940	0.940	1.223	1.040
Notation	Trp	Ser	Thr	Met	Asn	Gln	Cys
M	186	87	101	131	114	128	103
μ	1.548	0.724	0.841	1.090	0.949	1.065	0.857
Notation	Asp	Glu	Tyr	His	Lys	Arg	Average
M	115	129	163	137	128	156	120.15
μ	0.957	1.074	1.357	1.140	1.065	1.298	1

The dynamics simulation carried out showed that the torsional deformation of the three chain end-links can be an effective mechanism of acoustic soliton initiation in a helix macromolecule. The effectiveness of the initiation can exceed 50 %.

2.7.7 Interaction of Solitons with Molecular Chain Heterogeneities

All protein molecules consist of 20 types of amino acids, each with a different mass (see Table 2.4). Therefore, in an α-helix chain macromolecule, the mass of each link depends on its number. Let us consider the acoustic soliton dynamics in a chain with a random distribution of amino acid residues.

Values of the mass $\{M_k\}_{k=1}^{20}$ of amino acid residues are given in Table 2.4. The average mass value is

$$\overline{M} = \frac{1}{20} \sum_{k=1}^{20} M_k = 120.15 m_p \,,$$

where m_p is the proton mass. Introducing the dimensionless masses $\mu_k = M_k/\overline{M}$, the equations of motion (2.72)–(2.74) of the helix chain take the form

$$m_n \frac{d^2 \eta_n}{d\tau^2} = m_n (1 + \eta_n) \left(\frac{d\theta_n}{d\tau}\right)^2$$
$$- \sum_{j=1,2,3} \left\{ W_j (r_{j,n-j}) \left[1 + \eta_n - (1 + \eta_{n-j}) \cos(j\phi + \theta_n - \theta_{n-j}) \right] \right.$$
$$\left. + W_j (r_{jn}) \left[1 + \eta_n - (1 + \eta_{n+j}) \cos(j\phi + \theta_{n+j} - \theta_n) \right] \right\} \,,$$

$$(2.86)$$

$$m_n \frac{d^2\theta_n}{d\tau^2} = \frac{1}{1+\eta_n}\left\{ -2m_n \frac{d\eta_n}{d\tau}\frac{d\theta_n}{d\tau} \right.$$

$$+ \sum_{j=1,2,3}\left[W_j(r_{jn})(1+\eta_{n+j})\sin(j\phi + \theta_{n+j} - \theta_n) \right.$$

$$\left. - W_j(r_{j,n-j})(1+\eta_{n-j})\sin(j\phi + \theta_n - \theta_{n-j}) \right]\right\} ,$$

$$(2.87)$$

$$m_n \frac{d^2\beta_n}{d\tau^2} = \sum_j \left[W_j(r_{jn})(jh + \beta_{n+j} - \beta_n) - W_j(r_{j,n-j})(jh + \beta_n - \beta_{n-j}) \right] ,$$

$$(2.88)$$

where the mass m_n of the nth link can take any of the 20 values $\{\mu_k\}_{k=1}^{20}$ with equal probability.

Let us consider a chain of $N = 2{,}000$ links, where the first and the last 500 chain links have equal masses: $m_n = 1, n = 1,\ldots,500, n = 1{,}501,\ldots,2{,}000$. The masses of the links with numbers $500 < n \leq 1{,}500$ are randomly chosen from the 20 dimensionless values $\{\mu_k\}_{k=1}^{20}$. To model the passage through this inhomogeneous region of the chain by an acoustic soliton, we integrate the equations of motion (2.86)–(2.88) with initial conditions corresponding to the acoustic soliton centered at $n_c = 50$.

The example of the passage of the soliton through the inhomogeneous region of the chain is shown in Fig. 2.17. We observe that the helix chain inhomogeneities

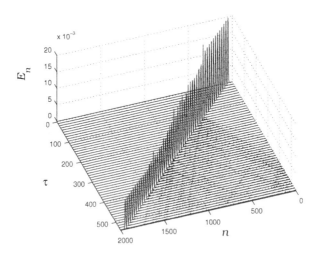

Fig. 2.17 Passage of a compression soliton through an inhomogeneous region of the helix chain. Dependence of energy E_n over the chain on time τ. Initial soliton velocity $s = 1.05s_1$, anharmonicity parameter $\gamma_3 = 1$, and transmission coefficient $p = 0.557$

do not cause the soliton to collapse. The soliton only emits phonons, thus leading to energy loss. We introduce the transmission coefficient $p = E_f/E_i$, where E_i is the initial energy of the soliton and E_f is its energy after passage through an inhomogeneous region of the helix chain. The value of p was estimated from the results of 100 independent computational experiments.

The dependences of the transmission coefficient p on the torsion velocity and the compression solitons are given in Tables 2.2 and 2.3, respectively. We observe that the energy loss grows steadily with the soliton velocity. The compression soliton is more stable with respect to the chain inhomogeneities than the torsion soliton. The simulation shows that the acoustic compression soliton can be an effective energy carrier in an α-helix macromolecule for a distance of up to 1,000 chain spacings.

2.8 Conclusion

The investigation undertaken shows that two types of acoustic solitons can exist in a helix chain: the soliton of torsion and the soliton of compression. The torsion soliton is a localized nonlinear packet of torsional phonons with velocity higher than the velocity of sound of torsional phonons in the helix chain. The soliton describes the motion of a localization region of helix untwisting along the chain (in the soliton localization region, the helix chain stretches and its radius decreases). The compression soliton is a localized nonlinear packet of longitudinal phonons with velocity higher than the longitudinal sound velocity in the helix chain (in the localization region of the soliton, the helix chain also slightly stretches and its radius increases). The existence of the torsion soliton is associated with the geometrical anharmonicity, while the compression soliton is defined by the physical anharmonicity (the molecular interaction anharmonicity) of the helix chain.

Solitons interact with each other as elastic particles. Their collision leads to reflection, followed by slight phonon emission. In a finite chain, the solitons can be formed as a result of the torsional deformation of three chain end-links. The effectiveness of such initiation of the soliton can exceed 50 %.

If the soliton interacts with inhomogeneities in the α-helix chain, this does not cause immediate soliton collapse. The soliton only emits phonons when it passes through the inhomogeneity, leading to energy loss. The compression soliton is more stable with respect to chain inhomogeneities than the torsion soliton. It can serve as an effective energy carrier for a distance of up to 1,000 chain spacings.

References

1. Fermi, E., Pasta, J., Ulam, S.: Studies of nonlinear problems, I. Los Alamos Report LA 1940 (1955)
2. Zabusky, N.J., Deem, G.S.: Dynamics of nonlinear lattices. Localized optical excitations, acoustic radiations and strong nonlinear behavior. J. Comput. Phys. **2**, 207 (1968)

3. Zabusky, N.J.: Nonlinear lattice dynamics and energy sharing. J. Phys. Soc. Jpn. Suppl. **26**, 196 (1969)
4. Zabusky, N.J., Kruskal, M.D.: Interaction of solitons in a collisionless plasma and the recurrence of initial states. Phys. Rev. Lett. **15**(6), 240 (1965)
5. Toda, M., Waddati, M.A.: Solitons and two solitons in an exponential lattice and related equation. J. Phys. Soc. Jpn. **34**, 18 (1973)
6. Eilbeck, J.C., Flesh, R.: Calculation of families of solitary waves on discrete lattices. Phys. Lett. A **149**, 200 (1990)
7. Duncan, D.B., Eilbeck, J.C., Fedderson, H., Wattis, J.A.D.: Solitons on lattices. Physica D **68**, 1 (1993)
8. Toda, M.: Theory of Nonlinear Lattices, 2nd edn. Springer, Berlin (1989)
9. Toda, M.: Vibration of a chain with nonlinear interaction. J. Phys. Soc. Jpn. **22**, 431 (1967)
10. Toda, M.: Wave propagation in anharmonic lattices. J. Phys. Soc. Jpn. **23**, 501 (1967)
11. Toda, M.: Wave in nonlinear lattices. Prog. Theor. Phys. Suppl. **45**, 174 (1970)
12. Toda, M.: Studies of a nonlinear lattice. Phys. Rep. C **18**, 1 (1975)
13. Toda, M.: Solitons and heat conduction. Phys. Scr. **20**, 424 (1979)
14. Mokross, F., Buttner, H.: Thermal conductivity in the diatomic Toda lattice. J. Phys. C Solid State Phys. **16**, 4539 (1983)
15. Jackson, E., Mistriotis, F.: Thermal conductivity of one- and two-dimensional lattices. J. Phys. C Condens. Matter. **1**, 1223 (1989)
16. Newell, A.C.: Solitons in Mathematics and Physics. Society for Industrial and Applied Mathematics, Philadelphia (1985)
17. Narayamurti, V., Varma, S.: Nonlinear propagation of heat pulses in solids. Phys. Rev. Lett. **25**, 1105 (1970)
18. Mertens, F.G., Buttner, H.: Modern Problems in Condensed Matter Sciences, vol. 17. North-Holland, Amsterdam (1986)
19. Gendelman, O.V., Manevich, L.I.: Exact soliton-like solutions in generalised dynamical models of quasi-one-dimensional crystal. Sov. Phys. JETP **102**, 511 (1992)
20. Bishop, A.R.: Solitons and physical perturbations. In: Lanngren, K. and Scott, A. (eds.) Solutions in Action. Academic Press, New York (1978)
21. Collins, M.A.: A quasicontinuum approximation for solitons in an atomic chain. Chem. Phys. Lett. **77**, 342 (1981)
22. Collins, M.A.: Solitons and nonlinear phenomena. Adv. Chem. Phys. **53**, 225 (1983)
23. Kosevich, A.M.: Theory of Crystal Lattice. Vishcha Shkola, Kharkov (1988)
24. Davydov, A.S.: Solitons in Molecular Systems. Naukova Dumka, Kiev (1988)
25. Zabusky, N.J.: Solitons and energy transport in nonlinear lattices. Comput. Phys. Commun. **5**, 1 (1973)
26. Olsen, O.H., Lomdhal, P.S., Kerr, W.C.: Localized excitations in a three-dimensional nonlinear model. Phys. Lett. A **136**, 402 (1989)
27. Christiansen, P.L., Lomdhal, P.S., Muto, V.: On a Toda lattice model with a transversal degree of freedom. Nonlinearity **4**, 477 (1990)
28. Muto, V., Lomdhal, P.S., Christiansen, P.L.: Two-dimensional discrete model for DNA dynamics: longitudinal wave propagation and denaturation. Phys. Rev. A **42**, 7452 (1990)
29. Lomdhal, P.S., Olsen, O.H., Samuelsen, M.R.: Transverse instabilities in a three-dimensional non-linear chain. Phys. Lett. A **152**, 343 (1991)
30. Turitsyn, S.K.: Blow-up in the Boussinesq equation. Phys. Rev. E **47**, R796 (1993)
31. Flytzanis, N., Savin, A.V., Zolotaryuk, Y.: Soliton dynamics in a thermalized molecular chain with transversal degree of freedom. Phys. Lett. A **193**, 148 (1994)
32. Zolotaryuk, Y., Savin, A.V.: Interaction of the acoustic soliton with transversal thermal vibrations of molecules in a chain. Ukr. Fiz. Zh. (Ukr. Phys. J.) **39**, 1051 (1994)
33. Savin, A.V., Zolotaryuk, Y.: The lattice soliton in a chain with orientational thermal vibrations. Phys. Lett. A **201**, 213 (1995)
34. Peyrard, M., Bishop, A.R.: Statistical mechanics of a nonlinear model for DNA denaturation. Phys. Rev. Lett. **62**, 2755 (1989)

35. Peyrard, M., Dauxois, T., Hoyet, H., Willis, C.R.: Biomolecular dynamics of DNA: statistical mechanics and dynamical models. Physica D **68**, 104 (1993)
36. Dauxois, T., Peyrard, M., Bishop, A.R.: Entropy-driven DNA denaturation. Phys. Rev. E **47**, R44 (1993)
37. Yakushevich, L.V.: Nonlinear Physics of DNA. Wiley, New York (1988)
38. Netrebko, N.V., Romanovsky, Y.A., Shidlovskaya, E.G.: Stochastic cluster dynamics of macromolecules. Izvestia Vuzov **2**, 26 (1994)
39. Romanovsky Yu.M.: Some problems of cluster dynamics of biological macromolecules. Lect. Notes Phys. **484**, 140 (1997)
40. Zolotaryuk, A.V., Christiansen, P.L., Savin, A.V.: Two-dimensional dynamics of a free molecular chain with a secondary structure. Phys. Rev. E **54**, 3881 (1996)
41. Manevich, L.I., Savin, A.V.: Solitons of tension in polyethylene molecules. Polym. Sci. Ser. A **38**, 789 (1996)
42. Christiansen, P.L., Savin, A.V., Zolotaryuk, A.V.: Soliton analysis in complex molecular systems: a zig-zag chain. J. Comput. Phys. **134**, 108 (1997)
43. Manevich, L.I., Savin, A.V.: Solitons in crystalline polyethylene: isolated chains in the transconformation. Phys. Rev. E **55**, 4713 (1997)
44. Savin, A.V., Manevich, L.I., Hristiansen, P.L., Zolotaruk, A.V.: Nonlinear dynamics of zigzag molecular chains. Physics-Uspekhi **42**, 245 (1999)
45. Christiansen, P.L., Zolotaryuk, A.V., Savin, A.V.: Solitons in an isolated helix chain. Phys. Rev. E **56**, 877 (1997)
46. Cadet, S.: Transverse envelope solitons in an atomic chain. Phys. Rev. Lett. **121**, 77 (1987)
47. Zolotaryuk, Y., Eilbeck, J.C., Savin, A.V.: Bound states of lattice solitons and their bifurcations. Physica D **108**, 81 (1997)
48. Savin, A.V., Zolotaryuk, Y., Eilbeck, J.C.: Moving kinks and nanopterons in the nonlinear Klein–Gordon lattice. Physica D **138**, 267 (2000)

Chapter 3
Topological Solitons

In general, it is not possible to isolate a molecular chain in physical systems. It is surrounded by other chains, which form its substrate. To consider the chain interaction with its environment, an additional potential $W(x)$ is introduced into the model, describing the interaction of a chain site with its substrate (see Fig. 3.1). If a chain in a system can have several steady states, the substrate potential must possess the same number of minima as well. Furthermore, if the equilibrium states are of equal energy (degenerate states), 'state transfer' can occur in the chain. A topological soliton (kink, antikink) describes the maximum efficient transition of the system from one equilibrium state to another [1, 2].

3.1 Solitons in a Chain with Substrate

We consider here a one-dimensional molecular chain arranged along the x-axis. As previously, let a be the chain spacing, M the mass of a single link, x_n the displacement of the nth site from its equilibrium position, and $V(r)$ the intermolecular potential. Then the Hamiltonian of the chain has the form

$$\mathscr{H} = \sum_{n=-\infty}^{+\infty} \left[\frac{1}{2} M \dot{x}_n^2 + V(r_n) + W(x_n) \right] , \tag{3.1}$$

where $r_n = x_{n+1} - x_n$ and $W(x)$ is the substrate potential. The following potentials are most frequently used as a substrate potential: the periodic sine–Gordon potential

$$W(x) = W_0 \frac{1}{2} \left(1 - \cos \frac{2\pi x}{a} \right) , \tag{3.2}$$

© Springer International Publishing Switzerland 2015
L.N. Lupichev et al., *Synergetics of Molecular Systems*, Springer Series
in Synergetics, DOI 10.1007/978-3-319-08195-3_3

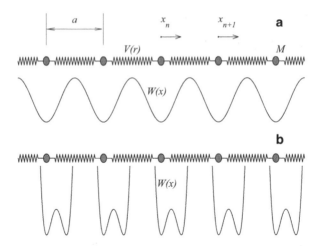

Fig. 3.1 Molecular chain (**a**) with the periodic (the Frenkel–Kontorova model) and (**b**) double-well (the ϕ-4 model) substrate potential $W(x)$

and the double-well ϕ-4 potential

$$W(x) = W_0 \left[1 - \left(\frac{x_n}{d} \right)^2 \right]^2 . \tag{3.3}$$

The model (3.1) with the periodic substrate potential (3.2) was first suggested by Frenkel and Kontorova [3,4] to describe the dislocation dynamics. The period of the substrate potential coincides with chain spacing a, and the potential amplitude W_0 corresponds to the bound energy between a chain link and the chain substrate. The chain has an infinite set of equivalent ground states $x_n \equiv 0, \pm a, \pm 2a, \dots$. A two-dimensional generalization of the Frenkel–Kontorova (FK) model was proposed in [5], where the molecular chain has the shape of a plane zigzag.

The chain with the substrate ϕ-4 potential (3.3) has only two ground states, viz., $x_n \equiv -d$ and $x_n \equiv d$, where $2d$ is the distance between them. This model is used in many areas of physics, and in particular in modeling proton transport along hydrogen bond chains [6–10]. In this case, each chain link can be in one of the two equivalent states: O–H$\cdots$O or O$\cdots$H–O, and the displacement of a proton H along the line of the hydrogen bond O–H$\cdots$O is described by a double-well potential.

For convenience of further calculation we introduce the following dimensionless variables:

- For the Frenkel–Kontorova model, time $\tau = t\sqrt{K/M}$, displacement $u_n = x_n/a$, energy $H = \mathcal{H}/Ka^2$.
- For the ϕ-4 model, displacement $u_n = x_n/2d$ and energy $H = \mathcal{H}/4Kd^2$.

Here $K = V''(0)$ is the rigidity of the intermolecular interaction. Then the chain is described by the dimensionless Hamiltonian

$$H = \sum_{n=-\infty}^{+\infty} \left[\frac{1}{2} u'^2_n + U(\rho_n) + \varepsilon Z(u_n) \right] , \qquad (3.4)$$

where the dimensionless intermolecular potential $U(\rho)$ is normalized by the condition (2.12), and $\rho_n = u_{n+1} - u_n$ are the relative displacements. The dimensionless substrate potential has two minimum states with the same energy, separated by a distance l from each other, and one maximum between them. For the Frenkel–Kontorova model, the potential is

$$Z(u) = \frac{1}{2}[1 - \cos(2\pi u)] , \qquad (3.5)$$

and for the ϕ-4 model it has the form

$$Z(u) = (1 - 4u^2)^2 . \qquad (3.6)$$

The dimensionless parameter ε describes the bond energy between the chain and its substrate.

It is also convenient to introduce the parameter $g = 1/\varepsilon$, describing the system's cooperativity. At $g = 0$, the motion of each particle does not link to the motion of neighboring particles and is defined only by the substrate potential. The greater the cooperativity coefficient g, the stronger the influence of neighboring particles. For the sine–Gordon potential, the cooperativity coefficient is $g = Ka^2/W_0$, while for the ϕ-4 potential (3.3), $g = 4Kd^2/W_0$.

3.1.1   Stationary State of Topological Soliton

In the system considered, the chain can be in at least two equivalent ground states. Let us consider the case when the first half of the chain is in one state and the second is in another. This results in the emergence of a topological defect in the system. It is localized on the boundary of two states, so it is often called a topological soliton.

To find a steady state of a topological soliton, the following minimum problem must be solved:

$$E = \sum_{n=-\infty}^{+\infty} \left[U(\rho_n) + \varepsilon Z(u_n) \right] \longrightarrow \min_{\dots, u_n, u_{n+1}, \dots} , \qquad (3.7)$$

with the boundary conditions

$$\lim_{n \to \pm\infty} u_n = u_{\pm\infty} , \qquad (3.8)$$

which correspond to these states. A topological soliton is called a kink if $u_{-\infty} <$ $u_{+\infty}$ and an antikink if $u_{-\infty} > u_{+\infty}$. The topological charge can also be defined as the difference between the limit displacements $q = u_{-\infty} - u_{+\infty}$.

The nonlinearity in this system results primarily from the substrate potential $Z(u)$. Therefore, the harmonic potential of the intermolecular interaction is commonly used:

$$U(\rho) = \frac{1}{2}\rho^2 . \tag{3.9}$$

As the harmonic potential is an even function, in a chain with this intermolecular interaction potential, the properties of a kink and antikink coincide. Let us study further a chain with this interaction potential. For the Frenkel–Kontorova model (3.4) and (3.5), it is sufficient to consider only a soliton with the topological charge $q = \pm 1$, i.e., take the limiting values $u_{-\infty} = 0$, $u_{+\infty} = -q$, and $u_{-\infty} = q/2$, $u_{+\infty} = -q/2$ for the ϕ-4 model.

It is not possible to obtain an explicit solution of the constrained minimum problem (3.7) and (3.8), but this problem can be solved numerically to any preassigned accuracy. To do this, a solution of the following minimum problem is sought using the conjugate gradient method [59]:

$$\mathcal{E} = \sum_{n=1}^{N-1} \left[gU(\rho_n) + Z(u_n) \right] \longrightarrow \min_{u_2,\dots,u_{N-1}} , \tag{3.10}$$

with the boundary conditions

$$u_1 = u_{-\infty} , \quad u_N = u_{+\infty} , \tag{3.11}$$

where $\mathcal{E} = E/\varepsilon = gE$ is the renormalized energy. The solution of this problem $\{u_n\}_{n=1}^{N}$ will change steadily as the link number n increases. Therefore, it is convenient to describe a steady state of the topological soliton by the position of its centre

$$\overline{n} = \sum_{n=1}^{N-1} \left(n + \frac{1}{2} \right) p_n$$

and the root-mean-square diameter

$$D = 1 + 2 \left[\sum_{n=1}^{N-1} \left(n + \frac{1}{2} - \overline{n} \right)^2 p_n \right]^{1/2} ,$$

where the sequence $p_n = |u_{n+1} - u_n|$ specifies the distribution of the deformation in the chain ($\sum_n p_n = 1$). In order to ensure that the boundary conditions (3.11) do not affect the shape of the soliton, it will suffice to solve the problem taking the number of chain links N to be ten times greater than the soliton width D.

The substrate potentials (3.5) and (3.6) are symmetrical with respect to a local minimum point. Therefore, the minimization problem will have two types of soliton solution: one solution with the half-integer centre of symmetry $\bar{n} = n_0 + 1/2$ (for the ϕ-4 model the displacements $u_{n_0+k} = -u_{n_0+1-k}$, $k = 0, 1, 2, \ldots$) and another solution with the integer centre $\bar{n} = n_0$ ($u_n n_0 = 0$, $u_{n_0+k} = -u_{n_0-k}$, $k = 1, 2, \ldots$). Only the solution with the half-integer centre is stable. In the absence of cooperativity ($g = 0$), the solution with the half-integer centre has the form

$$u_n = u_{-\infty}, \quad \text{for } n = n_0, n_0 - 1, n_0 - 2, \ldots,$$

$$u_n = u_{+\infty}, \quad \text{for } n = n_0 + 1, n_0 + 2, \ldots.$$

This is the ground state with energy $E = 0$ and width $D = 1$. The solution with the integer centre, viz.,

$$u_n = u_{-\infty}, \quad \text{for } n = n_0 - 1, n_0 - 2, \ldots, \quad u_{n_0} = \frac{1}{2}(u_{-\infty} + u_{+\infty}),$$

$$u_n = u_{+\infty}, \quad \text{for } n = n_0 + 1, n_0 + 2, \ldots,$$

is unstable. Its renormalized energy and width are $\mathscr{E} = 1$ and $D = 2$, respectively. The difference between these two states is $\Delta\mathscr{E} = 1$.

The solutions of the minimum problem (3.10) and (3.11) for the ϕ-4 model at $N = 100$ and $g = 10$ with the different types of symmetry are shown in Fig. 3.2a (left). The solution with the half-integer centre is always the energetic ground state for all values of the cooperativity coefficient g. Its energy $E_{1/2}$ is less than that of the steady state of the soliton with an integer centre. If in solving the minimum problem the displacement $u_{N/2}$ of a central particle is fixed, it is possible to find the dependence of the topological soliton energy E on the position of its centre $\bar{n}$. (For a monotonic change in the value $u_{N/2}$, the soliton centre changes. Using this, the dependence of $u_{N/2}$ on $\bar{n}$ can be obtained, and from this the dependence of $E(\bar{n})$.) As can be seen from Fig. 3.2b (left), the profile of the Peierls potential $E(\bar{n})$ is a sine-like periodic function with unit period. The minima of the Peierls potential are associated with the steady states of the soliton with a half-integer centre, while the maxima are associated with the states with an integer centre of symmetry. The amplitude of the Peierls potential corresponds to the difference between these states, viz., $\Delta E = E_0 - E_{1/2}$.

For $g = 0$, we have the amplitude of the Peierls potential $\Delta\mathscr{E} = g\Delta E = 1$, the energy of the ground state of the topological soliton $E_{1/2} = 0$, and the width $D = 1$. With an increase in the cooperativity parameter g, the renormalized energy $\mathscr{E}$ and the width of the soliton grow as $\sqrt{g}$ (see Fig. 3.2a, b right), and the amplitude of the Peierls potential tends exponentially to zero (see Fig. 3.2c right).

The Peierls potential profile resists the motion of the topological defect [11]. The defect is in the pinning state. If it is to start moving, it must overcome the potential

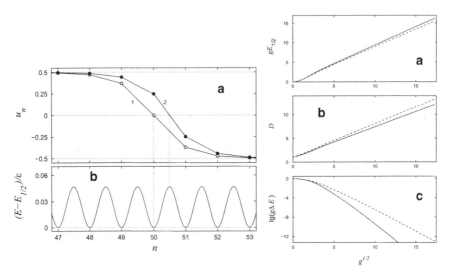

Fig. 3.2 *Left*: Steady state of the topological soliton (antikink) in the ϕ-4 chain, $\varepsilon = 0.1$ ($g = 10$) with the integer and half-integer centres of symmetry, *lines 1* and *2* (**a**), and the corresponding Peierls potential profile $E(n)$ (**b**). *Right*: Dependencies of the renormalized energy $\mathscr{E}_{1/2} = gE_{1/2}$ (**a**), width of the steady ground state of the topological soliton D (**b**), and the logarithm of the renormalized amplitude of the Peierls potential $\lg \Delta\mathscr{E} = \lg(g\Delta E)$ (**c**) on the square root of the cooperativity coefficient $\sqrt{g}$. The dependencies for the discrete ϕ-4 model and the sine–Gordon model are shown by *solid* and *dashed lines*, respectively

profile threshold, and the motion itself will be accompanied by phonon emission. With an increase in system cooperativity, the pinning energy tends exponentially to zero. As can be seen from Fig. 3.2c (right), the pinning actually disappears for $g > 100$ ($\varepsilon < 0.01$). For a strong cooperativity, the topological defect can already move as a solitary wave with a constant subsonic velocity. Therefore, the defect is often called the topological soliton.

3.1.2 Interaction of Topological Solitons

To describe soliton interactions, we will derive the potential of their pair interaction, i.e., the dependence of the energy of soliton steady states on the distance between their centres.

In the ϕ-4 model only the opposite-sign solitons (kink and antikink) can interact. In order to find the energy of a pair of opposite-sign solitons, the following constrained minimum problem must be solved:

$$
gE = \sum_{n=1}^{N-1} \left[\frac{1}{2} g(u_{n+1} - u_n)^2 + Z(u_n) \right] \longrightarrow \min_{u_2,\dots,u_{N-1}} : \ u_1 = u_N = -1/2 ,
$$

$$(3.12)$$

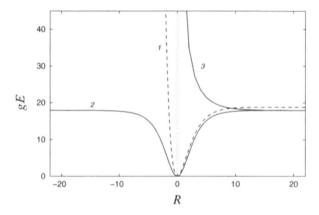

Fig. 3.3 Dependence of the renormalized energy gE of a pair of solitons on the distance between their centres R (cooperativity coefficient $g = 100$). *Line 1*: dependence for opposite-sign solitons in the ϕ-4 chain. *Lines 2* and *3*: dependence for opposite-sign and same-sign solitons in the sine–Gordon chain, respectively

for a fixed displacement of the central particle $u_{N/2}$. The solution of the problem (3.12) corresponds to the homogeneous state $u_n \equiv -1/2$ at $u_{N/2} = -1/2$ and noninteracting kink–antikink pair at $u_{N/2} = 1/2$.

Let $\{u_n\}_{n=1}^N$ be the solution of the problem (3.12) with energy E. The distance between solitons can be defined as

$$R = \sum_{n=1}^{N}(u_n + 1/2) .$$

The distance R depends continuously on $u_{N/2}$. We have $R < 0$ for $u_{N/2} < 0$, $R = 0$ for $u_{N/2} = 0$, and $R \to +\infty$ as $u_{N/2} \to 1/2 - 0$.

To be specific, let us take the cooperativity parameter $g = 100$ ($\varepsilon = 0.01$). The dependence of the renormalized energy gE of two opposite-sign solitons in the ϕ-4 chain on the distance R between them is shown in Fig. 3.3. As can be seen from this figure, the soliton interaction is attractive. The energy is $E = 0$ at $R = 0$, and it tends monotonically to the doubled energy of the isolated steady state as $R \to +\infty$, and to ∞ as $E \nearrow +\infty$. This form of the interaction potential is due to the inability of topological solitons in the ϕ-4 chain to pass through each other – they can only either recombine or repel one another other.

In the sine–Gordon model, both opposite-sign and same-sign solitons can interact. To find the energy of opposite-sign solitons, one must solve the constrained minimum problem (3.12) with boundary conditions $u_1 = u_N = 0$. The interaction potential is shown in Fig. 3.3. This potential is a finite symmetrical function with minimum at $R = 0$. Under these circumstances, opposite-sign solitons can already pass through each other. To find the energy of same-sign solitons, it is essential to solve the constrained minimum problem (3.12) with the boundary conditions $u_1 = 0$

and $u_N = 2$. Fixing the displacement of soliton centres, the dependence $E(R)$ can be obtained. As can be seen from Fig. 3.3, the energy tends steadily to infinity as $R \searrow 0$. Therefore, the interaction of same-sign solitons corresponds to their mutual repulsion.

3.1.3 Soliton Dynamics

The Hamiltonian (3.4) gives the equations of motion

$$u_n'' = u_{n+1} - 2u_n + u_{n-1} - \varepsilon Z_u(u_n) , \quad n = 0, \pm 1, \pm 2, \ldots . \tag{3.13}$$

For strong cooperativity, the relatively broad soliton justifies the use of the continuum approximation. We seek a soliton solution of (3.13) in the form of a solitary wave $u_n(\tau) = u(\xi)$, smoothly dependent on n, where $\xi = n - s\tau$ is the wave variable and s is the wave velocity. For the ϕ-4 and sine–Gordon models, the equations of motion in the continuum approximation have an explicit solution for $s < 1$. In general, it is convenient to search for a soliton solution numerically. For this purpose, one must move from a continuous derivative with respect to time to a discrete derivative with respect to n: $u_n'(\tau) = -s(u_{n+1} - u_n)$. Then, a soliton solution will correspond to the minimum of the discrete Lagrangian

$$L = \sum_n \left[\frac{1}{2}(1 - s^2)(u_{n+1} - u_n)^2 + \varepsilon Z(u_n) \right] .$$

To find a soliton form, one must solve the minimum problem

$$L = \sum_{n=1}^{N-1} \left[\frac{1}{2}(1 - s^2)(u_{n+1} - u_n)^2 + \varepsilon Z(u_n) \right] \longrightarrow \min_{u_2,\ldots,u_{N-1}} : u_1 = u_{-\infty}, \ u_N = u_{+\infty} .$$

$$\tag{3.14}$$

For strong cooperativity, the topological soliton has the subsonic velocity spectrum $0 \le s < 1$. When the velocity increases ($s \nearrow 1$), the soliton energy increases steadily ($E \nearrow \infty$), while its width decreases monotonically ($D \searrow 1$).

The interaction of opposite-sign solitons in the continuous ϕ-4 model has been treated in numerous studies [12–20]. The interaction of kink and antikink in their collision was first investigated numerically in [12]. It was shown that the soliton collision at velocity $s = 0.1$ leads to formation of a bound state (breather) which is a weakly decaying, autolocalized nonlinear vibration. An analytical expression describing the breather dynamics with appropriate accuracy was found in [13], and its half-life was estimated to be such that it could be referred to as a long-lived state. The breather forms at a collision velocity s not exceeding a certain threshold velocity s_c. For $s > s_c$, there is inelastic reflection of solitons, accompanied by phonon emission.

Further study [14–20] showed that the interaction of opposite-sign solitons exhibits a resonant behavior. In the value range $s < s_c$, there exist regions of velocities in which solitons are inelastically reflected. These regions interchange with regions of velocities for which the soliton collision leads to breather formation. An analysis of this phenomenon was carried out in [20]. For $s < s_c$, solitons converge and diverge several times before they completely diverge (the number of collision resonances). This allows us to consider resonance interaction of solitons at certain collision velocities. The resonance phenomenon is also observed in the case of the modified sine–Gordon equation [21] and double sine–Gordon equation [22].

For strong cooperativity of the intermolecular interaction, all these phenomena are observed in the discrete ϕ-4 model as well. To illustrate this, we consider a soliton collision in the ϕ-4 chain for the cooperativity coefficient $g = 100$ ($\varepsilon = 0.01$). We integrate the equations of motion (3.13) with $n = 1, 2, \ldots, N$, taking the initial condition which corresponds to a kink–antikink pair moving towards each other with velocity s, and introducing viscous friction at the chain ends to ensure absorption of emitting phonons (thereby excluding the influence of boundary conditions on the topological soliton dynamics). Inelastic reflection of solitons is shown in Fig. 3.4 (left). As can be clearly seen in this figure, at velocity $s = 0.3$, the soliton collision leads to their reflection, followed by excitation of internal vibrations and slight phonon emission. As a result of energy loss, their velocities decrease (after collision they already diverge with velocities $s = 0.16$).

Therefore, the soliton collision is accompanied by energy loss through excitation of the soliton's internal modes. As can be seen from the static interaction potential $E(R)$ of the solitons (see line 1 in Fig. 3.3), at small collision velocities, the rest of the energy is not sufficient to overcome the interaction energy of opposite-sign solitons. Indeed, for $s = 0.05$, the collision already causes the formation of a bound state of solitons, a low frequency breather. As Fig. 3.4 (right) illustrates, the collision itself is accompanied by energy loss through intensive phonon emission.

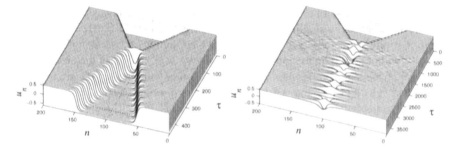

Fig. 3.4 *Left:* Nonelastic reflection of opposite-sign solitons in the ϕ-4 chain ($\varepsilon = 0.01$). The soliton velocity is $s = 0.3$ before the collision and $s = 0.16$ afterwards. Viscous friction at the chain ends was introduced in order to ensure absorption of emitting phonons. *Right:* Formation of the low frequency breather upon the collision of opposite-sign topological solitons in the ϕ-4 chain ($\varepsilon = 0.01$). The soliton velocity is $s = 0.05$. The chain length is $N = 200$. Viscous friction at the ends of the chain was introduced in order to ensure absorption of emitting phonons

After the collision, long-lived localized nonlinear vibrations form in the chain. The frequency of this vibration lies below the lower limit of the frequency spectrum of linear vibrations (phonons) in the chain and approaches this limit when the vibration damps. Low frequency breathers, such as topological solitons, can move along the chain with subsonic velocities.

Note that, in the ϕ-4 chain, in addition to the low-frequency breathers, there exist discrete high-frequency breathers – the highly localized vibrations of the chain. The frequencies of these nonlinear vibrations lie above the upper limit of the phonon frequency spectrum and their energy significantly exceeds the interwell barrier height of the ϕ-4 potential. In the ϕ-4 model, the discrete breather is essentially a localized high-frequency vibration of a single particle in the substrate potential with energy $E \gg \varepsilon$. The existence of the breathers is associated with positive anharmonicity of the substrate potential for the deformation $|u_n| > 1$. In the sine–Gordon model, for which the potential has a negative anharmonicity, only the low-frequency breathers exist.

The discrete breathers, or intrinsic localized modes, are localized, periodic stable vibrations of the nonlinear discrete system. Intensive investigation of the discrete breathers began in 1988 with the pioneering work of Sievers and Takeno [23]. The existence of discrete breathers is ensured by a theorem proven in [24, 25], as well as a large number of numerical studies (see the review [26]). Today, their role in the mechanisms of energy transfer and relaxation in molecular systems has become quite clear [27–29].

The continuous sine–Gordon model is a completely integrable system. All topological solitons have a subsonic band of velocities and elastically interact with each other without any changes to their shapes and velocities, as well as not emitting phonons. Opposite-sign solitons pass through one another, while same-sign solitons reflect from each other (this interaction scenario corresponds to the shape of the soliton interaction potential, see Fig. 3.3). There exist different discretizations of the continuous sine–Gordon model [30, 31] which are also integrable systems. However, the discrete sine–Gordon chain is not an integrable system for any values of the cooperativity coefficient [11]. In this case, the Peierls potential always exists.

Here, we choose the amplitude of the substrate parameter $\varepsilon = 0.01$, corresponding to a strong cooperativity $g = 100$, for which solitons move along the chain virtually without any phonon emission. At velocity $s = 0.5$, as in the continuous model, opposite-sign solitons pass through one another without any changes to their shapes and velocities (see Fig. 3.5 left). The system discreteness manifests itself in an inelastic interaction of solitons at small velocities. At velocity $s = 0.02$, soliton collision already leads to the formation of breathers (see Fig. 3.5 right). Same-sign solitons repel each other, so their collision causes elastic reflection (see Fig. 3.6).

Note that, in a thermalized chain, the properties of the soliton interaction can change. In fact, in the thermalized ϕ-4 chain, an indirect reflection of opposite-sign solitons occurs through the thermal phonons and the resulting interaction has longer range than the static attraction [32]. As a result, in the thermalized chain, opposite-sign solitons attract each other only at short range and repel one another at longer ranges. In a chain with weak cooperativity, the Peierls potential profile prevents the

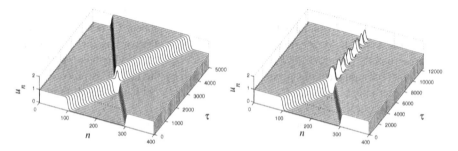

Fig. 3.5 *Left*: Collision of same-sign topological solitons in the sine–Gordon chain ($\varepsilon = 0.01$). The soliton velocity is $s = 0.5$ and the length of the chain $n = 400$. *Right*: Formation of a low-frequency breather as a result of the collision of opposite-sign solitons in the sine–Gordon chain ($\varepsilon = 0.01$). Velocity of the soliton $s = 0.02$ and length of chain $n = 400$

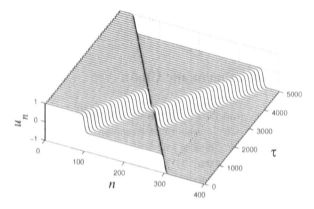

Fig. 3.6 Inelastic reflection of topological solitons of opposite signs in the sine–Gordon chain ($\varepsilon = 0.01$). Soliton velocity $s = 0.5$ and length of chain $N = 400$

solitons from approaching each other. In this case, the interaction of opposite-sign solitons can lead to the formation of soliton bound states [33].

3.1.4 Supersonic Regimes of Topological Soliton Motion

In the absence of anharmonicity of the site–site interaction, the topological solitons of opposite signs possess the same properties: they can only have a continuous supersonic velocity spectrum. With the appearance of cubic anharmonicity in the site–site interaction, the situation changes dramatically. The properties of the topological solitons of opposite signs become different. If the site–site interaction has positive anharmonicity, i.e., $\alpha = -U^{(3)}(0) > 0$, the negative soliton (kink) is narrower than the positive soliton (antikink). The pinning energy of the negative soliton is greater, while its mobility is less than that of the positive soliton. For strong cooperativity and increased anharmonicity, the velocity spectrum of the negative soliton becomes narrower and the spectrum even disappears when the velocity approaches a threshold, whereas the positive soliton always has the continuous subsonic velocity spectrum [34]. Note that the lack of symmetry in the properties of the solitons of opposite signs is entirely determined by the lack of symmetry in the site–site interaction, so the quartet anharmonicity does not cause any difference in the soliton properties.

The positive soliton can also travel at supersonic velocity [35]. The supersonic spectrum has a discrete structure [36, 47]. There exists only a finite number of supersonic velocity values $s_1 > s_2 > \ldots > s_N$, for which the antikink motion is not accompanied by phonon emission. The supersonic kink, corresponding to the nth velocity value s_n, is a bound state of n acoustic solitons, and in this case the sum of their amplitudes must coincide with the width of the substrate potential barrier. The number N of supersonic velocity values increases when the anharmonicity parameter of the site–site interaction is increased.

3.2   Solitons in an Anharmonic Chain

Let us consider how the properties of a topological soliton are modified when the anharmonicity of the site–site interaction is changed. The chain model (3.4) with a harmonic potential for the site–site interaction $U(\rho)$ and the periodic substrate potential (3.5) was first suggested by Frenkel and Kontorova [3, 4] to describe dislocation dynamics. A chain model with the FPU substrate potential (3.5) is a natural generalization of the Frenkel–Kontorova model.

3.2.1   Stationary Soliton State

The Hamiltonian of the chain can be written in the dimensionless form

$$H = \sum_{n=-\infty}^{+\infty} \left[\frac{1}{2} u_n'^2 + g U(\rho_n) + (1 - \cos u_n) \right] , \qquad (3.15)$$

where the prime denotes differentiation with respect to the dimensionless time $\tau = t\sqrt{2\pi^2 W_0/Ma^2}$, $g = a^2 K/2\pi^2 W_0$ is the dimensionless cooperativity parameter, $U(\rho)$ is the dimensionless potential of the site–site interaction, normalized according to the condition $d^2 U/d\rho^2|_{\rho=0} = 1$, and $\rho_n = u_{n+1} - u_n$ is the relative displacement. We take the potential in the form

$$U(\rho) = \frac{1}{2}\rho^2 - \frac{1}{3}\alpha\rho^3 + \frac{1}{4}\beta\rho^4 , \qquad (3.16)$$

where $\alpha \geq 0$ and $\beta \geq 0$ are the nonlinearity parameters. For $\beta = 0$, the potential has only the cubic anharmonic term (the FPU-α potential), while for $\alpha = 0$, it includes the quartet anharmonic term alone (the FPU-β potential).

The chain under consideration has infinitely many ground states $\{u_n = 2\pi n\}_{n=-\infty}^{+\infty}$. We focus on the case where the first half of the chain is in one state and the second is in another. As a result, a topological defect (soliton) emerges in the system, localized on the boundary of two states. To find the stationary state of the topological defect, one must solve the minimum problem

$$E = \sum_{n=-\infty}^{+\infty} \left[gU(\rho_n) + (1 - \cos u_n) \right] \longrightarrow \min_{\ldots, u_n, u_{n+1}, \ldots} , \qquad (3.17)$$

with the boundary conditions

$$\lim_{n \to \pm\infty} u_n = u_{\pm\infty} , \qquad (3.18)$$

which correspond to these two states. A topological defect (soliton) is called a kink if $u_{-\infty} > u_{+\infty}$ and an antikink if $u_{-\infty} < u_{+\infty}$. A kink is the soliton of chain compression and an antikink is the soliton of chain extension. The topological charge of a soliton can be defined as the difference between the limit displacements, viz., $q = (u_{+\infty} - u_{-\infty})/2\pi$. Thus, kink and antikink have negative $q = -1$ and positive $q = +1$ charges, respectively.

It is impossible to obtain an analytical solution of the minimum problem (3.17). A comprehensive review of analytical methods is given in the book by Braun and Kivshar [37, 38]. An analytical study can be conducted only by using approximate methods. For strong cooperativity ($g \gg 1$), the continuum approximation can be used, while in the case of weak cooperativity, the variational approach is applicable. Meanwhile, explicit results can be obtained only numerically by solving, to any given accuracy, the constrained minimum problem (3.17) with the boundary condition (3.18). Using the method of conjugate gradients [59], we seek a solution of the minimum problem

$$E = \sum_{n=1}^{N-1} \left[gU(\rho_n) + (1 - \cos u_n) \right] \longrightarrow \min_{u_2, \ldots, u_{N-1}} , \qquad (3.19)$$

with the boundary conditions

$$u_1 = u_{-\infty} , \qquad u_N = u_{+\infty} . \qquad (3.20)$$

The solution of this problem depends monotonically on n. Therefore, for the solution (the stationary state of a topological soliton), the position of the soliton centre can be determined as

$$\bar{n} = \sum_{n=1}^{N-1} \left(n + \frac{1}{2} \right) p_n ,$$

and its diameter as

$$D = 1 + 2 \left[\sum_{n=1}^{N-1} \left(n + \frac{1}{2} - \bar{n} \right)^2 p_n \right]^{1/2} ,$$

where the sequence $p_n = |u_{n+1} - u_n|/2\pi$ defines the distribution of chain deformation ($\sum_n p_n = 1$). In order to avoid the boundary conditions (3.20) influencing the soliton shape when we solve the minimum problem (3.19), it suffices to take the number of chain links N to be ten times the soliton width D.

Let us consider a chain with the FPU-β potential, i.e., with cubic nonlinearity parameter $\alpha = 0$. This potential is an even function, so the deformation energy will depend only on the deformation amplitude, but not the deformation sign. As a result, a kink and antikink will possess the same properties. To be specific, we consider here only the kink when $u_{-\infty} = 2\pi$ and $u_{+\infty} = 0$.

The periodic potential of the substrate is a symmetric function with respect to all of its maxima. Thus, the maximum problem has two types of soliton solutions: the solution with a half-integer centre of symmetry $\bar{n} = n_0 + 1/2$ and the solution with an integer centre $\bar{n} = n_0$ ($u_{n_0} = \pi$, $u_{n_0+k} = 2\pi - u_{n_0-k}$, $k = 1, 2, \ldots$). In the absence of cooperativity ($g = 0$), the solution with a half-integer center has the form

$$u_n = 2\pi , \quad \text{for } n = n_0, n_0 - 1, n_0 - 2, \ldots ,$$

$$u_n = 0 , \quad \text{for } n = n_0 + 1, n_0 + 2, \ldots .$$

This solution is the ground state with energy $E = 0$ and width $D = 1$. The solution with an integer centre, viz.,

$$u_n = 2\pi , \quad \text{for } n = n_0 - 1, n_0 - 2, \ldots , \quad u_{n_0} = \pi ,$$

$$u_n = 0 , \quad \text{for } n = n_0 + 1, n_0 + 2, \ldots ,$$

is unstable. Its energy and width are $E = 1$ and $D = 2$, respectively. The energy difference between these states is $\Delta E = 1$.

In the case of the FK model, i.e., when $\alpha = \beta = 0$, for any cooperativity $g \geq 0$, only the solution with a half-integer centre is stable. For the FPU model, this state can already be unstable and the solution with an integer centre can become a stable solution.

The solutions of the minimum problem (3.19) and (3.20) with different types of symmetry and $N = 100$, $g = 1$, $\alpha = 0$, and $\beta = 1$ are shown in Fig. 3.7. If in solving the minimum problem we fix the displacement of the central particle $u_{N/2}$,

Fig. 3.7 Stationary state of the topological soliton (kink) in the chain with $g = 1$, $\alpha = 0$, and $\beta = 1$. *Line 1*: half-integer centered soliton (energy $E = 10.042\,585\,94$ and width $D = 4.50$). *Line 2*: integer centered soliton, $\bar{n} = 51$ ($E = 10.042\,585\,89$ and $D = 4.50$)

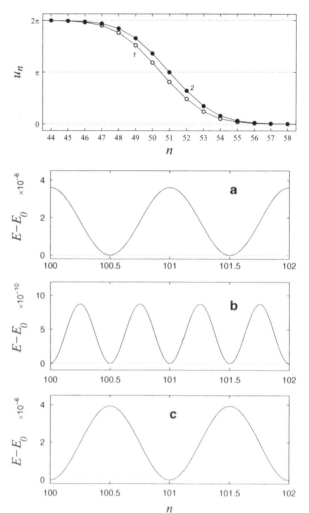

Fig. 3.8 Peierls potential profile $E(\bar{n}) - E_0$ for the kink in the chain with $\alpha = 0$ and $\beta = 2$ when the cooperativity parameter has value $g = 0.34$ (**a**), $g = 0.348\,333\,714$ (**b**), and $g = 0.36$ (**c**)

we can find the dependence of the energy of the topological soliton E on the position $\bar{n}$ of its centre. Indeed, with a monotonic change in the value of $u_{N/2}$, the soliton centre $\bar{n}$ also changes monotonically. Using this fact, we can obtain the dependence of $u_{N/2}$ on $\bar{n}$ and thus the dependence $E(\bar{n})$. As can be seen from Fig. 3.8, the Peierls potential $E(\bar{n}) - E_0$ is a sine-like function with unit period equal (E_0 is the kink ground state energy). The minima of the Peierls potential correspond to the stationary state of the kink, while the maxima correspond to its unstable state. When

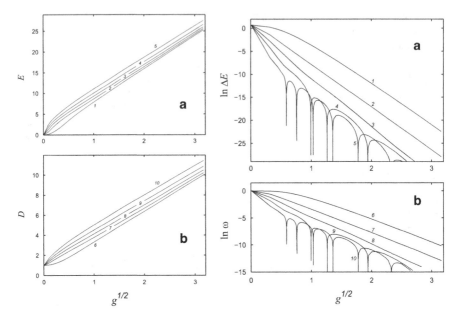

Fig. 3.9 *Left*: Dependence of the energy E (**a**) and width D (**b**) of the stationary kink on the square root of the cooperativity parameter $g^{1/2}$ for nonlinearity values $\alpha = 0$ and $\beta = 0$ (*lines 1* and *6*), 0.2 (*lines 2* and *7*), 0.5 (*lines 3* and *8*), 1 (*lines 4* and *9*), 2 (*lines 5* and *10*). *Right*: The dependence of the logarithm of the Peierls potential amplitude $\ln \Delta$ (**a**) and the logarithm of the lowest eigenfrequency of the kink $\ln \omega$ (**b**) on the square root of the cooperativity parameter $g^{1/2}$ for nonlinearity values $\alpha = 0$ and $\beta = 0$ (*lines 1* and *6*), 0.2 (*lines 2* and *7*), 0.5 (*lines 3* and *8*), 1 (*lines 4* and *9*), and 2 (*lines 5* and *10*)

the nonlinearity $\beta = 2$ and the cooperativity $g = 0.34$, only the kink with a half-integer centre is stable, and when $g = 0.36$, the kink with an integer centre becomes stable. At an intermediate value $g = 0.348\,333\,714$, both types of stationary states are stable and they have the same energy, although in this case the Peierls potential amplitude is $\Delta E = \max |E(\bar{n}) - E_0| > 0$.

Figure 3.9 (left) shows the dependence of the energy E and width D of the stationary kink on the cooperativity parameter g, calculated for various values of the nonlinearity parameter β. As can be seen in this figure, for all values of $\beta \geq 0$, the energy and width of the kink increase steadily as $\sqrt{g}$ with increasing g. The coefficients of proportionality

$$e_1 = \lim_{g \to \infty} \frac{E}{\sqrt{g}}, \qquad d_1 = \lim_{g \to \infty} \frac{D}{\sqrt{g}},$$

are virtually independent of the nonlinearity parameter β. Only the following terms in the asymptotic expansion depend on β:

$$e_2 = \lim_{g \to \infty} \left(E - e_1 \sqrt{g} \right), \qquad d_1 = \lim_{g \to \infty} \left(D - d_1 \sqrt{g} \right).$$

The energy and width of the topological soliton increase steadily with increasing nonlinearity.

The dependence of the pinning energy ΔE on the model parameters g and β is more complex (see Fig. 3.9 right). When $g = 0$, we have the amplitude of the Peierls potential $\Delta E = 1$, the ground state energy of the topological defect $E_{1/2} = 0$, and its width $D = 1$. Increasing the cooperativity parameter g, the amplitude of the Peierls potential tends exponentially to zero (see Fig. 3.9a right). As can be seen in this figure, for weak nonlinearity $\beta \leq 0.5$, the amplitude ΔE decreases monotonically. In this case, the stationary state with half-integer centre always remains as the (stable) ground state. The situation changes for stronger nonlinearity. In fact, the amplitude ΔE has local minima at $g = g_1$, g_2, and g_3 for $\beta = 1$, where

$$g_1 = 0.988\,654 < g_2 = 1.598\,386 < g_3 = 3.762\,3 ,$$

and minima at $g = g_1, \ldots, g_6$ for $\beta = 2$, where

$$g_1 = 0.348\,333\,714 < g_2 = 0.579\,470\,53 < g_3$$
$$= 1.065\,449\,51 < g_4 = 1.844\,469 < g_5 = 3.166 < g_6 = 5.44 .$$

Under these circumstances, the stationary state of the kink with a half-integer centre already becomes the ground state at $0 \leq g < g_1$, $g_2 < g < g_3$, $g_4 < g < g_5, \ldots$ and the state with an integer centre becomes the ground state at $g_1 < g < g_2$, $g_3 < g < g_4, \ldots$ At $g = g_k$, $k = 1, 2, \ldots$ both states are stable and have the same energy. As a result, a sharp drop in the amplitude of the Peierls potential, ΔE, take places and its period decreases by a factor 2 (see Fig. 3.8). However, the Peierls potential never disappears completely: $\Delta E > 0$ at all values $\beta \geq 0$ and $g \geq 0$.

The Peierls potential profile prevents the motion of the topological defect [11]. The defect is in the pinning (immobile) state. For it to start moving, it must overcome the potential threshold, and the motion itself is accompanied by phonon emission. If the initial kinetic energy does not exceed the potential amplitude, instead of straight motion, the defect will oscillate relative to a corresponding minimum of the Peierls potential. The frequency ω of this oscillation is the lowest eigenfrequency of the defect, which is directly proportional to the potential amplitude (see Fig. 3.9b right). For strong cooperativity $g > 10$, pinning practically vanishes, and the topological defect can move as a solitary wave with constant subsonic velocity. Therefore, the defect is often called a topological soliton.

In the FPU-α model ($\alpha > 0$), there is symmetry breaking between kink and antikink. This was observed for the first time by Milchev and Markov [39, 40] (see also [41]). In this case, the kink and antikink have different shapes, energies, and widths. This effect results from the fact that, when $\alpha > 0$, the interaction of neighboring particles in a kink localization region (in a region of local compression) is stronger than the interaction in an antikink localization region (in a region of local extension). The dependencies of the kink energy E and width D on the

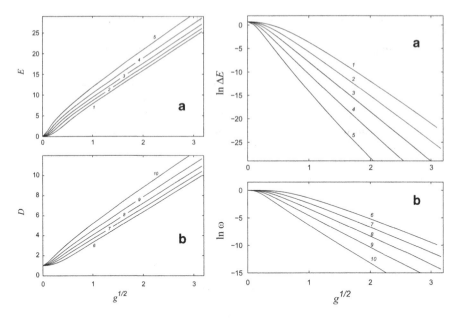

Fig. 3.10 *Left*: Dependence of the energy E (**a**) and width D (**b**) of the stationary kink on the square root of the cooperativity parameter $g^{1/2}$ for nonlinearity values $\beta = 0$ and $\alpha = 0$ (*lines 1 and 6*), 0.2 (*lines 2 and 7*), 0.5 (*lines 3 and 8*), 1 (*lines 4 and 9*), and 2 (*lines 5 and 10*). *Right*: Dependence of the logarithm of the Peierls potential amplitude $\ln \Delta$ (**a**) and the logarithm of the lowest eigenfrequency of the kink $\ln \omega$ (**b**) on the square root of the cooperativity parameter $g^{1/2}$ for nonlinearity values $\beta = 0$ and $\alpha = 0$ (*lines 1 and 6*), 0.2 (*lines 2 and 7*), 0.5 (*lines 3 and 8*), 1 (*lines 4 and 9*), and 2 (*lines 5 and 10*)

cooperativity parameter g are shown in Fig. 3.10 (left) for different values of the nonlinear parameters $\alpha > 0$ and $\beta = 0$. As can be clearly seen in this figure, the pinning energy ΔE and the frequency ω decrease exponentially with increasing values of $\sqrt{g}$. The greater the value of the nonlinearity parameter α, the faster it decreases. For all values of $g \geq 0$ and $\alpha \geq 0$, only the stationary state of the kink with a half-integer centre is stable.

Let us consider now the general case of the FPU model with parameters $\alpha = 1$ and $\beta = 1$. As can be seen in Fig. 3.11 (left), the kink energy and width are greater than those of the antikink. They increase proportionally to $\sqrt{g}$ with increasing cooperativity, and the proportionality coefficient in the case of kink is greater for the antikink. The dependencies of the pinning energy ΔE and the lowest eigenfrequency ω of the topological defect on the parameter g for the kink and antikink differ significantly (Fig. 3.11 right). The values of ΔE and ω for the kink decrease exponentially with increasing $\sqrt{g}$, and the ground state always corresponds to the kink with a half-integer centre. In the case of the antikink, ΔE and ω decrease exponentially much more slowly. This decrease is not monotonic. There are three local minima at $g = g_1, g_2,$ and g_3 ($g_1 = 0.342\,996 < g_2 = 0.821\,412 < g_3 = 2.834\,920\,6$). The

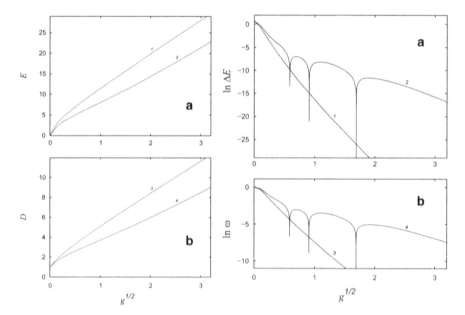

Fig. 3.11 *Left*: Dependence of the energy E (**a**) and width D (**b**) of the kink (*lines 1* and *3*) and antikink (*lines 2* and *4*) on the square root of the cooperativity parameter $g^{1/2}$ for nonlinearity values $\alpha = 1$ and $\beta = 1$. *Right*: Dependence of the logarithm of the Peierls potential amplitude $\ln \Delta$ (**a**) and the logarithm of the lowest eigenfrequency $\ln \omega$ (**b**) of the kink (*lines 1* and *3*) and antikink (*lines 2* and *4*) on the square root of the cooperativity parameter $g^{1/2}$ for nonlinearity values $\alpha = 1$ and $\beta = 1$

stationary state of the antikink with a half-integer centre is the ground state only if $0 \le g \le g_1$ and $g_2 \le g \le g_3$. If $g_1 \le g \le g_2$ and $g \ge g_3$, the ground state will be the state with an integer centre.

3.2.2 Vibrational Eigenmodes of a Topological Soliton

The symmetry between kink and antikink breaks once the cubic anharmonicity is taken into account [37, 42]. In this case, the topological soliton shape changes and additional vibrational eigenmodes can appear [43]. Let us consider these vibrational eigenmodes. For this purpose we find the frequency spectrum of the vibrational eigenmodes of a kink and antikink and analyse the stability of these localized vibrations.

The Hamiltonian (3.15) gives the equations of motion

$$\ddot{u}_n = gF(\rho_{n+1}) - gF(\rho_n) - \sin u_n , \quad n = 0, \pm 1, \pm 2, \dots , \tag{3.21}$$

where

$$F(\rho) = \frac{dU}{d\rho} = \rho - \alpha\rho^2 + \beta\rho^3 .$$

When $\alpha = \beta = 0$, (3.21) in the continuum approximation has the form of the sine–Gordon equation. For small displacements $|u_n| \ll 1$, one can neglect all anharmonic terms. Then (3.21) becomes linear:

$$\ddot{u}_n = g(u_{n+1} - 2u_n + u_{n-1}) - u_n , \quad n = 0, \pm 1, \pm 2, \dots , \tag{3.22}$$

and its solution can be represented as a sum of linear waves

$$u_n(t) = A \exp i(qn - \omega t) ,$$

where A and $q \in [-\pi, \pi]$ are the wave amplitude and wavenumber, respectively, and the wave frequency is

$$\omega(q) = \sqrt{2g(1 - \cos q) + 1} . \tag{3.23}$$

It follows from the dispersion equation (3.23) that the frequency spectrum of small amplitude vibrations (phonons) comprises a band $[\omega(0), \omega(\pi)]$, where $\omega(0) = 1$ and $\omega(\pi) = \sqrt{4g + 1}$ are the minimum and maximum frequencies, respectively.

3.2.3 Numerical Method for Finding the Vibrational Eigenmodes

In order to find a stationary state of the topological soliton, one must solve the minimum problem (3.19) with the boundary conditions (3.20), viz., $u_{-\infty} = 2\pi$ and $u_{+\infty} = 0$ for the kink and $u_{-\infty} = 0$ and $u_{+\infty} = 2\pi$ for the antikink.

Let $\{u_n^0\}_{n=1}^N$ be the solution of the problem (3.19), corresponding to a stationary topological soliton. To find its vibrational eigenmodes, we represent the solution of the equations of motion in the form of a small perturbation of the stationary topological soliton: $u_n(t) = u_n^0 + v_n(t)$, where the perturbation $|v_n| \ll 1$, $n = 1, \dots, N$. Substituting this relationship into the equations of motion of a finite chain of N links with fixed ends, we get the linear equations

$$\ddot{v}_n = b_{n-1}v_{n-1} + a_n v_n + b_n v_{n+1} , \quad n = 2, 3, \dots, N - 1 , \tag{3.24}$$

where

$$a_n = -gU''(u_{n+1}^0 - u_n^0) - gU''(u_n^0 - u_{n-1}^0) + \cos(u_n^0) , \quad b_n = gU''(u_{n+1}^0 - u_n^0) .$$

The search for vibrational eigenmodes of the stationary topological soliton is thereby reduced to finding the eigenvalues and eigenvectors of the symmetric tridiagonal matrix:

$$
B = \begin{pmatrix}
a_2 & b_2 & 0 & \cdots & 0 & 0 & 0 \\
b_2 & a_3 & b_3 & \cdots & 0 & 0 & 0 \\
\vdots & \vdots & \vdots & \ddots & \vdots & \vdots & \vdots \\
0 & 0 & 0 & \cdots & b_{N-3} & a_{N-2} & b_{N-2} \\
0 & 0 & 0 & \cdots & 0 & b_{N-2} & a_{N-1}
\end{pmatrix}.
$$

Let λ and $\Psi_\lambda = (\psi_2, \psi_3, \ldots, \psi_{N-1})^*$ be the eigenvalue and eigenvector corresponding to the matrix B ($\sum_n \psi_n^2 = 1$). Then the solution for the system of linear equations (3.24) has the form

$$
v_n(t) = -A\psi_n \exp(-i\omega t), \quad n = 2, 3, \ldots, N-1, \tag{3.25}
$$

where $A > 0$ and $\omega = \sqrt{-\lambda}$ are the amplitude and frequency of the vibration, respectively. Let us define the centre of the vibration as

$$
\bar{n} = \sum_{n=2}^{N-1} \psi_n^2,
$$

and its diameter

$$
D = 1 + 2 \left[\sum_{n=2}^{N-1} (n - \bar{n})^2 \psi_n^2 \right]^{1/2}.
$$

We treat the vibration as localized if its diameter $D < N/4$. The localized vibrations correspond to the vibrational eigenmodes of the soliton and the nonlocalized vibrations are referred to as phonons of the chain.

3.2.4 Vibrational Eigenmodes of a Soliton in the FPU-β Model

In the FK model, i.e., when $\alpha = 0$ and $\beta = 0$, the topological solitons have only the low-frequency vibrational eigenmodes. In addition to the trivial vibration, corresponding to the translation mode (vibration of the soliton centre around the minimum of the Peierls profile), in the narrow interval of the cooperativity parameter $0.29 \le g \le 1.05$, a high-frequency vibration exists near the low limit of the frequency spectrum (see Fig. 3.12). This vibration was investigated in detail in [44]. When anharmonicity $\beta > 0$ appears, this interval quickly begins to get narrower (see Fig. 3.12), and when $\beta \approx 0.007$, it completely vanishes (for nonlinearity parameter

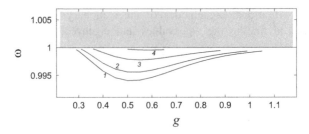

Fig. 3.12 Dependence of the second eigenfrequency of the kink on the cooperativity parameter g for $\alpha = 0$ and $\beta = 0, 0.001, 0.003,$ and 0.004 (*lines 1–4*). The *shaded area* shows the phonon frequency spectrum

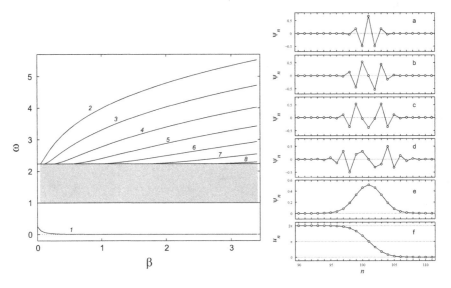

Fig. 3.13 *Left*: Dependence of the eigenfrequencies ω of the stationary topological soliton on the parameter of site–site interaction nonlinearity β for cooperativity $g = 1$ in the FPU chain with $\alpha = 0$. The *shaded area* shows the phonon frequency spectrum. *Line 1* represents the dependence of the lowest eigenfrequency, *lines 2–8* show the dependence of the rest of eigenfrequencies. *Right*: Profiles of the vibrational eigenmodes with frequencies $\omega = 3.98$ (**a**), 3.30 (**b**), 2.77 (**c**), 2.40 (**d**), and 0.00045 (**e**). Profile of the corresponding kink (**f**) in the FPU model with parameters $g = 1$, $\alpha = 0$, and $\beta = 1$

$\beta \geq 0.007$, there is only a single eigenfrequency in the low-frequency region $0 \leq \omega < 1$ which corresponds to the translational mode). With increasing nonlinearity, the high-frequency vibration eigenmodes appear. These lie above the phonon frequency spectrum. Both the number of these eigenfrequencies and their values increase steadily as the parameter β increases, while the lowest eigenvalue tends to zero (see Fig. 3.13 left). The profile of the vibration eigenmodes (corresponding to eigenvectors ψ_n) and the profile of the kink are shown in Fig. 3.13 (right). The width of the vibration decreases monotonically with increasing frequency. The number

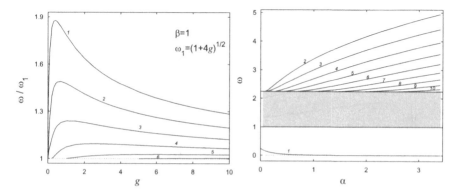

Fig. 3.14 *Left*: Ratio of the high eigenfrequencies ω of the topological soliton to the maximum frequency ω_1 of the phonon spectrum as a function of the cooperativity parameter g in the FPU chain with parameters $\alpha = 0$ and $\beta = 1$. *Right*: Dependence of the eigenfrequencies ω of the stationary kink on the parameter α of site–site interaction nonlinearity, for cooperativity $g = 1$ in the FPU chain with $\beta = 0$. The *shaded area* shows the phonon frequency spectrum. *Line 1* shows the dependence of the lowest eigenfrequency and *lines 2–10* depict the dependencies for the rest of the eigenfrequencies

of eigenfrequencies also increases with increasing cooperativity parameter g (see Fig. 3.14 left).

3.2.5   Vibrational Eigenmodes of a Kink in the FPU-α Model

In the FPU chain with cubic anharmonicity, $\alpha > 0$ and $\beta = 0$, the properties of the kink and antikink differ significantly. For $\beta = 0$ and $\alpha > 0$, the site–site interaction potential (3.16) is no longer a function bounded from below (the point $\rho = 0$ here is only a local minimum). When extended, this chain may break, so the solitons of extension, i.e., antikinks, can cause chain breakdown. In this connection, we consider only the soliton of compression (kinks). In the case of both the FPU-α and FPU-β models in a range of frequencies lying below the phonon frequency spectrum, the stationary kink can have two eigenfrequencies only for very weak nonlinearity $\alpha < 0.39$. When $\alpha \geq 0.39$, in this range, the kink has only the trivial (translational) mode. The frequency of this mode is proportional to the amplitude of the Peierls potential and so tends exponentially to zero with increasing cooperativity parameter g. An increase in the nonlinearity parameter leads to the appearance of the high-frequency vibrational eigenmodes with frequencies lying above the phonon frequency spectrum. With increasing α, the number of these frequencies rises, and the frequencies themselves increase steadily (see Fig. 3.14 right). The number of eigenfrequencies also increases with rising cooperativity parameter g (Fig. 3.15 left).

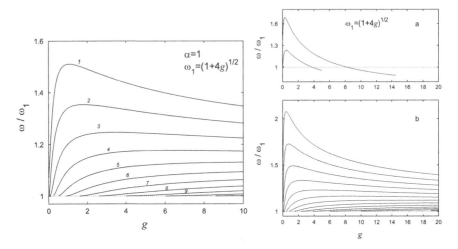

Fig. 3.15 *Left*: Ratio of the eigenfrequencies ω of the topological soliton to the maximum frequency ω_1 of the phonon spectrum as a function of the cooperativity parameter g in the FPU chain with parameters $\alpha = 1$ and $\beta = 0$. *Right*: Ratio of the high eigenfrequencies ω of the antikink (**a**) and kink (**b**) to the maximum frequency ω_1 of the phonon spectrum as a function of the cooperativity parameter g in the FPU chain with parameters $\alpha = 1$ and $\beta = 1$

3.2.6 Soliton Vibrational Eigenmodes in the Mixed FPU-α-β Model

Here we consider the FPU chain with nonlinearity parameters $\alpha = 1$ and $\beta = 1$. The site–site interaction potential (3.16) is bounded from below and has only a single minimum at the point $\rho = 0$. Opposite-sign topological solitons can exist simultaneously in this chain. The properties of the kink and antikink differ from each other. The kink always has a broader profile, with a large number of high-frequency vibrational eigenmodes. The number of vibrational eigenmodes increases steadily with increasing cooperativity parameter g (see Fig. 3.15b right). For weak cooperativity, the antikink also has high-frequency vibrational eigenmodes, but their number decreases with increasing g (see Fig. 3.15a right). With increasing cooperativity, these frequencies can even overlap with the phonon spectrum. For strong cooperativity $g > 14.5$, these frequencies vanish and the antikink (as opposed to the kink) has only a single trivial low-frequency translational mode.

3.2.7 Modeling Vibrational Eigenmodes

Numerical modeling of the vibrational eigenmodes of topological solitons shows that they are exactly linear. To check this, we modeled the dynamics of a finite

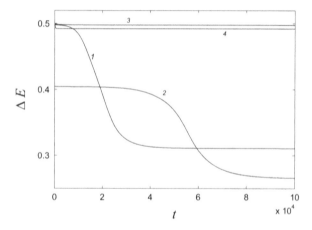

Fig. 3.16 Energy eigenmode of the kink $\Delta E = E - E_0$ (E and E_0 are the energies of the excited and stationary soliton states, respectively) with frequencies $\omega = 4.493\,662\,069\,0$, $3.851\,170\,574\,6$, and $3.323\,743\,239\,1$ (*lines 1–3*) and antikink with the frequency $\omega = 3.456\,200\,056\,9$ (*line 4*) as a function of time t in the FPU chain with $\alpha = 1$, $\beta = 1$, and cooperativity parameter $g = 1$

chain with fixed ends. The following equations were integrated numerically:

$$\ddot{u}_n = g F(\rho_{n+1}) - g F(\rho_n) - \sin u_n , \quad n = 2, 3, \ldots, N - 1 , \tag{3.26}$$

with initial condition

$$u_n(0) = u_n^0 , \quad \dot{u}_n(0) = A\psi_n , \quad n = 1, 2, \ldots, N, \tag{3.27}$$

where $u_1 = u_1^0$ and $u_N = u_N^0$, N is the number of chain links, $\{u_n^0\}_{n=1}^N$ is the stationary soliton profile which was obtained by solving the minimum problem (3.19), $\{\psi_n\}_{n=1}^N$ is the normalized eigenvector corresponding to the eigenfrequency of the topological soliton, and $A > 0$ is the amplitude of the vibration. (The energy of the excited soliton (3.27) will be equal to $E_0 + A^2/2$, where E_0 is the energy of its stationary state.) To model the dynamics of the excited state of the soliton in an infinite chain, we introduce viscous friction at the ends of the chain in order to ensure absorption of phonons emitted by the soliton.

Numerical integration of (3.26) has shown that, at a sufficiently high amplitude A, the vibration is accompanied by phonon emission which almost vanishes upon reaching a certain excitation energy threshold (see Fig. 3.16). Under these circumstances, the emission is already absent and the oscillation frequency does not depend on the amplitude, i.e., the vibration is linear. Thus, an increase in the amplitude of vibrational eigenmodes does not lead to the formation of nonlinear vibrations. The excess energy is spent on phonon emission until the vibration becomes completely linear.

The contribution of the vibrational eigenmodes of the topological soliton to the frequency spectrum of thermal vibrations can be estimated by considering the

Langevin equation

$$\ddot{u}_n = gF(\rho_{n+1}) - gF(\rho_n) - \sin u_n + \xi_n - \gamma \dot{u}_n , \quad n = 2, 3, \ldots, N-1 , \quad (3.28)$$

with initial condition

$$u_n(0) = u_n^0 , \quad \dot{u}_n(0) = 0 , \quad n = 1, 2, \ldots, N , \quad (3.29)$$

where $\{u_n^0\}_{n=1}^n$ is the profile of the stationary topological soliton, $\gamma = 1/t_r$ is the friction coefficient, and t_r is the relaxation time. The random forces ξ_n describing the interaction of the chain sites with a thermal bath have normal distribution and their correlation functions are

$$\langle \xi_n(t_1) \xi_m(t_2) \rangle = 2\gamma T \delta_{nm} \delta(t_1 - t_2) ,$$

where T is the dimensionless temperature of the thermal bath.

The equations of motion (3.28) with the initial conditions (3.29) were integrated numerically by the standard fourth-order Runge–Kutta method with a constant integration step Δt [45]. In the numerical procedure, the lagged Fibonacci random number generator [46] was used and the δ function has the form $\delta(t) = 0$, if $|t| > \Delta t/2$ and $\delta(t) = 1/\Delta t$, if $|t| \leq \Delta t/2$, i.e., the step of numerical integration corresponds to the correlation time of the random force. To use the Langevin equation, this correlation time is assumed to be $\Delta t \ll t_r$. We thus chose the integration step $\Delta t = 0.05$ and the relaxation time $t_r = 10$. During the time $t_0 = 10t_r$, the system comes to equilibrium with the thermal bath and the point

$$\{u_n(t_0), \dot{u}_n(t_0)\}_{n=1}^N \quad (3.30)$$

gives a random realization of the thermalized state of the chain. To analyse the dynamics of the thermalized chain further, one must switch off the interaction with the thermal bath, i.e., the equations of motion (3.26) should be integrated with the initial condition (3.30).

The frequency spectral density $p(\omega)$ for the thermal vibration of the chain ($N = 200$) with the kink as a function of temperature T is shown in Fig. 3.17 (left). In the chain with nonlinearity parameters $\alpha = 1$, $\beta = 1$, and cooperativity parameter $g = 1$, the kink has eight vibrational eigenmodes (see Fig. 3.15b right). As can be seen in Fig. 3.17c (left), at temperature $T = 0.01$, each eigenfrequency is related to the clearly pronounced peak in the frequency spectral density. At the higher temperature $T = 0.1$, these local maxima are still present in the frequency spectral density, but they exhibit tailing (see Fig. 3.17b left). At the temperature $T = 1$, they already vanish (see Fig. 3.17a left). The right edge of the frequency spectrum shifts to the right. This is associated with the appearance of high-frequency breathers in the chain. A similar situation also takes place for the antikinks (see Fig. 3.17 right). Here, the maxima corresponding to eigenfrequencies are pronounced at the

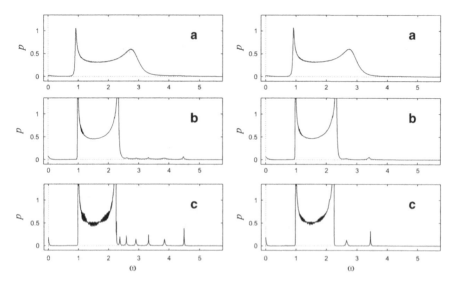

Fig. 3.17 *Left*: Frequency spectral density $p(\omega)$ for the thermal vibration of the FPU chain with the kink (under the fixed-end boundary condition, $N = 200$) at temperatures $T = 1$ (**a**), $T = 0.1$ (**b**), and $T = 0.01$ (**c**) (chain parameters $\alpha = 1$, $\beta = 1$, and $g = 1$). *Right*: Frequency spectral density $p(\omega)$ for the thermal vibration of the FPU chain with the antikink (under the fixed-end boundary condition, $N = 200$) at temperatures $T = 1$ (**a**), $T = 0.1$ (**b**), and $T = 0.01$ (**c**) (chain parameters $\alpha = 1$, $\beta = 1$, and $g = 1$)

temperature $T = 0.01$, poorly expressed at $T = 0.01$, and completely absent at $T = 1$. Therefore, the vibrational eigenmodes of topological solitons manifest themselves in the frequency spectrum only at low temperatures, when all vibrations are almost linear. At high temperatures, when the vibrations become essentially nonlinear, the vibrational eigenmodes no longer manifest themselves in the thermal vibration spectrum, which again highlights their linear character.

3.2.8 Supersonic Soliton Motion

In the absence of the site–site interaction anharmonicity, the opposite-sign topological solitons possess the same properties – they have a continuous subsonic velocity spectrum. Static interaction of opposite-sign topological solitons leads to their attraction, but in discrete chains, the pinning of solitons can lead to stable bound states [33]. In a thermalized chain with the ϕ-4 potential, the indirect repulsive interaction of opposite-sign solitons occurs via phonons. This is longer-ranged than the static interaction [32]. With the appearance of anharmonicity in the site–site interaction, the situation changes drastically. The properties of opposite-sign topological solitons now become different.

As shown in [35], with increasing negative anharmonicity, the velocity spectrum of the negative soliton becomes narrow, and when it reaches a certain threshold, it even vanishes (the soliton has only the zero velocity component), whereas the positive soliton always has a continuous subsonic velocity spectrum. It was shown in [34] that anharmonicity can lead to the positive topological soliton having a single supersonic velocity value. The structure of the supersonic velocity spectrum of the positive topological soliton in the ϕ-4 model with negative cubic anharmonicity of the site–site interaction was analysed in detail in [36]. There it was shown that the supersonic spectrum has a discrete structure. There exists only a finite number of supersonic velocity values $s_1 > s_2 > \ldots > s_N$, for which the kink motion is not accompanied by phonon emission. The supersonic kink corresponding to the nth velocity value s_n is a bound state of n acoustic (nontopological) solitons. For other supersonic velocity values, the kink motion is always accompanied by phonon emission. In the Frenkel–Kontorova model the situation becomes more complex. Here, anharmonicity also leads to the appearance of the discrete supersonic velocity spectrum of the topological soliton, but anharmonicity can also bring about association of solitons of the same signs. In this case, supersonic topological solitons with multiple charge emerge [47].

The Model

The Hamiltonian of a bistable molecular chain has the form

$$\mathscr{H} = \sum_n \left[\frac{1}{2} m \dot{u}_n^2 + \mathscr{U}(u_{n+1} - u_n) + \mathscr{V}(u_n) \right] , \tag{3.31}$$

where m is the mass of a chain link, u_n is the displacement of the nth link from its equilibrium position, $\mathscr{U}(\rho)$ is the site–site interaction potential, and $\mathscr{V}(u)$ is the symmetric two-well potential describing the interaction of chain sites with its substrate. For the ϕ-4 model with cubic anharmonicity, the potentials are

$$\mathscr{U}(\rho) = \frac{1}{2} \kappa \rho^2 - \frac{1}{3} \gamma \rho^3 , \quad \mathscr{V}(u) = \varepsilon \left[(u/l)^2 - 1 \right]^2 ,$$

where κ and $\gamma > 0$ are the stiffness and anharmonicity of the site–site interaction potential and ε and $2l$ are the height and width of the barrier of the two-site potential, respectively.

The Hamiltonian (3.31) gives the equations of motion

$$m \ddot{u}_n = \mathscr{F}(u_{n+1} - u_n) - \mathscr{F}(u_n - u_{n-1}) - \mathscr{G}(u_n) , \quad n = 0, \pm 1, \pm 2, \ldots , \tag{3.32}$$

where

$$\mathscr{F}(\rho) = \frac{d}{d\rho} \mathscr{U}(\rho) = \kappa \rho - \gamma \rho^2 , \quad \mathscr{G}(u) = \frac{d}{du} \mathscr{V}(u) = 4\varepsilon u \frac{(u/l)^2 - 1}{l^2} .$$

For convenience, we introduce dimensionless variables: time $\tau = t\sqrt{\kappa/m}$, displacement $x_n = u_n/l$, and energy $H = \mathcal{H}/\kappa l^2$. Then the Hamiltonian of the system (3.31) takes the form

$$H = \sum_n \left[\frac{1}{2} x'^2_n + U(r_n) + V(x_n) \right] , \tag{3.33}$$

where the prime denotes differentiation with respect to the dimensionless time τ and $r_n = x_{n+1} - x_n$ is the relative displacement. The potentials are

$$U(r) = \frac{1}{2} r^2 - \frac{1}{3}\beta r^3 , \qquad V(x) = g(x^2 - 1)^2 ,$$

where $\beta = \gamma l/\kappa > 0$ is the dimensionless anharmonicity parameter and $g = \varepsilon/\kappa l^2 \geq 0$ is the dimensionless height of the two-well potential barrier. The equations of motion (3.32) have the form

$$x''_n = F(r_n) - F(r_{n-1}) - G(x_n) , \quad n = 0, \pm 1, \pm 2, \ldots , \tag{3.34}$$

where

$$F(r) = \frac{d}{dr} U(r) = r - \beta r^2 , \qquad G(x) = \frac{d}{dx} V(x) = 4gx(x^2 - 1) .$$

The Continuum Approximation

We assume that (3.34) has a soliton solution $x_n(\tau) = x(\xi) = x(n - st)$ which smoothly depends on the number of chain sites n, i.e., the solution is suggested to have the form of a solitary wave of constant shape with asymptotic behavior

$$x_n \longrightarrow \mp 1 \, (\pm 1) \quad \text{for } n \to \pm\infty , \tag{3.35}$$

for the positive (negative) soliton. Here, $\xi = n - st$ and s are the wave variable and soliton velocity, respectively. The positive topological soliton describes the transition of the chain from one equilibrium state $x_n = +1$ to another $x_n = -1$, while the negative topological soliton describes the reverse transition from the state $x_n = -1$ to $x_n = +1$. The compression or extension of the chain occurs in the localization region of the positive or negative soliton, respectively.

Without considering the dispersion of long-wavelength phonons, the equations of motion (3.34) in the continuum approximation reduce to the differential equation

$$(1 - s^2)x_{\xi\xi} - 2\beta x_\xi x_{\xi\xi} - 4gx(x^2 - 1) = 0 . \tag{3.36}$$

This equation can be integrated, thus allowing a complete study in this approximation of the properties of topological solitons in an anharmonic chain [34]. Indeed, letting $\varphi = x_\xi$, (3.36) takes the form

$$\left(1 - s^2 - 2\beta\varphi\right)d\varphi = 4gx(x^2 - 1)d\xi \; . \tag{3.37}$$

Multiplying (3.37) by φ, then integrating and taking into account the boundary conditions (3.35), we obtain

$$\left[\frac{1}{2}(1 - s^2) - \frac{2}{3}\beta\varphi\right]\varphi^2 = g(x^2 - 1)^2 \; . \tag{3.38}$$

For the positive soliton ($\varphi \leq 0$), one can obtain from (3.38) the continuous dependence $\varphi = \varphi(x)$, $-1 \leq x \leq 1$ only if $|s| \leq 1$, while for the negative soliton ($\varphi \geq 0$) the continuous dependence can be obtained only if $|s| < s_-$, where the velocities are

$$s_- = \begin{cases} \sqrt{1 - (24\beta^2 g)^{1/3}} & \text{for } 24\beta^2 g \leq 1 \; , \\ 0 & \text{for } 24\beta^2 g > 1 \; . \end{cases}$$

Thus, using the continuum approximation without considering the dispersion of long-wavelength phonons shows that the positive topological soliton always has only the continuous subsonic velocity spectrum $0 \leq s \leq 1$, while the negative topological soliton has the continuous spectrum $0 \leq s \leq s_-$.

Accounting for the dispersion of long-wavelength phonons, the equations of motion (3.34) in the continuum approximation reduce to the differential equation

$$(1 - s^2)x_{\xi\xi} + \frac{1}{12}x_{\xi\xi\xi\xi} - 2\beta x_\xi x_{\xi\xi} - 4gx(x^2 - 1) = 0 \; , \tag{3.39}$$

which cannot generally be integrated analytically. It was shown in [35] that, for the fixed velocity

$$s = s'_1(\beta) = \sqrt{1 + \frac{4}{3}\beta^2 - \frac{1}{2\beta^2}g} \; , \tag{3.40}$$

(3.39) has the soliton solution

$$x(\xi) = -\tanh\frac{\mu\xi}{2} \; , \tag{3.41}$$

with the inverse width $\mu = 4\beta$. It follows from (3.40) that, for a small height g of the two-well potential barrier, the positive topological soliton (3.41) has the supersonic velocity $s = s'_1(\beta) > 1$.

In the limit $g \to 0$, after a single integration, (3.39) becomes the Boussinesq equation

$$\frac{1}{12}\varphi_{\xi\xi} + (1 - s^2)\varphi - \beta\varphi^2 = 0 ,$$

which defines the supersonic acoustic soliton

$$\varphi(\xi) = -\frac{a}{\cosh^2(\mu\xi)}$$

in the one-dimensional lattice with cubic anharmonicity, where $a = 3(s^2 - 1)/2\beta$, $\mu = \sqrt{3(s^2 - 1)}$ is the inverse width, and $s > 1$ is the velocity. The boundary condition (3.35), conserved in the passage to the limit $g \to 0$, allows the acoustic soliton to have only the single velocity

$$s = \sqrt{1 + \frac{4}{3}\beta^2} . \tag{3.42}$$

In fact, the total chain compression, which we shall henceforth refer to as the acoustic soliton amplitude, is

$$R(s) = x(+\infty) - x(-\infty) = \int_{-\infty}^{+\infty} \varphi(\xi)d\xi = -\frac{\sqrt{3(s^2 - 1)}}{\beta} = -2 ,$$

which leads to (3.42). Note also that (3.42) follows from (3.40) in the passage to the limit $g \to 0$.

On the other hand, for $g = 0$, the boundary conditions (3.35) will also be fulfilled when several identical acoustic solitons are present in the system. In this case, the velocity s_N of the N-soliton state can be found from the equation $NR(s_N) = -2$, and is equal to

$$s_N^\circ(\beta) = \sqrt{1 + \frac{4}{3}\left(\frac{\beta}{N}\right)^2} . \tag{3.43}$$

Thus, in the limit $g \to 0$, the positive topological soliton has the infinite discrete supersonic spectrum $\{s_N^\circ\}_{N=1}^\infty$ with the sound velocity $s = 1$ as the limit point. It is impossible to find an analytical N-soliton solution of (3.39) at $N \geq 2$ and $g > 0$, so we seek it numerically.

Numerical Methods for Finding Supersonic Soliton States

We assume once again that the equations of motion of the chain (3.34) have a solution $x_n(\tau) = x(n - s\tau)$ which depends smoothly on n and satisfies the asymptotic behavior (3.35). Then if we replace the second time derivative by its discrete analogue

$$x''_n = s^2 \frac{d^2}{dn^2} x = \frac{1}{12} s^2 \left[16(x_{n+1} - 2x_n + x_{n-1}) - (x_{n+2} - 2x_n + x_{n-2}) \right],$$

the differential equations (3.34) are transformed into completely discrete equations

$$-\frac{1}{12} s^2 \left[16(x_{n+1} - 2x_n + x_{n-1}) - (x_{n+2} - 2x_n + x_{n-2}) \right]$$
$$+ F(r_n) - F(r_{n-1}) - G(x_n) = 0, \quad n = 0, \pm1, \pm2, \dots, \qquad (3.44)$$

which coincide with (3.39) in the continuum approximation. The discrete equations (3.44) determine an extremum of the Lagrangian

$$L_s = \sum_n \left\{ -\frac{1}{24} s^2 \left[16(x_{n+1} - x_n)^2 - (x_{n+2} - x_n)^2 \right] + U(x_{n+1} - x_n) + V(x_n) \right\}.$$

Therefore, a soliton solution of (3.39) can be sought numerically as the extremum of the Lagrangian L_s. The supersonic topological soliton corresponds to the saddle point of the Lagrangian and so can be found by numerically solving the constrained minimum problem

$$F_s = \frac{1}{2} \sum_{n=3}^{M-2} \left(\frac{\partial L_s}{\partial x_n} \right)^2 \longrightarrow \min_{x_3, \dots, x_{M-2}} : \quad x_1 = x_2 = +1, \quad x_{M-1} = x_M = -1.$$
$$(3.45)$$

The boundary conditions should not affect the soliton shape. To satisfy this requirement, it suffices to take the number of sites M to be ten times the soliton width.

In order to solve the minimum problem (3.45) numerically, we used the conjugate gradient method [59] with $M = 400$. By solving this problem, we can find all the soliton solutions of (3.39). The minimum point $\{x_n^\circ\}_{n=1}^{M}$ of the functional F_s corresponds to a soliton solution only if x_n° depends smoothly on n and the actual minimum value $F_s(x_1^\circ, \dots, x_M^\circ) \approx 0$. The absence of such minima unambiguously indicates that (3.44), and hence also (3.39), do not have soliton solutions for a given velocity.

Supersonic Soliton States

Here we take the value of the substrate parameter to be $g = 0.001$. Numerical solution of the minimum problem (3.45) shows that the positive soliton always has the continuous subsonic velocity spectrum $0 \leq s \leq s_0 < 1$. The dependence of the

Fig. 3.18 Upper edge of the continuous velocity spectrum s_0 (*thick line 1*), the supersonic velocities $s_1, s_2, \ldots, s_7$ (*thin lines 2, 3, $\ldots$, 8*), s'_1 (*dashed line 9*), and $s_1^{\circ}, s_2^{\circ}, \ldots, s_5^{\circ}$ (*dotted lines 10, 11, $\ldots$, 14*) as a function of the anharmonicity parameter β of the chain

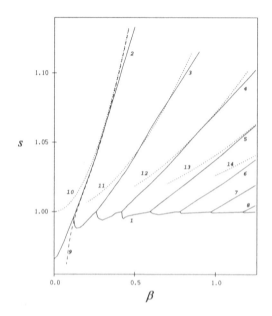

upper edge of the continuous spectrum on the anharmonicity parameter β is shown in Fig. 3.18. For $\beta = 0$, we have $s_0 = 0.966$, and with increasing anharmonicity the upper edge of the spectrum tends to the sound velocity: $s_0 \to 1$ as $\beta \to \infty$.

In addition to the continuous subsonic spectrum, the soliton has the finite discrete supersonic velocity spectrum $\{s = s_n\}_{n=1}^{N}$, where $s_1 > \ldots > s_N > 1$ (for the other values $s > 1$ the problem (3.45) does not have soliton solutions). The number of permissible supersonic values of the velocity N increases steadily with increasing anharmonicity parameter β. There exists a sequence of values of β tending to infinity, viz.,

$$0 < \beta_1 < \beta_2 < \ldots < \beta_n < \ldots \, ,$$

at which the number N increases by 1. Thus, for $0 \le \beta < \beta_1$, we have $N = 0$ (the topological soliton does not have supersonic states), and for $\beta_n \le \beta < \beta_{n+1}$, we have $N = n$ (the topological soliton has n supersonic states, $n = 1, 2, \ldots$). For the value $g = 0.001$ used here, the critical values of the anharmonicity parameter are $\beta_1 = 0.12$, $\beta_2 = 0.25$, $\beta_3 = 0.42$, $\beta_4 = 0.59$, $\beta_5 = 0.78$, $\beta_6 = 0.97$, and $\beta_7 = 1.17$.

The supersonic velocities s_n increase steadily with increasing anharmonicity parameter. The dependencies of the supersonic values $s_1, s_2, \ldots,$ and s_7 on the parameter β is shown in Fig. 3.18. As can be seen from this figure, the dependence $s_1(\beta)$ obtained by numerical solution of the problem (3.45) coincides completely with the dependence $s'_1(\beta)$ obtained analytically. The corresponding curves 2 and 9 in Fig. 3.18 differ only at velocities $s > 1.07$ when the soliton gets narrow and the continuum approximation we used ceases to be correct.

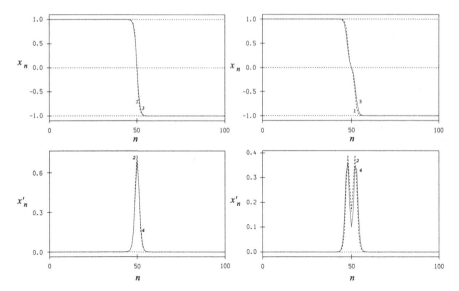

Fig. 3.19 *Left*: Profile of the one-soliton supersonic kink state found in the numerical solution of the problem (3.45) at the initial time $\tau = 0$ (*dashed lines 1* and *2*) and time $\tau = \tau_e = 92,542.1$ after the passage of $N = 10^5$ chain links (*solid lines 3* and *4*). The anharmonicity parameter is $\beta = 0.4$, the barrier height $g = 0.001$, the initial velocity $s_1 = 1.095$, and the final velocity $\bar{s}_1 = 1.079$. *Right*: Profile of the two-soliton supersonic kink state. All notations are the same as for the plots *on the left* and $\beta = 0.9$, $g = 0.001$, $s_2 = 1.115$, $\bar{s}_2 = 1.095$, $N = 5 \times 10^4$, and $\tau_e = 45,530.0$

As can be seen in Fig. 3.18, the functions $s_n(\beta)$ and $s_n^{\circ}(\beta)$ behave equivalently as $\beta \to \infty$, i.e., $s_n(\beta)/s_n^{\circ}(\beta) \to 1$. In the limit, the supersonic state of the topological soliton (kink), which has velocity s_n, decays into identical acoustic solitons which we will call an n-soliton state. From the analysis above, we can also conclude that the discreteness of the supersonic velocity spectrum is only due to the boundary conditions (3.35), i.e., the two-well nature of the potential $V(x)$, but not its specific form.

A typical profile of the one-soliton supersonic state of the kink is shown in Fig. 3.19 (left). Considering the relative displacement of chain links $r_n = x_{n+1} - x_n$, which is proportional to the velocities x'_n ($x'_n \approx -s r_n$), the kink is seen to have a one-hump profile. The one-soliton state can be considered as an acoustic soliton satisfying the boundary conditions (3.35). Typical profiles of the two-, three-, and four-soliton supersonic states of the kink are shown in Figs. 3.19 (right) and 3.20 (left and right). The corresponding kinks are characterized by the two-, three-, and four-hump profiles according to their relative displacements. This shows that these supersonic states are the bound states of two, three, or four acoustic solitons, respectively. We will show in the next section that if the substrate is taken away, i.e., $g = 0$, the n-soliton state decays into n acoustic solitons.

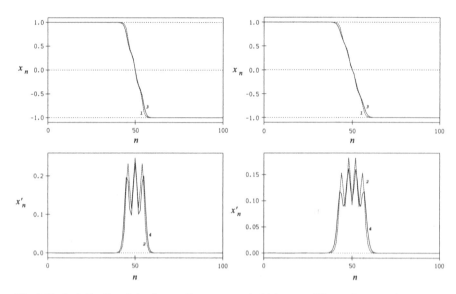

Fig. 3.20 *Left*: Profile of the three-soliton supersonic kink state. All notations are the same as in Fig. 3.19 (left) and $\beta = 1.1$, $g = 0.001$, $s_3 = 1.083$, $\bar{s}_3 = 1.065$, $N = 5 \times 10^4$, and $\tau_e = 46{,}823.0$. *Right*: Profile of the four-soliton supersonic kink state. All notations are the same as in Fig. 3.19 (left) and $\beta = 1.2$, $g = 0.001$, $s_4 = 1.058$, $\bar{s}_2 = 1.041$, $N = 5 \times 10^4$, and $\tau_e = 47{,}944.8$

Modeling the Supersonic Dynamics

Let us consider the dynamics of supersonic states of the topological soliton (kink) in a finite chain of L links with free ends. The dynamics of this chain are given by the equations of motion

$$x''_1 = F(r_1) - G(x_1),$$

$$\vdots$$

$$x''_n = F(r_n) - F(r_{n-1}) - G(x_n), \tag{3.46}$$

$$\vdots$$

$$x''_L = -F(r_{L-1}) - G(x_L),$$

with the energy integral

$$H = \sum_{n=1}^{L} \left[\frac{1}{2} x'^2_n + V(x_n) \right] + \sum_{n=1}^{L-1} U(r_n). \tag{3.47}$$

We take the number of chain sites L equal to $M + 100$, where M is the number of sites used in the solution of the minimum problem (3.45). The soliton solution of the problem (3.45), $\{x_n^\circ\}_{n=1}^M$, with velocity s satisfies the initial conditions

$$x_n(0) = x_n^\circ , \quad \text{for} \quad n = 1, 2, \ldots, M ,$$

$$x_n(0) = x_M^\circ , \quad \text{for} \quad n = M + 1, \ldots, L ,$$

$$x'_n(0) = -\frac{1}{2}s\left[x_{n+1}(0) - x_{n-1}(0)\right] , \quad \text{for} \quad n = 2, \ldots, L - 1 , \quad (3.48)$$

$$x'_1(0) = 0 ,$$

$$x'_L(0) = 0 .$$

The soliton centre $\{x_n(\tau)\}_{n=1}^L$ is conveniently defined as the point at which the broken line sequentially linking the points (n, x_n) intersects the n-axis. At the initial time, the soliton is centered at $m = M/2$. To model the soliton dynamics in an infinite chain, we shift the soliton through 100 links to the left as soon as it passes through 100 chain links, i.e., when its centre reaches the site $M/2 + 100$. This procedure leads to the change of variables

$$x_n = x_{100+n} , \quad x'_n = x'_{100+n} , \quad \text{for} \quad n = 1, \ldots, M ,$$
$$x_n = x_L , \quad x'_n = 0 , \quad \text{for} \quad n = M + 1, \ldots, L .$$

This numerical method for topological soliton dynamics simulation allows us to avoid the integration of a high dimensional system of equations. This method is especially efficient in the analysis of supersonic soliton dynamics. Using this method, the nonsoliton subsonic component of the initial condition is cut off as a result of the soliton shift.

Numerical simulation of the topological soliton dynamics confirmed the discreteness of the supersonic velocity spectrum. As can be seen from Figs. 3.19 and 3.20, the initial condition (3.48), corresponding the n-soliton supersonic kink state ($s = s_n > 1$), leads to the formation of a supersonic kink of unchanged shape moving along the discrete chain with constant supersonic velocity $s = \bar{s}_n < s_n$.

The initial condition (3.48), obtained in the continuum approximation, is not exact for a soliton in the discrete chain. The discreteness of the chain leads to the difference between the actual $\bar{s}_n$ and calculated s_n values of the velocity. The motion of the supersonic kink is always accompanied by phonon emission as long as the kink velocity $s > \bar{s}_n$. Phonon emission causes the kink to slow down. At the velocity $s = \bar{s}_n$, the emission disappears and the kink motion gets completely stabilized. The kink now begins to move with this constant velocity, and its shape does not change. Note that the final velocity $s = \bar{s}_n$ does not change with small variations in the initial velocity and shape of the kink. Such stability unambiguously points to the discreteness of the supersonic velocity spectrum of the kink in an anharmonic chain.

In order to understand the structure of the kink supersonic state, we consider its dynamics in a chain without the substrate ($g = 0$). For this purpose, we integrate the equation of motion (3.46) for $g = 0$ and $L = 1,100$. We take the initial conditions

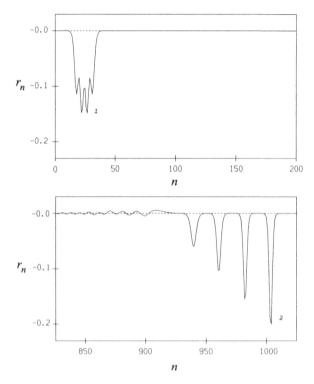

Fig. 3.21 Decay of the four-soliton supersonic kink state ($\beta = 1.2$, $g = 0.001$, and $\bar{s}_4 = 1.041$) into four acoustic solitons and a subsonic phonon tail in the free chain ($g = 0$). *Lines 1* and *2* show the relative displacement $r_n = x_{n+1} - x_n$ of the chain links at the initial time $\tau = 0$ and time $\tau = 900$, respectively. The chain length $L = 1,100$

corresponding to the supersonic n-soliton state of the kink ($n = 1, \ldots, 5$) in a discrete chain with the substrate ($g = 0.001$). In this case, numerical integration showed that exactly n uncoupled acoustic solitons and a subsonic phonon tail are formed from the kink (see Fig. 3.21). This allows us to conclude that the supersonic n-soliton state of the kink is really a bound state of n acoustic solitons, which is only stable at the velocity $s = \bar{s}_n$.

The dependence of the velocity $\bar{s}_n$ of the n-soliton supersonic state of the kink in the discrete chain on the anharmonicity parameter β is shown in Fig. 3.22. The velocity value $\bar{s}_n$ agrees with the calculated value s_n only near the sound velocity. In this case, a sufficiently large kink width justifies the use of the continuum approximation in the previous section. Despite this, the numerical simulation performed for the supersonic kink dynamics suggests that the conclusion about the discreteness of the supersonic velocity spectrum also remains true for higher velocities when the kink width becomes commensurate with the chain spacing. However, in this case, the derivation of an accurate value for the velocity requires the use of finer methods which take into account the chain discreteness [48].

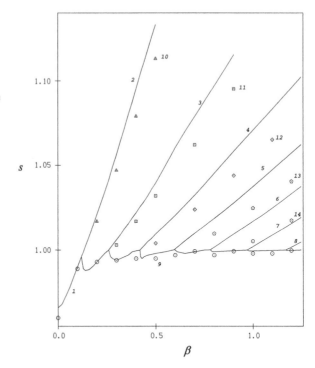

Fig. 3.22 Dependencies of the velocities $s_0, s_1, \ldots, s_7$ (*lines 1, 2, \ldots, 8*) on the anharmonicity parameter β. The markers *9, 10, \ldots, 14* give the velocity values $\bar{s}_0, \bar{s}_1, \ldots, \bar{s}_5$ obtained in the numerical simulation of the soliton dynamics

At a subsonic velocity $s < 1$, the positive topological soliton has a sufficiently large width, so solution of the minimum problem (3.45) allows one to find the soliton shape to high accuracy (see Fig. 3.23).

Numerical integration of the equations of motion (3.46) was carried out by the standard fourth order Runge–Kutta method with a constant integration step [45]. The accuracy of the numerical integration was controlled by checking the conservation of the energy integral (3.47). Using the step value $\Delta\tau = 0.05$, the energy is conserved up to five digits.

3.2.9   *Conclusion*

The anharmonicity of the site–site interaction leads to a significant change in the properties of topological solitons. In the FPU chain with sinusoidal substrate potential, the properties of the stationary states of topological solitons depend substantially on the relationship between the nonlinearity and cooperativity parameters. In contrast to the Frenkel–Kontorova model, states with both half-integer and integer centres of symmetry can be stable. The energy of soliton pinning can have deep local minima, but it is nonzero for all parameter values. The anharmonicity also leads to the appearance of localized vibrational eigenmodes:

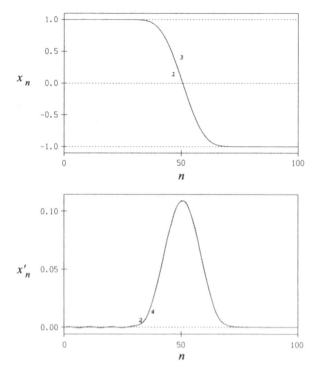

Fig. 3.23 Profiles of the subsonic topological soliton at the initial time $\tau = 0$ (*dashed lines 1* and *2*) and time $\tau = 10{,}103.0$ after the soliton passage of $N = 10^4$ chain links (*solid lines 3* and *4*). Initial velocity $s = 0.99$, $\beta = 1.0$, $g = 0.001$, and velocity $\bar{s} = 0.9898$

in addition to the low-frequency translational mode, the vibrational eigenmodes appear with frequencies lying above the phonon frequency spectrum. The number of these high-frequency vibrational eigenmodes increases with increasing values of the nonlinearity and cooperativity parameters. These vibrations are linear and give a clearly pronounced peak in the frequency spectral density for the thermal vibrations of a chain with topological solitons at low temperatures, when all vibrations are almost linear. Additionally, anharmonicity can cause supersonic motion of the topological soliton: there are a finite number of the velocity values $s_1 > s_2 > \cdots > s_N > 1$ for which the supersonic motion of the soliton is not accompanied by phonon emission. The supersonic kink corresponding to the nth velocity value s_n is the bound state of n acoustic solitons and the sum of their amplitudes should coincide with the width of the two-well potential barrier. The number N of supersonic velocity values rises with increasing chain anharmonicity. In the continuum approximation, the numerical method was suggested to obtain the shape and velocity of a supersonic state of the topological soliton.

3.3   Solitons in a Quasi-One-Dimensional Crystal

Topological solitons represent one of the most important types of localized nonlinear excitations in periodic nonlinear media. They can be identified with the point structural defects in polymer crystals, which play a significant role in heat transfer and other physical processes related to chain mobility, such as relaxation processes, diffusion of chains between the crystal and amorphous phases, and phase transitions [49]. Generally, topological solitons are investigated in the approximation of a one-dimensional chain interacting with a substrate (via a periodic interaction potential). Even within this approximation, there were found to be deviations from the behavior predicted by the continuous models (particularly, the Frenkel–Kontorova model), which result from the pinning phenomenon (the soliton slowing down due to dispersion).

The physical mechanism of this interaction has been analysed in [11], where the role of resonant relationships between the soliton and phonon characteristics was elucidated. A series of works conducted later [50–52] was devoted to analysis of the soliton slowdown through the Cherenkov radiation mechanism. In view of the application of these models to polymer crystals, the question arises concerning the role of the chains, which are close to strongly excited chains, in the dynamics of point structural defects. To solve this problem, the simplest model of a polymer crystal was suggested in [53–55]. This involves a system of linear chains with strong intrachain and weak interchain interactions. In the modeling, the existence of point structure defects such as vacancies or interstitials (areas of local extension and compression with no break in the intrachain bonds) was revealed. Furthermore, the dependence of the defect dynamics on the interchain interaction was investigated. It was shown that, at a weak bond between the chains, the defects represent smooth profile topological solitons with smooth profile peculiar to a one-dimensional chain in a periodic substrate potential. The spectrum of possible velocities was also obtained. When the interchain interaction increases, there are no soliton-like solutions in the form of waves traveling with a constant velocity. Defect motion is always accompanied by emission with a series of typical properties that are absent in the Frenkel–Kontorova model.

3.3.1   Modelling a Quasi-One-Dimensional Molecular System

In the modeling of topological solitons discussed above we assumed that the chain environment forms an immobile substrate for the chain which interacts with it via the substrate potential $V(u)$. This model does not allow a correct description of the interaction between a molecular chain and its environment. To do this, one must consider the mobility of neighboring chains.

Let us consider the two-dimensional model of the quasi-one-dimensional molecular system represented in Fig. 3.24. The system is made up of parallel linear chains

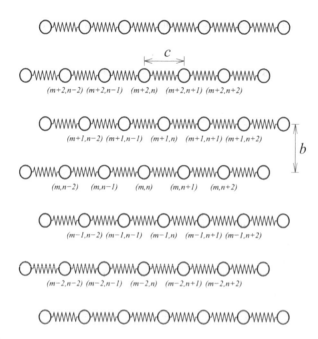

Fig. 3.24 Schematic representation of the two-dimensional model of a quasi-one-dimensional molecular system

of massive particles. The structure of the system is defined by two periods: the longitudinal chain spacing c and transverse chain spacing b. If we number the links and the particles of the chain with indices m and n, respectively, then in an equilibrium position the particle (m, n) has coordinates

$$x_{m,n} = \frac{n + [1 + (-1)^m]}{4} c , \quad y_{m,n} = mb , \quad m, n = 0, \pm 1, \pm 2, \dots .$$

We assume that the chain links can move along the x-axis, i.e., their transverse coordinates $y_{m,n}$ do not change. We denote the particle mass by M and the rigidity of a spring connecting the particles by K. We then change to a system of units in which the constants $\sqrt{M/K}$ and Kc_1^2 are the units of time and energy, respectively, with c_1 a unit of length which will be defined below. We also denote the longitudinal displacements of particles from their equilibrium positions by $u_{m,n}$. Then the Hamiltonian of the system takes the form

$$H = \sum_{m,n} \left[\frac{1}{2} \dot{u}_{m,n}^2 + \frac{1}{2} (u_{m,n+1} - u_{m,n} - \delta)^2 + \sum_{k=1}^{+\infty} \sum_{j=-\infty}^{+\infty} U(r_{m,n;k,j}) \right] , \quad (3.49)$$

where the dot denotes differentiation with respect to time and the compression of the isolated chain spacing which occurs during the formation of a quasi-one-dimensional crystal is specified by the parameter $\delta = c_0 - c$, with c_0 the isolated chain spacing. The potential $U(r_{m,n;k,j})$ describes the interaction of the nth particle in the mth chain with the $(n+j)$th particle in the $m+k$ chain ($k \neq 0$). The distance between the particles is

$$r_{m,n;k,j} = \left\{ \left[(j + d_{m,k})c + u_{m+k,n+j} - u_{m,n} \right]^2 + (kb)^2 \right\}^{1/2} ,$$

where $d_{m,k} = (-1)^m \left[(-1)^k - 1 \right] / 4$.

The parameter δ can be either positive or negative depending on the potential $U(r_{m,n;k,j})$. Most often, this potential is chosen so that it causes repulsion at short range and attraction of particles at long range. The parameter δ is negative if the attraction decreases sufficiently slowly when the distance between the particles increases. Then the attraction of the chain to non-nearest neighbors is so strong that nearest neighbors repel each other and the chains are extended during crystal formation. In the case of a sufficiently rapid decrease in particle attraction at infinity, the nearest chains attract each other weakly and $\delta > 0$.

We describe the particle interaction between different chains by the Lennard-Jones potential

$$U(r) = \varepsilon \left(\frac{r_0}{r} \right)^6 \left[\left(\frac{r_0}{r} \right)^6 - 2 \right] f(r) , \qquad (3.50)$$

where the dimensionless parameter ε specifies the ratio of energies of inter- and intrachain interactions and r_0 is the equilibrium distance between the interacting particles. The truncation function is

$$f(r) = \frac{1}{2} \left\{ 1 - \tanh \left[\mu (r - R_0) \right] \right\} ,$$

which is introduced for the convenience of numerical calculations. It allows one to avoid taking into account interactions between particles separated by a distance greater than $r > R_0$, where $R_0 \gg r_0$ is the truncation radius and the parameter $\mu > 0$ describes the smoothness of the truncation function. Figure 3.25 shows profiles of the interaction potential $U(r)$ and truncation function $f(r)$ for the values $R_0 = 20$ and $\mu = 2$ used for the truncation radius and smoothness parameter, respectively.

To find the equilibrium values of the lattice periods b and c, one must solve the minimum problem

$$E(b,c) = \frac{1}{2}(c_0 - c)^2 + \sum_{k=1}^{+\infty} \sum_{j=-\infty}^{+\infty} U(r_{jk}) \longrightarrow \min_{b,c} , \qquad (3.51)$$

where the distance between particles is

$$r_{j,k} = \left[(kb)^2 + (j + \Delta_k)^2 c^2 \right]^{1/2} ,$$

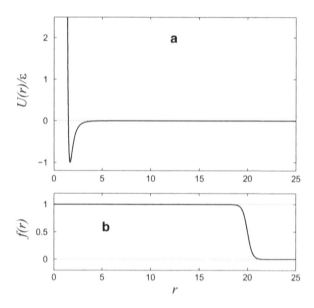

Fig. 3.25 Profiles of the interaction potential $U(r)$ (**a**) and the truncation function $f(r)$ (**b**)

and parameter $\Delta_k = \left[1 - (-1)^k\right]/4$. The solution to this problem (3.51) satisfies the system of equations

$$E_c = c - c_0 + c \sum_{k=1}^{+\infty} \sum_{j=-\infty}^{+\infty} U'(r_{jk})(j + \Delta_k)^2/r_{jk} = 0 \, , \tag{3.52}$$

$$E_b = b \sum_{k=1}^{+\infty} \sum_{j=-\infty}^{+\infty} U'(r_{jk})k^2/r_{jk} = 0 \, , \tag{3.53}$$

where $U'(r) = dU/dr$. We assume that, in the ground state, the longitudinal period is $c = 1$, i.e., the crystal longitudinal period c_1 is taken as a unit of length. Substituting this value into (3.53), we obtain an equation which uniquely defines the value of the longitudinal period b, viz.,

$$g(b) = \sum_{k=1}^{+\infty} \sum_{j=-\infty}^{+\infty} U'(r_{jk})k^2/r_{jk} = 0 \, .$$

Now the distance r_{jk} is determined by the relation $r_{jk} = \left[(kb)^2 + (j + \Delta_k)^2\right]^{1/2}$. After obtaining the equilibrium value of the period b from (3.52), the value of the parameter δ is immediately found to be

$$\delta = c_0 - 1 = \sum_{k=1}^{+\infty} \sum_{j=-\infty}^{+\infty} U'(r_{jk})(j + \Delta_k)^2 / r_{jk} \,.$$

For these parameter values b and δ, the ground state of the system will be characterized by zero relative displacements $\{u_{m,n} = 0\}_{m,n=-\infty}^{+\infty}$. The energy of the ground state is equal to

$$e_0 = \frac{1}{2}\delta^2 + \sum_{k=1}^{+\infty} \sum_{j=-\infty}^{+\infty} U(r_{jk}) \,.$$

Therefore, thanks to the judicious choice of variables, there remain two dimensionless parameters in the problem which define the properties of the system dynamics: the ratio of characteristic inter- and intrachain distances r_0 and the ratio of characteristic inter- and intrachain interaction energies ε. The value $0.5 < r_0 < \infty$ influences the form of the substrate potential generated by the neighbors of a given chain. It was shown in [56] that, for $0.5 < r_0 < 0.82$, the substrate potential near a given particle is generated mainly by a pair of nearest atoms belonging to the two nearest neighboring chains ('local' interaction of chains), and its form is significantly different from sinusoidal. For $0.91 < r_0 < \infty$, neighboring chains create a substrate potential which is close to a sinusoidal form with good accuracy. In this case, the substrate near a given particle is formed by many particles (more than four) belonging to the neighboring chain ('collective' interaction of chains).

It is the latter case that is of interest when investigating the dynamics of soliton-like localized excitations in a two-dimensional system, because this leads to the Frenkel–Kontorova model in the limit of immobile neighbors. Here, we choose an equilibrium distance r_0 for the Lennard-Jones potential which corresponds to a polyethylene crystal in the 'united atoms' model [57, 58]. In this model, the CH_2-group is substituted for particles with a total mass of 14 amu, and parameters of the Lennard-Jones potential between these particles are chosen such that the density of the crystal with united atoms is close to the density of the polyethylene crystal. As the chain period and equilibrium distance for the Lennard-Jones potential are equal to 2.54 and 4.265 Å, respectively, we obtain the value $r_0 = 1.67$. The interaction properties in a three-dimensional crystal composed of zigzag chains differ from those in a planar system of linear chains. Therefore, in our model we consider the value of ε corresponding to polyethylene at which the width of a static soliton coincides with that of a static soliton obtained in the model developed in [57, 58]. This condition appears to give the value $\varepsilon = 0.0007$. We also analyse two cases with a stronger interchain interaction, viz., $\varepsilon = 0.007$ and 0.07.

3.3.2 Immobile Neighbor Approximation

In the immobile neighbor approximation, the dynamics of only a single chain in the system are considered, while the rest of the chains are assumed to be immobile and form the substrate potential for the mobile chain. Let u_n be the displacement of the nth particle in the mobile chain from its equilibrium position. Then the Hamiltonian of the chain has the form

$$H = \sum_n \left[\frac{1}{2} \dot{u}_n^2 + \frac{1}{2}(u_{n+1} - u_n - \delta)^2 + V(u_n) \right] , \tag{3.54}$$

where the substrate potential is

$$V(u) = 2 \sum_{k=1}^{+\infty} \sum_{j=-\infty}^{+\infty} U(r_{jk}) . \tag{3.55}$$

The substrate potential (3.55) is a periodic function with period equal to the chain spacing. The potential shape depends only on the dimensionless equilibrium distance r_0 of the Lennard-Jones potential (3.50), and the energy of interaction ε gives the amplitude of the potential.

For the value $r_0 = 1.67$ used in the calculation, the substrate potential $V(u)$ coincides with the sinusoidal potential $V_0 + 0.3513\varepsilon \sin^2(\pi u)$ to 0.1 % accuracy, where $V_0 = \min V(u)$ is the minimum value of the substrate potential. Therefore, the Hamiltonian (3.54) can be written in the same form as the Hamiltonian for the Frenkel–Kontorova model up to a difference of zero-energy reference level:

$$H = \sum_n \left[\frac{1}{2} \dot{u}_n^2 + \frac{1}{2}(u_{n+1} - u_n)^2 + \epsilon \sin^2(\pi u) \right] , \tag{3.56}$$

where $\epsilon = 0.3513\varepsilon$ is the substrate potential amplitude. Thus, in the immobile neighbor approximation, the two-dimensional model of a quasi-one-dimensional molecular crystal reduces to the well-studied Frenkel–Kontorova model with the dimensionless Hamiltonian (3.56).

3.3.3 Dispersion of Low-Amplitude Waves

The Hamiltonian of the quasi-one-dimensional crystal (3.49) gives the equations of motion

$$\ddot{u}_{m,n} = -\frac{\partial H}{\partial u_{m,n}} , \quad n, m = 0, \pm 1, \pm 2, \dots . \tag{3.57}$$

For a low-amplitude displacement $|u_{m,n}| \ll 1$, one can replace the nonlinear equations (3.57) by the system of linear equations

$$\ddot{u}_{2m+1,n} = u_{2m+1,n+1} - 2u_{2m+1,n} + u_{2m+1,n-1}$$

$$+ \sum_{i=1}^{+\infty} \sum_{j=-\infty}^{+\infty} \left[K_{2i-1,j}(u_{2m+2i,n+j} + u_{2m-2i+2,n+j} - 2u_{2m+1,n}) \right.$$

$$\left. + K_{2i,j}(u_{2m+2i+1,n+j} + u_{2m-2i+1,n+j} - 2u_{2m+1,n}) \right],$$

$$(3.58)$$

$$\ddot{u}_{2m,n} = u_{2m,n+1} - 2u_{2m,n} + u_{2m,n-1}$$

$$+ \sum_{i=1}^{+\infty} \sum_{j=-\infty}^{+\infty} \left[K_{2i-1,j-1}(u_{2m+2i-1,n+j} + u_{2m-2i+1,n+j} - 2u_{2m,n}) \right.$$

$$\left. + K_{2i,j}(u_{2m+2i,n+j} + u_{2m-2i,n+j} - 2u_{2m,n}) \right], \quad (3.59)$$

for $n, m = 0, \pm 1, \pm 2, \ldots$, where

$$K_{2i-1,j} = U''(r_{2i-1,j})\frac{(j + 1/2)^2}{r_{2i-1,j}^2} + U'(r_{2i-1,j})\frac{(2i - 1)^2 b^2}{r_{2i-1,j}^3},$$

$$K_{2i,j} = U''(r_{2i,j})\frac{j^2}{r_{2i,j}^2} + U'(r_{2i,j})\frac{(2ib)^2}{r_{2i,j}^3},$$

for $i = 1, 2, \ldots, j = 0, \pm 1, \pm 2, \ldots$, and the primes denote derivatives. The intermolecular distances are

$$r_{2i-1,j} = \left[(j + 1/2)^2 + (2i - 1)^2 b^2 \right]^{1/2},$$

$$r_{2i,j} = \left[j^2 + (2ib)^2 \right]^{1/2}.$$

It can be readily seen that $r_{2i-1,-j-1} = r_{2i-1,j}$ and $r_{2i,-j} = r_{2i,j}$, whence the same equalities will hold for the corresponding rigidity coefficients, viz.,

$$K_{2i-1,-j-1} = K_{2i-1,j}, \quad K_{2i,-j} = K_{2i,j}. \quad (3.60)$$

Using the equalities (3.60), the linearized equations of motion (3.58) and (3.59) can be rewritten in the more convenient form

$$\ddot{u}_{2m+1,n} = u_{2m+1,n+1} - 2u_{2m+1,n} + u_{2m+1,n-1}$$

$$+ \sum_{i=1}^{+\infty} \sum_{j=0}^{+\infty} K_{2i-1,j} \left(u_{2m+2i,n+j} + u_{2m-2i+2,n+j} \right.$$

$$\left. + u_{2m+2i,n-j-1} + u_{2m-2i+2,n-j-1} - 4u_{2m+1,n} \right)$$

$$+ \sum_{i=1}^{+\infty} \left[K_{2i,0} \left(u_{2m+2i+1,n} + u_{2m-2i+1,n} - 2u_{2m+1,n} \right) \right.$$

$$+ \sum_{j=1}^{+\infty} K_{2i,j} \left(u_{2m+2i+1,n+j} + u_{2m-2i+1,n+j} + u_{2m+2i+1,n-j} \right.$$

$$\left. \left. + u_{2m-2i+1,n-j} - 4u_{2m+1,n} \right) \right] , \qquad (3.61)$$

$$\ddot{u}_{2m,n} = u_{2m,n+1} - 2u_{2m,n} + u_{2m,n-1}$$

$$+ \sum_{i=1}^{+\infty} \sum_{j=0}^{+\infty} K_{2i-1,j} \left(u_{2m+2i-1,n+j+1} + u_{2m-2i+1,n+j+1} \right.$$

$$\left. + u_{2m+2i-1,n-j} + u_{2m-2i+1,n-j} - 4u_{2m,n} \right)$$

$$+ \sum_{i=1}^{+\infty} \left[K_{2i,0} \left(u_{2m+2i,n} + u_{2m-2i,n} - 2u_{2m,n} \right) \right.$$

$$+ \sum_{j=1}^{+\infty} K_{2i,j} \left(u_{2m+2i,n+j} + u_{2m-2i,n+j} + u_{2m+2i,n-j} \right.$$

$$\left. \left. + u_{2m-2i,n-j} - 4u_{2m,n} \right) \right] , \qquad (3.62)$$

for $n, m = 0, \pm 1, \pm 2, \ldots$. In order to obtain the dispersion equation, we substitute the solution in the form of a traveling wave into (3.61) and (3.62):

$$u_{2m,n} = A \exp i(q_1 n + q_2 2m - \omega t) ,$$
$$u_{2m+1,n} = A \exp i \left[q_1 (n - 1/2) + q_2 (2m + 1) - \omega t \right] , \qquad (3.63)$$

for $n = 0, \pm 1, \pm 2, \ldots$, and $m = 0, \pm 1, \pm 2, \ldots$, where $A \ll 1$ and $q_1, q_2 \in [0, \pi]$ are the amplitude and wavenumbers, respectively.

Substituting the solution (3.63) into (3.61) and (3.62) gives the same dispersion equation:

$$\omega^2(q_1, q_2) = 2(1 - \cos q_1)$$
$$+ 4 \sum_{i=1}^{+\infty} \left\{ \sum_{j=0}^{+\infty} K_{2i-1,j} \left[1 - \cos q_1 (j + 1/2) \cos q_2 (2i - 1) \right] \right.$$
$$\left. + \frac{1}{2} K_{2i,0} (1 - \cos 2i q_2) + \sum_{j=1}^{+\infty} K_{2i,j} (1 - \cos j q_1 \cos 2i q_2) \right\}.$$
$$(3.64)$$

It can be shown that, in a numerical simulation, it suffices to consider only the interaction of the chain with a finite number L (limited above and below) of the nearest neighboring chains. Therefore, the infinite sums describing the chain interaction must be substituted for the finite sums. This leads to the Hamiltonian of the system (3.49) to take the form

$$H = \sum_{m,n} \left[\frac{1}{2} \dot{u}_{m,n}^2 + \frac{1}{2} (u_{m,n+1} - u_{m,n} - \delta)^2 + \sum_{k=1}^{L} \sum_{j=-\infty}^{+\infty} U(r_{m,n;k,j}) \right]. \quad (3.65)$$

Similarly, in (3.58)–(3.64), the infinite sum $\sum_{i=1}^{+\infty}$ over index i must be substituted for the finite sum $\sum_{i=1}^{L}$.

If one confines attention to only the nearest neighboring chains, the dispersion equation (3.64) has the particularly simple form

$$\omega^2(q_1, q_2) = 2(1 - \cos q_1) + 4 \sum_{j=0}^{+\infty} K_{1,j} \left[1 - \cos q_1 (j + 1/2) \cos q_2 \right]. \quad (3.66)$$

The dispersion equations (3.64) and (3.66) give the two-dimensional dispersion surface $\omega = \Omega(q_1, q_2)$ which is shown in Fig. 3.26 (left) for the three values of the interchain interaction parameters ε. When the wavenumbers are varied over the ranges $0 < q_1 < \pi$ and $0 < q_2 < \pi$, the dispersion surface behaves like a smooth function which increases steadily as the variables q_1 and q_2 increase. With an increase in the interaction parameter ε, the slope of the surface decreases monotonically over the variable q_2 and vanishes in the limit $\varepsilon \to +0$:

$$\lim_{\varepsilon \to 0} \Omega(q_1, q_2) = \Omega_0(q_1),$$

where $\Omega_0(q_1) = 2 \sin(q_1/2)$ specifies the dispersion curve of an isolated molecular chain. Note that, in the immobile neighbor approximation, in the framework of the

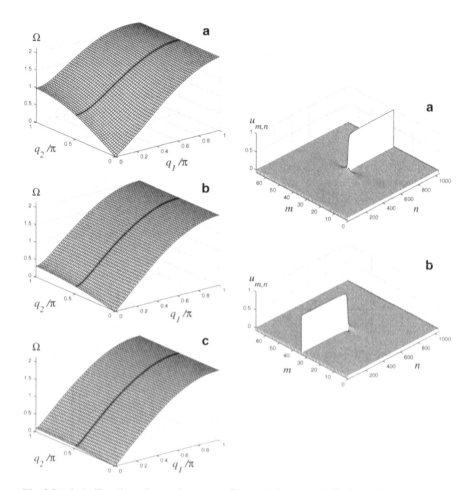

Fig. 3.26 *Left*: The dispersion surface $\omega = \Omega(q_1, q_2)$ for $r_0 = 1.67$, $L = 1$, and $\varepsilon = 0.07$ (**a**), $\varepsilon = 0.007$ (**b**), $\varepsilon = 0.0007$ (**c**). For comparison, the dispersion curve for the corresponding Frenkel–Kontorova model (the immobile neighbor approximation) $\omega = \Omega_1(q_1) = \Omega(q_1, \pi/2)$ is shown in all the figures. *Right*: Profiles of the positive (**a**) and negative (**b**) topological solitons in a quasi-one-dimensional crystal at $\varepsilon = 0.007$ and $L = 1$. The dependence of the displacement $u_{m,n}$ on the number of the link n in the chain m is shown

Frenkel–Kontorova model (3.56), the dispersion curve has only the optical branch:

$$\Omega_1(q_1) = \sqrt{\kappa + 4\sin^2(q_1/2)}, \qquad (3.67)$$

where $\kappa = 2\pi^2\epsilon$ is the rigidity of the substrate potential. This one-dimensional curve always lies strictly on a two-dimensional dispersion surface at $q_2 = \pi/2$: $\Omega(q_1, \pi/2) = \Omega_1(q_1)$ (see Fig. 3.26 left).

Let us find the velocities

$$s_x = \lim_{q_1 \to 0} \Omega(q_1, 0)/q_1$$

and

$$s_y = \lim_{q_2 \to 0} \Omega(0, q_2)/q_2$$

of longitudinal and transverse long-wavelength phonons, the maximum velocities of longitudinal and transverse sound. Using (3.64) and (3.66), it can be easily shown that the velocity of longitudinal sound is

$$
s_x = \left\{ 1 + 2 \sum_{i=1}^{+\infty} \left[\sum_{j=0}^{+\infty} (j + 1/2)^2 K_{2i-1,j} + \sum_{j=1}^{+\infty} j^2 K_{2i,j} \right] \right\}^{1/2}
$$

$$
= \left\{ 1 + \sum_{i=1}^{+\infty} \sum_{j=-\infty}^{+\infty} \left[(j + 1/2)^2 K_{2i-1,j} + j^2 K_{2i,j} \right] \right\}^{1/2}, \tag{3.68}
$$

and the velocity of transverse sound is

$$
s_y = \left\{ \sum_{i=1}^{+\infty} \left[2 \sum_{j=0}^{+\infty} (2i - 1)^2 K_{2i-1,j} + (2i)^2 K_{2i,0} + 2 \sum_{j=1}^{+\infty} (2i)^2 K_{2i,j} \right] \right\}^{1/2}
$$

$$
= \left(\sum_{i=1}^{+\infty} \sum_{j=-\infty}^{+\infty} i^2 K_{i,j} \right)^{1/2}, \tag{3.69}
$$

where we have used (3.60) for the rigidity coefficients.

When we take into account the interaction of only the nearest neighboring chains, (3.68) and (3.69) are significantly simplified:

$$
s_x = \left[1 + \sum_{j=-\infty}^{+\infty} (j + 1/2)^2 K_{1,j} \right]^{1/2},
$$

$$
s_y = \left(\sum_{j=-\infty}^{+\infty} K_{1,j} \right)^{1/2}.
$$

The dependencies of s_x and s_y on the energy of interchain interaction ε and the number of interacting neighboring chains L are given in Tables 3.1 and 3.2. As the tables show, the longitudinal sound velocity is always greater than the transverse.

Table 3.1 Dependence of the longitudinal sound velocity s_x on the parameters ε and L

ε	$L = 1$	$L = 2$	$L = 4$	$L = 6$	$L = 8$
0.07	0.752958	0.736010	0.733030	0.732789	0.732739
0.007	0.978108	0.976817	0.976593	0.976575	0.976571
0.0007	0.997832	0.997706	0.997684	0.997682	0.997682

Table 3.2 Dependence of the transverse sound velocity s_y on the parameters ε and L

ε	$L = 1$	$L = 2$	$L = 4$	$L = 6$	$L = 8$
0.07	0.491599	0.524050	0.528294	0.528671	0.528745
0.007	0.155457	0.165719	0.167054	0.167173	0.167196
0.0007	0.049160	0.052405	0.052829	0.052865	0.052872

This phenomenon results from the prevalence of the intrachain interaction over the interchain interaction. The velocities are commensurable only for the energy of longitudinal interaction $\varepsilon = 0.07$. For greater values of ε, the interchain interaction becomes comparable with the intrachain interaction and the system ceases to be quasi-one-dimensional. If the interchain interaction is decreased, the longitudinal sound velocity decreases monotonically, while the transverse sound velocity increases. As $\varepsilon \searrow 0$, we have $s_y \searrow 0$ and $s_x \nearrow 1$. Tables 3.1 and 3.2 also show that the velocities change slightly with increasing number of interacting neighboring chains L. It allows us to conclude that even with the power-law decrease in the Lennard-Jones potential, the interaction between nearest neighbors is predominant.

3.3.4   Stationary States of Topological Solitons

Here we consider a finite segment of a one-dimensional crystal with $1 \le n \le N$ and $1 \le m \le M$. The end particles of the segment are assumed to be immobile. The potential energy of the segment (soliton energy) is defined by the equation

$$E = \sum_{m=1}^{M-1} \sum_{n=1}^{N-1} \left[\frac{1}{2}(u_{m,n+1} - u_{m,n} - \delta)^2 - e_0 + \sum_{k=1}^{L} \sum_{j} U(r_{m,n;k,j}) \right]. \qquad (3.70)$$

To find the stationary state of the topological soliton, it is necessary to solve the constrained minimum problem

$$E \xrightarrow[u_{m,n}]{} \min : \quad \begin{cases} u_{m,1} = 0, & u_{m,N} = 0, & m = 1, 2, \ldots, M, \\ u_{1,n} = 0, & u_{M,n} = 0, & n = 1, 2, \ldots, N, \end{cases} \qquad (3.71)$$

with the boundary conditions

$$u_{M/2+1,1} = 0 , \quad u_{M/2+1,N} = 1 ,$$

for the positive soliton (kink) and

$$u_{M/2+1,1} = 1 , \quad u_{M/2+1,N} = 0 ,$$

for the negative soliton (antikink).

Let $\{u_{m,n}\}_{n=1,m=1}^{N,M}$ be a solution of the minimum problem (3.71) with the boundary conditions corresponding to the topological soliton. Then the coordinates of the stationary soliton centre can be obtained as

$$\overline{m} = \sum_{m=1}^{M} \sum_{n=1}^{N} m p_{m,n} , \qquad \overline{n} = \sum_{m=1}^{M} \sum_{n=1}^{N} n p_{m,n} ,$$

along with its transverse and longitudinal diameters

$$D_y = 1 + 2 \left[\sum_{m=1}^{M} \sum_{n=1}^{N} (m - \overline{m})^2 p_{m,n} \right]^{1/2} , \quad D_x = 1 + 2 \left[\sum_{m=1}^{M} \sum_{n=1}^{N} (n - \overline{n})^2 p_{m,n} \right]^{1/2} ,$$

where the two-dimensional sequence

$$p_{m,n} = \frac{1}{S} \left[(u_{m,n+1} - u_{m,n})^2 + (u_{m,n} - u_{m,n-1})^2 \right] ,$$

with

$$S = \sum_{m=1}^{M} \sum_{n=1}^{N} \left[(u_{m,n+1} - u_{m,n})^2 + (u_{m,n} - u_{m,n-1})^2 \right] ,$$

specifies the energy distribution of longitudinal chain deformation over the crystal.

The minimum problem (3.71) was solved numerically using the conjugate gradient method [59]. The crystal segment with $N = 1{,}000$ and $M = 67$ was adapted for the calculation. A soliton-like profile with centre between the $N/2$th and $(N/2 + 1)$th particles in the $(M/2 + 1)$th chain was used as the initial point:

$$u_{m,n} = 0 , \quad m = 1, 2, \ldots, N , \quad n = 1, 2, \ldots, M ,$$
$$u_{M/2+1,n} = 1 , \quad n = N/2 + 1, N/2 + 2, \ldots, N , \quad \text{for the kink}$$
$$u_{M/2+1,n} = 1 , \quad n = 1, 2, \ldots, N/2 , \quad \text{for the antikink.}$$

Views of the positive and negative soliton solutions are shown in Fig. 3.26 (right). For the positive soliton (see Fig. 3.26a right) in the central chain ($m = M/2 +$

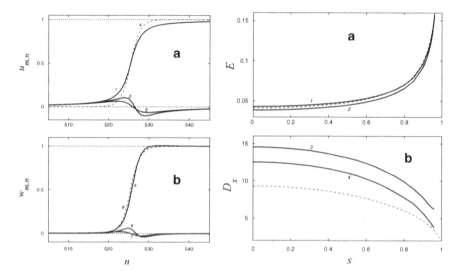

Fig. 3.27 *Left*: Dependencies of the displacement $u_{m,n}$ (**a**) and relative displacement $w_{m,n} = u_{m,n} - u_{m-1,n}$ (**b**) on the chain link number n for the chains in a region of localization of the positive topological soliton. The dependencies are shown for the central chain with the number $m = M/2 + 1$ (*lines 1* and *5*) and two neighboring chains with $m = M/2 + 2$ (*lines 2* and *6*) and $m = M/2 + 3$ (*lines 3* and *7*). For comparison, kink profile obtained in the approximation of immobile neighbor chains is shown (*dashed lines 4* and *8*). $\varepsilon = 0.07$, $L = 1$. *Right*: Dependencies of the energy E (**a**) and longitudinal diameter D_x (**b**) of the topological soliton on its velocity s in the case of the antikink (*lines 1* and *3*) and kink (*lines 2* and *4*) in a quasi-one-dimensional crystal with $\varepsilon = 0.007$. For comparison, the dependencies obtained in the approximation of immobile neighbor chains are shown (*dashed lines 4* and *8*)

1), the displacement has the form of a smooth antikink. In the range of soliton localization, extension of the central chain occurs, followed by local compression of neighboring chains. For the negative soliton (see Fig. 3.26b right), the relative displacements of the central chain already have the form of a kink. In the range of antisoliton localization, there is compression of the central chain, accompanied by local extension of neighboring chains.

Figure 3.27 (left) compares the profiles of the topological soliton and soliton obtained in the immobile neighbor approximation. As can be seen from the figure, although the derivatives of the curves, taken at the kink centre, coincide for the quasi-one-dimensional and Frenkel–Kontorova models, the decay character of the curves at infinity differs qualitatively. In the case of mobile neighbors, the decay is not exponential, but follows a power law, as does the perturbation decay of neighboring chains. However, the curve corresponding to the displacement of atoms in the central chain relative to atoms in the neighboring chain appears to be well approximated by the Frenkel–Kontorova model (see Fig. 3.27b left). In the region of soliton localization, the nearest neighboring chains adjust to the topological defect of the central chain. For this reason, the topological soliton in a quasi-one-dimensional crystal has not only the longitudinal width D_x but also the transverse width $1 < D_y < D_x$.

If in solving the minimum problem (3.71) we fix the particle displacement, taken at the soliton centre $u_{M/2+1,N/2}$, and then vary it, we can obtain the dependence of the stationary state energy E on the position of its centre $\bar{n}$, i.e., find the Peierls potential profile $E(M/2 + 1, \bar{n})$. The Peierls potential is a periodic function with period equal to the longitudinal crystal spacing. The minima and maxima of the potential correspond to the ground and metastable states of the stationary topological soliton, respectively. The amplitude of the Peierls potential gives the soliton pinning energy ΔE.

The dependencies of the energy E, diameters D_x, D_y, and pinning energy ΔE of the topological soliton on the interchain interaction energy ε and soliton sign are listed in Table 3.3. The difference in the values of the kink and antikink parameters is defined by the fact that the chains are slightly compressed when they associate in a crystal, so that asymmetry in extension and compression appears as a result of nonlinearity of the interchain interactions. As can be seen from Table 3.3, the soliton is weakly pinned only for the strong interchain interaction $\varepsilon = 0.07$. For the weak interactions $\varepsilon = 0.007$ or $\varepsilon = 0.0007$, the pinning actually vanishes ($\Delta E < 10^{-9}$).

Table 3.4 gives the dependencies of the energy E and diameters D_x and D_y of the positive soliton on the number of interacting neighboring chains L for the interaction energy $\varepsilon = 0.007$. The soliton energy and shape remain practically unchanged as the chain number increases. Therefore, for both the nonlinear and linear waves in the quasi-one-dimensional crystal, the interaction with only two nearest neighbors is dominant. Thus, for the analysis of soliton dynamics, it is

Table 3.3 Dependencies of the energy E, diameters D_x and D_y, and pinning energy ΔE of the topological soliton on the interchain interaction energy ε

ε	Type	E	D_x	D_y	ΔE
	Antikink	0.15117	6.48	3.62	3.1×10^{-5}
0.07	Kink	0.10516	3.90	1.49	1.8×10^{-5}
	FK model	0.10574	3.76	1.00	1.0×10^{-5}
	Antikink	0.04302	14.5	2.55	0
0.007	Kink	0.03851	12.6	2.18	0
	FK model	0.04115	9.29	1.00	0
	Antikink	0.01310	40.3	2.27	0
0.0007	Kink	0.01266	37.8	2.14	0
	FK model	0.01376	27.1	1.00	0

Table 3.4 Dependencies of the energy E and diameters D_x and D_y of the positive soliton on the number of interacting neighboring chains L for the interaction energy $\varepsilon = 0.007$

L	E	D_x	D_y
1	0.043017	12.61	2.18
2	0.045279	11.97	2.16
4	0.045657	11.74	2.14
6	0.045691	11.71	2.12
8	0.045700	11.68	2.11

sufficient to take $L = 1$, i.e., assume that only two nearest neighboring chains interact.

3.3.5 Topological Soliton Dynamics

For weak interaction between neighboring chains, viz., $\varepsilon = 0.007$ and 0.0007, pinning is practically absent. Therefore, soliton solutions are expected to exist in the form of a traveling wave $u_{m,n}(t) = u(m, n - st)$, where s is the wave velocity and the wave shape depends smoothly on the chain link number n. Then the second derivative with respect to time can be replaced by the second discrete derivative

$$\ddot{u}_{m,n} = s^2 \frac{\partial^2}{\partial n^2} u(m, n - st) = s^2 (u_{m,n+1} - 2u_{m,n} + u_{m,n-1}) .$$

Substituting this expression into the equations of motion (3.57), we obtain the set of discrete equations

$$s^2 (u_{m,n+1} - 2u_{m,n} + u_{m,n-1}) = -\frac{\partial H}{\partial u_{m,n}} , \qquad n, m = 0, \pm 1, \pm 2, \dots .$$

The solution to the discrete equations (3.72) corresponds to the minimum of the functional

$$F = \sum_{m=1}^{M-1} \sum_{n=1}^{N-1} \left[\frac{1}{2} (u_{m,n+1} - u_{m,n} - \delta)^2 - \frac{1}{2} s^2 (u_{m,n+1} - u_{m,n})^2 \right.$$

$$\left. + \sum_{k=1}^{L} \sum_{j} U(r_{m,n;k,j}) - e_0 \right] .$$

Therefore, the form of the moving soliton (like the form of the stationary soliton) can be sought numerically as the solution of the minimum problem

$$F \longrightarrow \min_{u_{m,n}} . \tag{3.72}$$

If $\{u_{m,n}^0\}$ is a solution to the minimum problem (3.72), the soliton energy is

$$E = \sum_{m=1}^{M-1} \sum_{n=1}^{N-1} \left[\frac{1}{2} (u_{m,n+1}^0 - u_{m,n}^0 - \delta)^2 \right.$$

$$\left. + \frac{1}{2} s^2 (u_{m,n+1}^0 - u_{m,n}^0)^2 + \sum_{k=1}^{L} \sum_{j} U(r_{m,n;k,j}) - e_0 \right] .$$

and the boundary conditions for the equations of motion at $t = 0$ are

$$u_{m,n}(0) = u_{m,n}^0 , \quad \dot{u}_{m,n}(0) = -s(u_{m,n+1}^0 - u_{m,n}^0) , \quad n = 1, 2, \ldots, N ,$$

$$m = 1, 2, \ldots, M . \tag{3.73}$$

Numerical solution of the problem (3.72) has shown that the topological soliton has the subsonic spectrum of velocities $0 < s < s_x$. The dependencies of the energy E and longitudinal width D_x of the soliton on its velocity s are shown in Fig. 3.27 (right). As in the Frenkel–Kontorova model, the soliton energy increases steadily with increasing velocity ($E \nearrow \infty$ as $s \nearrow s_x$), and the longitudinal diameter decreases monotonically ($D_x \searrow 1$ as $s \nearrow s_x$).

We also modeled the soliton dynamics. Numerical integration of the equations of motion

$$\ddot{u}_{m,n} = -\frac{\partial H}{\partial u_{m,n}} , \quad n = 1, 2, \ldots, N , \quad m = 1, 2, \ldots M . \tag{3.74}$$

with the initial condition (3.73) has shown that, for weak interchain interaction, viz., $\varepsilon = 0.007$ and 0.0007, the motion of topological solitons in the quasi-one-dimensional crystal is not accompanied by phonon emission. The solitons retain their initial forms and velocities over the whole spectrum of velocities $0 < s < s_x$.

The situation changes for the stronger interchain interaction $\varepsilon = 0.07$. The transverse diameter of the kinks increases, while the longitudinal diameter decreases to sizes at which discreteness effects become significant (see Table 3.3). It turns out that the pinning energy of the kinks in the quasi-one-dimensional crystal is two to three times greater than that predicted by the corresponding Frenkel–Kontorova model (in the immobile neighbor approximation). Therefore, the discreteness effects in the crystal are expected to be more pronounced. The smooth soliton solutions do not exist in this case. The kink motion is always accompanied by phonon emission (Fig. 3.28 left). As a result, its velocity decreases monotonically. Numerical simulation of the kink dynamics allows us to find the dependence of its velocity s on time t. The function $s(t)$ is a monotonically decreasing function, so it is possible to obtain the dependence of the kink slowdown $|ds/dt|$ on its velocity s (see Fig. 3.28 right). For comparison, the same dependence as obtained in the corresponding Frenkel–Kontorova model (all chains except one are fixed) is also shown in Fig. 3.28 (right). As expected, the slowdown in the quasi-one-dimensional crystal at any velocity turns out to be stronger than that in the corresponding Frenkel–Kontorova model.

Phonon emission by a moving kink was investigated in the discrete Frenkel–Kontorova model [11]. Numerical simulation has shown that, during the motion, the kink radiates a phonon of a certain mode if it resonates with this phonon mode. The physical condition for this resonance can be stated as follows: moving along the chain together with the phonon mode, the kink 'sees' all particles oscillating in

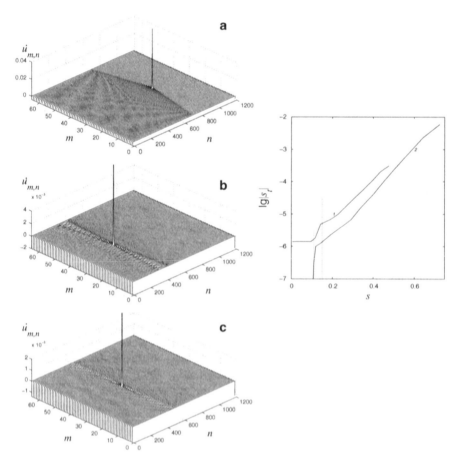

Fig. 3.28 *Left*: Phonon emission during the motion of the weakly pinned antikink in the quasi-one-dimensional crystal ($\varepsilon = 0.07$). Distribution of velocities $\dot{u}_{m,n}$ in the crystal with the condition of fixed ends. The antikink velocities are $s = 0.254$ (**a**), $s = 0.137$ (**b**), and $s = 0.093$ (**c**). *Right*: Dependence of the kink slowdown on its velocity s in the one-dimensional crystal with $\varepsilon = 0.07$ (*line 1*) and in the corresponding Frenkel–Kontorova model (*line 2*). *Dashed lines* show the critical values of velocities $s_1 = \Omega_1(0)/2\pi = \Omega(0, \pi/2)/2\pi = 0.1109$ and $s_2 = \Omega(0, \pi)/2\pi = 0.1509$

phase with each other, i.e.,

$$q - \frac{\omega(q)}{s} = -2\pi n , \quad n = 0, 1, 2 \ldots ,$$

or

$$\frac{\omega(q)}{q + 2\pi n} = s , \quad n = 0, 1, 2 \ldots .$$

Fig. 3.29 Dispersion curve
for the discrete
Frenkel–Kontorova model
(the approximation of
immobile neighboring chains)
$\omega = \Omega_1(q_1)$ (2.19). The
points where it intersects the
straight line with slope equal
to the kink velocity
correspond to the phonon
modes radiated by the kink.
These modes are also shown
by the *filled circles* in the first
Brillouin zone $-\pi < q_1 < \pi$

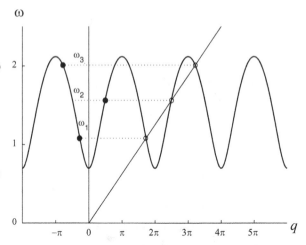

Graphically, the different modes in which the kink can radiate are easy to represent
on the dispersion curve $\omega = \omega(q)$, shown in all Brillouin zones, by drawing a
straight line with slope equal to the kink velocity. The intersections of the straight
line with the curve $\omega(q)$ correspond to the resonant modes of the kink.

The dispersion curve corresponding to the discrete Frenkel–Kontorova model
$\omega = \Omega_1(q_1)$ (see (3.67)) is depicted in Fig. 3.29. In this case, the kink radiates
forward at a frequency ω_2 and back at frequencies ω_1 and ω_3. The farther the intersection point is from the first Brillouin zone, the weaker, as a rule, the association
of the kink with this mode and the lower the radiation level corresponding to this
mode. Even at high discreteness, the emission into the radiation modes lying outside
the region of $0 \le q \le 2\pi$ is very low. In the case shown in Fig. 3.29, only the
back radiation at frequency ω_1 is noticeable. When the kink velocity decreases to
$s < s_1 = \Omega_1(2\pi)/2\pi$, the phonon emission falls drastically and the velocity of the
soliton becomes almost constant.

For $\varepsilon = 0.07$, the critical value of the velocity is $s_1 = 0.1109$. Numerical
simulation of the soliton dynamics in the immobile neighbor approximation cor-
roborates this result. At $s > s_1$, the soliton velocity decreases monotonically and
the logarithm of the slowdown $\lg |s_t|$ decreases proportionally with the velocity s
(Fig. 3.28 right). After reaching the value of s_1, the soliton velocity stops changing.
The soliton radiates phonons and 'descends' the dispersion surface with decreasing
velocity, following the curve

$$\frac{\Omega(q_1, \pi/2)}{q_1} = \frac{\Omega_1(q_1)}{q_1} = s , \qquad (3.75)$$

up to the breakup velocity s_1. In Fig. 3.30, the dependence of the kink's phonon
emission frequency on its velocity (line 2), obtained in numerical simulation, is
compared with the curve obtained by substituting $q_1(s)$ as given by (3.75) into the

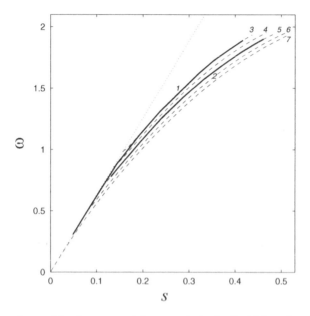

Fig. 3.30 Dependence of the frequency of phonon emission by the kink on its velocity for the interchain interaction value $\varepsilon = 0.07$. *Lines 1* and *2* show the dependencies for the quasi-one-dimensional crystal (all chains are mobile) and for immobile neighboring chains (the Frenkel–Kontorova model), respectively. *Dashed lines* give the theoretical dependencies $\omega = \Omega(q(s), q_2)$, where $q(s)$ is obtained from the equation $\Omega(q, q_2) - qs = 0$ with $q_2 = 0, \pi/3, \pi/2, 2\pi/3$, and π (*lines 3–7*). The *dotted line* is asymptotic to *line 1* for low velocities, $\omega = 2\pi s$

dispersion equation (line 5)

$$\omega(s) = \Omega(q_1(s), \pi/2) . \tag{3.76}$$

As can be seen, the curves coincide exactly.

In a quasi-one-dimensional crystal, a similar analysis of the soliton radiation frequencies leads to the resonance condition

$$\frac{\Omega(q_1, q_2)}{q_1} = s , \tag{3.77}$$

which now defines, not a single mode, but the set of modes shown in Fig. 3.31, represented by constant velocity curves for several values of the soliton velocity. For the soliton velocity $s = 0.254$ lying within the wavenumber region, where resonance with the mode can lead to significant radiation into this mode, only back radiation is possible. The level line of the function (3.77) lies to the left of the straight line $q_1 = 2\pi$, and correspondingly to the left of the straight line $q_1 = 0$ upon transferring to the first Brillouin zone. Indeed, only the back radiation can be seen in Fig. 3.28a (left). For the soliton velocity $s = 0.137$, both forward

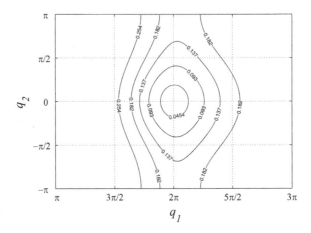

Fig. 3.31 Level lines of the function $s = \Omega(q_1, q_2)/q_1$ for $s = 0.0454, 0.093, 0.137, 0.182$, and 0.254 ($\varepsilon = 0.07$)

and backward radiation are possible, as observed in the numerical experiment (Fig. 3.28b left).

Since the soliton width is very small, the soliton radiates mainly into a mode with the maximum possible value of q_2 ($0 \leq q_2 \leq \pi$). Clearly, the optimal value is $q_2 = \pi$ (particles in neighboring chains oscillate out of phase, as happens in the motion within the soliton centre, and the best resonance with the mode is thereby reached). However, when the soliton velocity decreases, this condition ceases to be fulfilled (see Fig. 3.31). Then the soliton radiates at the frequency

$$\omega = \Omega\big(q_1^{\max}(s), q_2^{\max}(s)\big) \,, \tag{3.78}$$

where $q_2^{\max}(s)$ is the maximum value of q_2 for a given kink velocity s lying on the level line of the function (3.77) in the range $0 < q_2 < \pi$, and $q_1^{\max}(s)$ is the corresponding value of q_1. As can be seen in Fig. 3.31, at small soliton velocities, the maximum of q_2 is reached at $q_1 \approx 2\pi$. Therefore, the dependence of the frequency of phonon emission by the soliton on its velocity should tend to the straight line $\omega = 2\pi s$, and this is indeed what is observed in the numerical simulation (see Fig. 3.30). Thus, in a quasi-one-dimensional crystal, the soliton radiates when it 'descends' the dispersion surface, first along the line $q_2 = \pi$ (q_1 changes from π to a value close to zero) and then along the line $q_1 \approx 0$ (q_2 decreases from π to zero). Interestingly, the energy emitted by the soliton changes sharply at two points (see Fig. 3.28 right): first at the transition from the line $q_2 = \pi$ to the line $q_1 \approx 0$ (the soliton is no longer in resonance with the radiation mode), and secondly when crossing the point ($q_1 \approx 0$, $q_2 = \pi/2$). At velocities lower than the second point, the soliton radiates after the passage of one chain spacing, and the radiation energy does not depend on its velocity.

As we have seen, the soliton spectrum in a quasi-one-dimensional molecular crystal with weak interchain interaction ($\varepsilon = 0.007$ or 0.0007) is bounded above by the velocity of longitudinal sound $0 < s < s_x$. For these velocities, the plane $\omega = q_1 s$ does not intersect the dispersion surface $\omega = \Omega(q_1, q_2)$ in the wavenumber range $0 < q_1 < 0.8\pi$ (the plane is tangent to the dispersion surface). For wavenumbers $q_1 \approx \pi$, the group velocity of phonons is close to zero, so the energy transfer into these modes is highly ineffective. At soliton velocities higher than the velocity of longitudinal sound, the intersection of the aforementioned plane with the dispersion surface already occurs in the wavenumber range where $0 < q_1 < \pi/2$ and q_2 can take the value of π. In this case, effective emission of corresponding phonons is possible. Indeed, we observed that, in this range of velocities, the solitons cannot move without emission even for weak interchain interaction. On the other hand, because of the small discreteness of the system, this resonance does not lead to noticeable emission. This is true despite the fact that the plane $\omega = q_1 s$ at velocities $0 < s < s_x$ intersects the dispersion surface in the range of wavenumbers $q_1 \approx 2\pi$, as well as in the case of a strong interchain interaction $\varepsilon = 0.07$.

3.3.6 Interaction of Topological Solitons

The Frenkel–Kontorova model is only able to describe the interaction of topological solitons located on the same molecular chain. The two-dimensional model of a quasi-one-dimensional crystal already allows one to describe the interaction of topological solitons located on different molecular chains [53, 60]. Let us consider a soliton interaction in the framework of our model with interchain interaction intensity $\varepsilon = 0.007$. First, we compare the interaction of solitons located on the same chain of the quasi-one-dimensional crystal with their interaction obtained in the FK model. Numerical simulation of the collision of same-sign solitons has shown that the two models lead to an almost identical description of their interaction. At all velocities, the solitons repel one another as elastic particles, almost without phonon emission. The situation changes dramatically when we consider the interaction of opposite-sign solitons.

In the FK model with $\varepsilon = 0.007$, the chain discreteness manifests itself very weakly, so the soliton dynamics is well described by the continuous sine–Gordon equation. In this model, solitons interact with each other as elastic particles, i.e., opposite-sign solitons pass through each other without emitting phonons or changing their velocities. In the weakly discrete FK model considered above, this interaction of solitons occurs at all velocities $s \leq 0.01$ (see Fig. 3.32a left). The chain discreteness begins to manifest itself only at very low velocities $s < 0.01$. At the velocity $s = 0.005$, soliton collision causes them to recombine, resulting in the formation of a long-lived, localized nonlinear vibration, that is, a low-frequency breather (a bound state of two opposite-sign solitons) (see Fig. 3.32b left).

In the two-dimensional molecular crystal, the collision of opposite-sign solitons always occurs inelastically. Solitons can pass through each other only at high

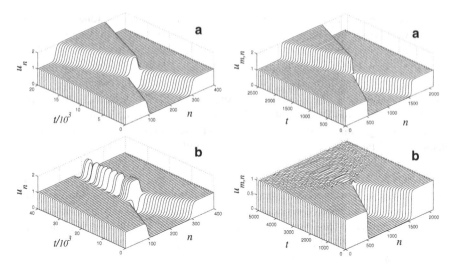

Fig. 3.32 *Left*: Collision of two opposite-sign solitons in the discrete Frenkel–Kontorova model ($\varepsilon = 0.007$) at the velocities $s_1 = -s_2 = 0.01$ (**a**) and $s_1 = -s_2 = 0.005$ (**b**). The distribution of the longitudinal displacement u_n over the chain is shown as a function of time t. *Right*: Collision of two opposite-sign solitons on the same chain in a quasi-one-dimensional system ($N = 2,100$, $M = 63$, $m = 32$, $\varepsilon = 0.007$) at the velocities $s_1 = -s_2 = 0.5$ (**a**) and $s_1 = -s_2 = 0.25$ (**b**). The distribution of the longitudinal displacement $u_{m,n}$ over the chain ($m = 32$) is shown as a function of time t

velocities ($s \geq 0.5$). In this case, a noticeable part of the energy is expended in radiation into the neighboring chains, and this decreases the soliton velocity after the collision (see Fig. 3.32a right). At lower velocities $s \leq 0.25$, solitons always recombine (see Fig. 3.32b right). This scenario of soliton recombination in the two-dimensional model differs essentially from the recombination scenario in the discrete FK model. In this case, localized breather-like excitations are not formed. All the energy of colliding solitons is radiated into the neighboring chains. This allows us to conclude that low-frequency breathers cannot exist in the quasi-one-dimensional system, i.e., the low-frequency breathers in the FK model are an artifact of the immobile neighbor approximation.

To test this conclusion, a numerical simulation of the low-frequency breather dynamics was carried out in the immobile neighbor approximation. The results show that the breathers have frequencies lying within the lowest gap in the phonon spectrum $0 < \omega < \Omega_1(0)$. For values of the intermolecular interaction energy $\epsilon = 0.07, 0.007$, and 0.0007, the low-frequency breather is a long-lived localized nonlinear vibration (see Fig. 3.33a left). The picture changes drastically if we let neighboring chains move. As there is no low-frequency gap in the phonon spectrum of the quasi-one-dimensional crystal, the breather begins to emit phonons intensively into the neighboring chains and collapses almost immediately (see Fig. 3.33b left). For all values of ϵ, the breather completely collapses during one

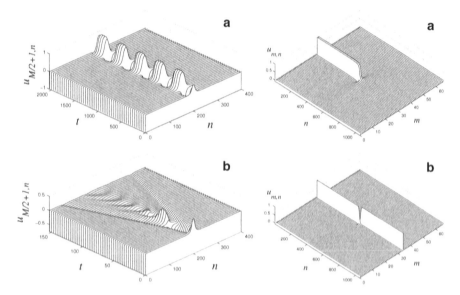

Fig. 3.33 *Left*: Low-frequency breather in the quasi-one-dimensional crystal ($M = 51$, $N = 400$) in the case of immobile neighboring chains (**a**) and its degradation when link motion in neighboring chains is possible (**b**). The displacement $u_{26,n}$ of the middle chain links is shown. Breather frequency $\omega_b = 0.0157$ and intermolecular interaction energy $\epsilon = 0.007$. *Right*: Bound state of a pair of kinks (**a**) and opposite-sign topological solitons (**b**) located on neighboring chains $m_2 = m_1 + 1$ ($\varepsilon = 0.007$). The distribution of longitudinal displacements $u_{m,n}$ is shown

oscillation period and its energy is uniformly distributed over all chains in the system.

Let us consider the interaction of solitons located on different chains with numbers $m_1 \neq m_2$. To find the bound state of the stationary solitons, one must solve the minimum problem (3.71) with a corresponding initial point. It is energetically favorable for solitons to form a bound state. Typical forms of the bound state of same- and opposite-sign solitons are shown in Fig. 3.33 (right).

In order to find the interaction potential between two solitons located on different chains m_1 and m_2 (i.e., the dependence of the energy E of a pair of solitons on the longitudinal distance R between their centres), one must fix the positions of their centres n_1 and n_2 when solving the minimum problem. The interaction potential of two same-sign solitons is shown in Fig. 3.34a. It follows from the form of the potential that, regardless of the distance between chains, the interaction of same-sign solitons located on these chains corresponds to weak repulsion at large distances and strong attraction at small distances. To form the bound state, the defects must overcome an energy barrier of height $\Delta E_1 = \max\left[E(R) - 2E_0\right]$, where E_0 is the energy of a single soliton. The depth $\Delta E_0 = 2E_0 - E(0)$ of the interaction potential well corresponds to the binding energy of the solitons.

Numerical simulation of the same-sign soliton dynamics in neighboring chains confirms the form obtained for the interaction potential. If the kinetic energy of

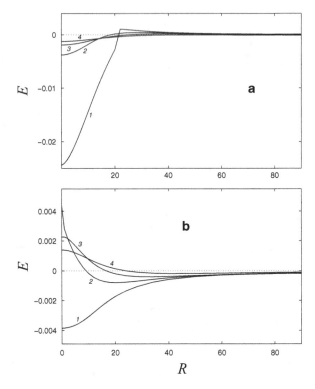

Fig. 3.34 Interaction potential of a pair of same-sign solitons (kink + kink) (**a**) and opposite-sign solitons (kink + antikink) (**b**) located on the chains m_1 and m_2, where $m_2 = m_1 + 1, m_1 + 2,$ $m_1 + 3$, and $m_1 + 4$ (*lines 1–4*). Parameter $\varepsilon = 0.007$

solitons is lower than the energy barrier ΔE_1, their collision always leads to repulsion between them. If the energy insignificantly exceeds the barrier, the solitons overcome the barrier and form the bound state. (The energy gain ΔE_0 due to the formation of the bound state is released by phonon emission.) Finally, solitons pass each other if their kinetic energy substantially exceeds the barrier ΔE_1. In this case, the solitons slow down and part of the kinetic energy is released by the phonon emission.

The interaction potential of two opposite-sign solitons is shown in Fig. 3.34b. The interaction of two solitons located on different chains always corresponds to their attraction. If the defects located on the chains are separated from each other by a distance of two or more transverse lattice spacings, their interaction corresponds to weak attraction at large distances and strong repulsion at small ones. Numerical simulation of the dynamics of solitons moving towards each other has completely confirmed this feature of their interaction.

3.3.7 Formation of Topological Solitons in a Thermalized Lattice

Consider a finite rectangular segment of a quasi-one-dimensional crystal $\Lambda = \{1 \leq n \leq N, 1 \leq m \leq M\}$ with periodic boundary conditions. We describe the dynamics of the thermalized segment by the Langevin equation

$$\ddot{u}_{mn} = -\frac{\partial H}{\partial u_{mn}} + \xi_{mn} - \gamma \dot{u}_{m,n} , \quad (m,n) \in \Lambda , \tag{3.79}$$

where the Hamiltonian H is given by (3.62), $\gamma = 1/t_r$ is the relaxation coefficient, and ξ_{mn} is the random force describing the interaction of the lattice site (m, n) with a thermal bath, which has a normal distribution with correlation function

$$\langle \xi_{mn}(t_1)\xi_{kl}(t_2)\rangle = 2\gamma T \delta_{mk}\delta_{nl}\delta(t_1 - t_2) , \tag{3.80}$$

where T is the dimensionless temperature of the thermal bath.

The equations of motion (3.79) were integrated numerically by the standard fourth-order Runge–Kutta method with a constant step of integration Δt [45]. In the numerical procedure, the lagged Fibonacci random number generator [46] was used and the δ function has the form $\delta(t) = 0$, if $|t| > \Delta t/2$ and $\delta(t) = 1/\Delta t$, if $|t| \leq \Delta t/2$, i.e., the step of numerical integration corresponds to the correlation time of the random force. For the Langevin equation to be used, one must have $\Delta t \ll t_r$. Therefore, we take the step $\Delta t = 0.1$ and the relaxation time $t_r = 10$. During the time $t = 10t_r$, the system comes to equilibrium with the thermal bath. The dynamics of the thermalized system can then be analysed.

The degree of nonlinearity of the thermal vibrations characterizes the heat capacity of the system, $c = \langle H \rangle/T$, where $\langle H \rangle$ is the average system energy. A typical dependence of the heat capacity of the system is shown in Fig. 3.35 (left).

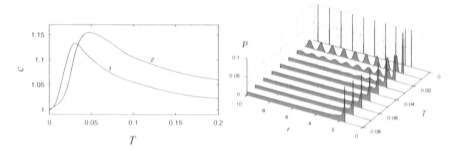

Fig. 3.35 *Left:* Dependence of the heat capacity c of the quasi-one-dimensional crystal (*line 1*) and the Frenkel–Kontorova chain (*line 2*) on temperature T. $\varepsilon = 0.07$. *Right:* Dependence of the distribution density p of the distance between the sites in neighboring chains on the temperature T of the quasi-one-dimensional crystal

At low temperatures $T \ll \varepsilon$, the heat capacity of the system is $c = 1$ (only the linear phonon modes are thermalized). With increasing temperature, the heat capacity increases steadily up to a maximum value $c_m > 1$ at T_m. A further increase in temperature leads to a monotonic decrease in the heat capacity. As $T \nearrow \infty$, the heat capacity tends to 1, $c \searrow 1$. At $\varepsilon = 0.07$, the maximum value of the heat capacity $c_m = 1.14$ is reached at temperature $T_m = 0.03$.

An increase in the heat capacity of the system at $T < T_m$ is associated with the accumulation of topological defects in the chain. The defects in the thermalized chain are produced by opposite-sign pairs and move along the chain as Brownian particles. The maximum concentration of defects is reached at temperature T_m.

To study the correlation in the motion of neighboring chains, we consider the distance r between the sites in neighboring chains as a random variable. At zero temperature, r can assume only a discrete set of values $r_k = \left[b^2 + (k + 1/2)^2 \right]^{1/2}$, $k = 0, 1, 2, \ldots$. Its distribution function $p(r)$ is a sum of delta-functions centered on the points r_k. Thermal vibrations cause spreading of this discrete set of values. The dependence of $p(r)$ on temperature T is shown in Fig. 3.35 (right). With increasing temperature, the local maxima of the distribution function become smaller and wider. The local maxima begin to disappear at temperatures $T > T_m$. There remains only one first maximum. The longitudinal order of the molecular arrangement in neighboring chains disappears and the concept of topological defect thus loses its meaning. The chains are no longer sensitive to each other's structure and begin to interact as solid rods. This manifests itself more and more as the temperature increases. As a result, the heat capacity decreases monotonically with increasing temperature.

The temperature dependence of the heat capacity of the Frenkel–Kontorova chain has the same form as in the quasi-one-dimensional crystal model, but the maximum is reached at a higher temperature $T_m = 0.05$ and $c_m = 1.16$ (see Fig. 3.35 left). This shift is associated with the fact that the formation of a defect pair modeled in the immobile neighbor approximation requires a higher energy in comparison to the case where motion of the sites in neighboring chains is possible. At temperatures $T > T_m$, the chain detaches from the substrate. As a consequence, with increasing temperature, the heat capacity decreases monotonically and tends to 1, i.e., $C \searrow 1$ as $T \longrightarrow \infty$.

The model used above cannot be used to model crystal melting. However, one can consider the loss of longitudinal order in the arrangement of sites in neighboring chains, resulting from the accumulation of topological defects, as crystal premelting. This creates the necessary conditions for further mutual detachment of the chains.

3.3.8 Conclusion

We have seen that, in the simplest nonlinear anisotropic medium, namely, a system of weakly bound molecular chains, there exist nonlinear localized excitations of soliton type with a topological charge. Their properties are determined by the intensity of the interchain interaction. At a low intensity of the interchain interaction, corresponding to a polyethylene crystal, when the sound velocity along the chain is significantly higher than in the transverse direction, these soliton-like excitations behave like the solitons in the Frenkel–Kontorova model (without changing their shape and velocity) at zero temperature. In this case, the system of bound molecular chains is a quasi-one-dimensional system. When the intensity of the interchain interaction increases so much that the transverse sound velocity becomes comparable with the longitudinal sound velocity, discreteness effects begin to appear. The solitons cannot maintain a constant velocity and shape, but slow down, emitting phonons.

References

1. Davydov, A.S.: Solitons in Molecular Systems. Naukova Dumka, Kiev (1988)
2. Krumhansl, J.A., Schrieffer, J.R.: Dynamics and statistical mechanics of one-dimensional model Hamiltonian for structural phase transition. Phys. Rev. B **11**, 3535 (1975)
3. Frenkel, Y.I., Kontorova, T.: On the theory of plastic deformation and twinning. I. Zh. Eksp. Teor. Fiz. **8**, 89 (1938)
4. Kontorova, T., Frenkel, Y.I.: On the theory of plastic deformation and twinning. II. Zh. Eksp. Teor. Fiz. **8**, 1340 (1938)
5. Christiansen, P.L., Savin, A.V., Zolotaryuk, A.V.: Zig-zag version of the Frenkel–Kontorova model. Phys. Rev. B **54**, 12892 (1996)
6. Weiner, J.H., Askar, A.: Proton migration in hydrogen-bonded chains. Nature **226**, 842 (1970)
7. Antonchenko, V.Ya., Davydov, A.S., Zolotaryuk, A.V.: Solitons and proton motion in ice-like structures. Phys. Stat. Sol. (b) **115**, 631 (1983)
8. Zolotaryuk, A.V., Spatschek, K.H., Laedke, E.W.: Stability of activation-barrier-lowering solitons. Phys. Lett. A **101**, 517 (1984)
9. Kashimori, Y., Kikuchi, I., Nishimoto, K.: The solitonic mechanism for proton transport in a hydrogen-bonded chain. J. Chem. Phys. **77**, 1104 (1982)
10. Yomosa, S.: Dynamics of the proton in one-dimensional hydrogen-bonded systems. J. Phys. Soc. Jpn. **51**, 3318 (1982)
11. Peyrard, M., Kruskal, M.D.: Kink dynamics in the highly discrete sine–Gordon system. Physica D **14**, 88 (1984)
12. Kudryavtsev, A.E.: Soliton-like solutions for a Higgs scalar field. JETP Lett. (USSR) (Engl. Transl.) **22**, 82 (1975)
13. Getmanov, B.S.: Bound states of soliton in the ϕ_2^4 field-theory model. Sov. J. Exp. Theor. Phys. Lett. **24**, 291 (1976)
14. Sigiyma, T.: Kink–antikink collisions in the two-dimensional ϕ-4 model. Prog. Theor. Phys. **61**, 1550 (1979)
15. Aubry, S.: An unified approach to the interpretation of displacive and order–disorder systems. II. Displacive systems. J. Chem. Phys. **64**, 318 (1976)
16. Ablowitz, M.J., Kruskal, M.D., Ladik, J.F.: Solitary wave collisions. SIAM J. Appl. Math. **36**, 428 (1979)
17. Klein, R., Hasenfratz, W., Theodorakopoulos, N., Wünderlich, W.: The kink–phonon and kink–kink interaction in the ϕ^4 model. Ferroelectrics **26**, 721 (1980)
18. Moshir, M.: Soliton–antisoliton scattering and capture in $\lambda\phi^4$ theory. Nucl. Phys. B **185**(2), 318 (1981)

19. Wingate, C.A.: Numerical research for ϕ^4 breather mode. SIAM J. Appl. Math. **43**(1), 120 (1983)
20. Campbell, D.K., Schonfeld, J.F., Wingate, C.A.: Resonance structure in kink–antikink interactions in ϕ^4 theory. Physica D **9**, 1 (1983)
21. Peyrard, M., Campbell, D.K.: Kink–antikink interactions in a modified sine–Gordon model. Physica D **9**, 33 (1983)
22. Campbell, D.K., Peyrard, M., Sodano, P.: Kink–antikink interactions in the double sine–Gordon equation. Physica D **19**, 165 (1986)
23. Sievers, A.J., Takeno, S.: Intrinsic localized modes in anharmonic crystals. Phys. Rev. Lett. **61**, 970 (1988)
24. MacKay, R.S., Aubry, S.: Proof of existence of breathers for time-reversible or Hamiltonian networks of weakly coupled oscillators. Nonlinearity **7**, 1623 (1994)
25. Aubry, S.: Breathers in nonlinear lattices: existence, linear stability and quantization. Physica D **103**, 201 (1997)
26. Flach, S., Willis, C.R.: Discrete breathers. Phys. Rep. D **295**, 181 (1998)
27. Kopidakis, G., Aubry, S.: Discrete breathers and delocalization in nonlinear disordered systems. Phys. Rev. Lett. **84**, 3236 (2000)
28. Kopidakis, G., Aubry, S., Tsironis, G.P.: Targeted energy transfer through discrete breathers in nonlinear systems. Phys. Rev. Lett. **87**, 165501 (2001)
29. Leitner, D.M.: Vibrational energy transfer in helices. Phys. Rev. Lett. **87**, 188102 (2001)
30. Orfanidis, S.J.: Discrete sine–Gordon equations. Phys. Rev. D **18**, 3823 (1978)
31. Pilloni, L., Levi, D.: The inverse scattering transform for solving the discrete sine–Gordon equation. Phys. Lett. A **92**, 5 (1982)
32. Kolbysheva, O.P., Sadreev, A.F.: The coupling of solitons via phonons in the ϕ^4–ϕ^2 model. Zh. Eksp. Teor. Fiz. **100**, 1262 (1991)
33. Beloshapkin, V.V., Berman, G.P., Tretyakov, A.G.: Stability of space structure and soliton–antisoliton pair annihilation in the discrete ϕ-4 model. Sov. Phys. JETP **68**, 410 (1989)
34. Milchev, A.: Breakup threshold of solitons in systems with nonconvex interactions. Phys. Rev. B **42**, 6727 (1990)
35. Kosevich, A.M., Kovalev, A.S.: The supersonic motion of a crowdion. The one-dimensional model with nonlinear interaction between the nearest neighbours. Solid State Commun. **12**, 763 (1973)
36. Savin, A.V.: Supersonic regimes of motion of a topological soliton. JETP **108**, 608 (1995)
37. Braun, O.M., Kivshar, Y.S.: The Frenkel–Kontorova Model. Springer, Berlin/Heidelberg (2004)
38. Braun, O.M., Kivshar, Yu.S.: Model' Frenkelya-Kontrovoi: Kontseptsii, metody, prilozheniya. Fizmatlit, M., Moscow (2008)
39. Milchev, A., Markov, I.: The effect of anharmonicity in epitaxial interfaces. I. Substrate-induced dissociation of finite epitaxial islands. Surf. Sci. **136**, 503 (1984)
40. Markov, I., Milchev, A.: Theory of epitaxy in Frank–van der Merwe model with anharmonic interactions. Thin Solid Films **126**, 83 (1985)
41. Braun, O.M., Zhang, F., Kivshar, Yu.S., Vazquez, L.: Kinks in the Klein–Gordon model with anharmonic interatomic interactions: a variational approach. Phys. Lett. A **157**, 241 (1991)
42. Milchev, A., Fraggis, Th., Pnevmatikos, St.: Formation of cracks from kinks in a Frenkel–Kontorova model with anharmonic interactions. Phys. Rev. B **45**, 10348 (1992)
43. Zhang, F.: Kink internal modes in discrete nonlinear chains. Phys. Rev. E **54**, 4325 (1996)
44. Braun, O.M., Kivshar, Y.S., Peyrard, M.: Kink's internal modes in the Frenkel–Kontorova model. Phys. Rev. E **56**, 6050 (1997)
45. Press, W.H., Teukolsky, S.A., Vetterling, W.T., Flannery, B.P.: Numerical Recipes in Fortran 77. The Art of Scientific Computing. Press Syndicate of the University of Cambridge, Cambridge/New York (1992)
46. Petersen, W.P.: Lagged Fibonacci random number generators for the NEC SX-3. Int. J. High Speed Comput. **6**, 387 (1993)

47. Zolotaryuk, Y., Eilbeck, J.C., Savin, A.V.: Bound states of lattice solitons and their bifurcations. Physica D **108**, 81 (1997)
48. Duncan, D.B., Eilbeck, J.C., Fedderson, H., Wattis, J.A.D.: Solitons on lattices. Physica D **68**, 1 (1993)
49. Manevich, L.I.: Solitons in polymer physics. Vysokomol. Soedin. **43**, 117 (2001)
50. Kurin, V.V., Yulin, A.V.: Radiation of linear waves by solitons in a Josephson transmission line with dispersion. Phys. Rev. B **55**, 11659 (1977)
51. Goldobin, E., Wallraff, A., Ustinov, A.V.: Cherenkov radiation from fluxon in a stack of coupled long Josephson junctions. J. Low Temp. Phys. **119**, 589 (2000)
52. Cattuto, C., Costantini, G., Guidi, T., Marchesoni, F.: Driven kinks in discrete chains: phonon damping. Phys. Rev. E **63**, 046611 (2001)
53. Christiansen, P.L., Savin, A.V., Zolotaryuk, A.V.: Topological solitons and dislocations in two- and three-dimensional anisotropic crystals. Phys. Rev. B **57**, 13564 (1998)
54. Zubova, E.A., Savin, A.V., Manevich, L.I.: Dynamics of quasi-1D topological soliton in 2D strongly anisotropic crystal. Physica D **211**, 294 (2005)
55. Savin, A.V., Zubova, E.A., Manevich, L.I.: Dynamics of topological solitons in a system of weakly coupled chains. Vysokomol. Soedin. Seriya A **47**, 637 (2005)
56. Zubova, E.A.: On the applicability of the Frenkel–Kontorova model to describing the dynamics of vacancies in a polymeric crystal chain. J. Exp. Theor. Phys. **93**, 895 (2001)
57. Zubova, E.A., Manevich, L.I., Balabaev, N.K.: Vacancy mobility in polymer crystals. J. Exp. Theor. Phys. **88**, 586 (1999)
58. Balabaev, N.K., Gendel'man, O.V., Mazo, M.A., Manevich, L.I.: Molecular dynamics simulation of essentially nonlinear vibrations in crystal polyethylene. Zh. Fiz. Khim. **69**, 26 (1995)
59. Fletcher, R., Reeves, C.M.: Function minimization by conjugate gradients. Comput. J. **7**, 149 (1964)
60. Savin, A.V., Khalack, J.M., Christiansen, P.L., Zolotaryuk, A.V.: Twisted topological solitons and dislocations in a polymer crystal. Phys. Rev. B **65**, 054106 (2002)

Chapter 4
Localized Nonlinear Vibrations

The description of various phenomena in solid state physics, the physics of waves and vibrations, biophysics, and engineering is based on models of (quasi-one-dimensional) molecular lattice dynamics. The study of linear quasi-periodic structures has allowed the explanation of many properties of systems consisting of a large number of interacting molecules and atoms. The introduction of the concept of collective excitations, describing the coordinated (coherent) motion of a large amount of particles has been especially fruitful. In the majority of cases, the exact quantitative results, obtained in the harmonic approximation, are in excellent agreement with observed properties of real systems.

However, there exist many significant physical phenomena which cannot be explained in the linear approach. These phenomena are determined by higher order terms in the power series expansion of the interaction energy of the molecular system components, taken in the vicinity of its equilibrium state, which are commonly neglected. In this chapter we will consider localized nonlinear vibrations, which are referred to as spatial-temporal structures. They are stable spatial-temporal formations in a dispersive medium which emerge in the absence of dissipation due to the development of instability and its subsequent stabilization through dispersion compensation by nonlinearity of these vibrations. If any, or even one, of their parameters are changed, these systems become unstable. In this case the instability may not be due to dissipation, but may result instead from effects associated with the dispersion. Structures formed this way can be called dispersive structures (by analogy with dissipative structures).

© Springer International Publishing Switzerland 2015
L.N. Lupichev et al., *Synergetics of Molecular Systems*, Springer Series in Synergetics, DOI 10.1007/978-3-319-08195-3_4

4.1 A Nonlinear Oscillator

Let us consider first the properties of an isolated nonlinear oscillator which can be
generally described by the dimensionless Hamiltonian

$$H = \frac{1}{2}\dot{u}^2 + U(u) , \tag{4.1}$$

where u is the coordinate of oscillation, the dot denotes differentiation with respect
to the dimensionless time t, the potential of the oscillator is normalized by the
conditions $U(0) = 0$, $U'(0) = 0$, and $U''(0) = 1$, and the dimensionless frequency
of the oscillator is equal to 1. Here, we restrict ourselves to cubic and quartic
anharmonicity. Then the nonlinear oscillator potential takes the form

$$U(u) = \frac{1}{2}u^2 + \frac{1}{3}\alpha u^3 + \frac{1}{4}\beta u^4 , \tag{4.2}$$

where $\alpha = \mathrm{d}^3 U/\mathrm{d}u^3|_{u=0}$ and $\beta = \mathrm{d}^4 U/\mathrm{d}u^4|_{u=0}$ are the cubic and quartic
anharmonicity parameters, respectively. Typical forms of the potential are shown in
Fig. 4.1 (left). The potential is a single-well if $\alpha^2 < 4\beta$, or a double-well if $\alpha^2 > 4\beta$
(when $\alpha^2 = 9\beta/2$, the double-well potential is a symmetric potential).

The equations of motion corresponding to the nonlinear oscillator (4.1) and (4.2)
have the form

$$\ddot{u} = -U'(u) = -u - \alpha u^2 - \beta u^3 , \tag{4.3}$$

with the integral of motion

$$\frac{1}{2}\dot{u}^2 + U(u) = E , \tag{4.4}$$

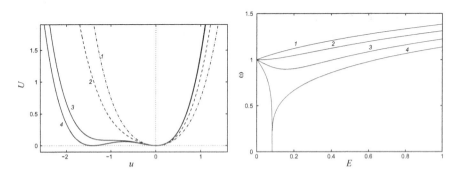

Fig. 4.1 *Left*: Profile of the substrate potential $U(u) = u^2/2 + \alpha u^3/3 + \beta u^4/4$ at $\beta = 1$ and
$\alpha = 0$ (*line 1*), $\alpha = 1$ (*line 2*), $\alpha = 2$ (*line 3*), and $\alpha = 3/\sqrt{2}$ (*line 4*). *Right*: Dependence of the
oscillation frequency ω of the nonlinear oscillator on the oscillation energy E at $\beta = 1$ and $\alpha = 0$
(*line 1*), $\alpha = 1$ (*line 2*), $\alpha = 1.5$ (*line 3*), and $\alpha = 2$ (*line 4*)

where E is the oscillation energy of the nonlinear oscillator. The energy conservation law (4.4) can be written in the form

$$du = \pm\sqrt{2[E - U(u)]}dt \, , \tag{4.5}$$

which can be integrated directly to give

$$\int_{u_0}^{u(t)} \frac{du}{\sqrt{2[E - U(u)]}} = t \, , \tag{4.6}$$

where $u_0 = u(0)$. The oscillation period can be readily found from the integral equation (4.6):

$$T = \sqrt{2} \int_{u_{\min}}^{u_{\max}} \frac{du}{\sqrt{E - U(u)}} \, , \tag{4.7}$$

where $u_{\min}$ and $u_{\max}$ are the minimum and maximum amplitudes of oscillation, i.e., the minimum and maximum solutions of the equation $U(u) = E$ with energy $E > 0$. The oscillation frequency is $\omega = 2\pi/T$.

The theoretical analysis of nonlinear oscillator dynamics carried out by Landau and Lifshitz [1] has shown that, in the second-order approximation of perturbation theory, the nonlinearity leads to a frequency shift by the amount

$$\Delta\omega = \left(\frac{3\beta}{8\omega_0} - \frac{5\alpha^2}{12\omega_0^3} \right) a^2 \, ,$$

where ω_0 is the oscillation frequency in the linear approximation and a is the oscillation amplitude. Thus, the frequency shift is directly proportional to the square of the oscillation amplitude and the shift direction depends on the relationship between the nonlinearity parameters α and β. The nonlinearity causes an increase in the oscillation frequency (hard anharmonicity) for $\alpha^2 < 9\beta\omega_0^2/10$ and a decrease for $\alpha^2 > 9\beta\omega_0^2/10$ (soft anharmonicity). In our case, the frequency is $\omega_0 = 1$. Here, we take $\beta = 1$, so that the nonlinear oscillator is soft for $\alpha^2 > 9/10$, i.e., for $|\alpha| > \sqrt{9/10} = 0.9487$, and otherwise hard.

The oscillation frequency of a linear oscillator does not depend on the oscillation energy (amplitude). When nonlinearity is considered, this gives rise to a strong dependence of the oscillation frequency ω on its energy E (see Fig. 4.1 right). For $\beta = 1$ and $\alpha = 0$, the oscillation frequency increases steadily with increasing the energy. For $\alpha = 1$ and $\alpha = 1.5$, the frequency decreases at low energy and increases monotonically at high energy, i.e., the oscillator possesses either soft anharmonicity for small displacements or hard anharmonicity for large ones. This is especially apparent for $\beta = 1$ and $\alpha = 2$, when the oscillation period (4.7) tends to infinity at the energy $E = 1/12$.

4.2   A Chain of Nonlinear Oscillators

The Hamiltonian of a chain of nonlinear oscillators has the form

$$H = \sum_n \left[\frac{1}{2}\dot{u}_n^2 + \frac{1}{2}g(u_{n+1} - u_n)^2 + U(u_n) \right] , \tag{4.8}$$

where g is the dimensionless parameter of system cooperativity. The Hamiltonian (4.8) leads to the equations of motion

$$\ddot{u}_n = k(u_{n+1} - 2u_n + u_{n-1}) - U'(u_n) , \quad n = 0, \pm 1, \pm 2, \dots . \tag{4.9}$$

For low-amplitude displacements $|u_n| \ll 1$, the equation of motion can be written as the linear system

$$\ddot{u}_n = k(u_{n+1} - 2u_n + u_{n-1}) - u_n , \quad n = 0, \pm 1, \pm 2, \dots , \tag{4.10}$$

with the dispersion equation

$$\omega(q) = \sqrt{1 + 2g(1 - \cos q)} ,$$

where $q \in [0, \pi]$ is the wavenumber. The phonon frequency spectrum consists of the bands $[\omega_0, \omega_m]$, where $\omega_0 = \omega(0) = 1$ and $\omega_m = \omega(\pi) = \sqrt{1 + 4g}$ are the minimum and maximum frequencies, respectively. Here, we take the cooperativity parameter $g = 1$, so the maximum frequency is $\omega_m = \sqrt{5} = 2.2361$.

Self-localization of vibrations in a one-dimensional anharmonic chain was investigated analytically by Kosevich and Kozlov [2]. Using an asymptotic method, it has been shown that, in the limit of low-amplitude displacements, the nonlinearity of the substrate potential (4.2) causes the appearance of localized eigenmode vibrations (breathers). The frequencies of these self-localized vibrations can lie both below and above the vibration frequency band of the corresponding harmonic chain. Here, the magnitude of the splitting of vibration frequencies from the edge of the phonon spectrum is a small parameter, so an asymptotic expansion of the solution of the nonlinear equations can be obtained. In this approach, the solutions of the equations can be found only in the continuum approximation.

Low-amplitude, low-frequency breathers (with frequencies $\omega < \omega_0$) can exist only when $\alpha^2 > 9\beta/10$ [2], i.e., in the case where an isolated oscillator possesses soft anharmonicity. In a discrete chain, these breathers are not the exact solutions, but behave as long-lived localized vibrations which are always accompanied by phonon emission [3]. The low-amplitude, high-frequency breathers (with frequencies $\omega > \omega_m$) can exist only when the following inequality is satisfied [2]:

$$\alpha^2 < \frac{3 - 12(1 + 4g)}{6 - 16(1 + 4g)}\beta . \tag{4.11}$$

For weak cooperativity $g \ll 1$, the inequality (4.11) takes the form $\alpha^2 < 9\beta/10$, i.e., low-amplitude, high-frequency breathers can exist in the chain only in the case of hard anharmonicity of an isolated oscillator.

In a chain with hard anharmonicity, there can exist high-amplitude breathers, which are the exact solutions in a discrete chain [3]. As shown above, the isolated oscillators (4.1) with the nonlinear potential (4.2) always possess hard anharmonicity for large displacements. It will be shown below that this anharmonicity always leads to the existence of high-amplitude, high-frequency breathers.

4.3 Numerical Method for Finding Breathers

To find a discrete breather with frequency ω, one must solve numerically the nonlinear equations

$$\mathbf{F(X)} = \mathbf{X} , \tag{4.12}$$

where $\mathbf{X} = \{u_n(0), \dot{u}_n(0)\}_{n=1}^{N}$ is the $2N$-dimensional vector representing the initial condition for the equations of motion (4.9) with $n = 1, 2, \ldots, N$, and the vector $\mathbf{F(X)} = \{u_n(T), \dot{u}_n(T)\}_{n=1}^{N}$ gives the chain position at time $t = T$, where $T = 2\pi/\omega$ is the nonlinear vibration period. Equation (4.12) always has the zero trivial solution $X \equiv 0$. In order to find a non-trivial solution, one must choose the right initial point for a sequence of iterations. For this purpose, it is convenient to use the anti-continuum approximation with $g = 0$ for $\omega \gg \omega_m$. Then the resulting solution is used as the initial point in solving (4.12) with a low frequency ω and so on. One can thereby find a solution for all frequencies at which a discrete breather can exist.

For definiteness, we set the cooperativity $g = 1$ and the nonlinearity $\beta = 1$. Numerical solution of (4.12) has shown that there can exist several types of localized vibrations (breathers). Each breather is described by its energy E, diameter D, centre m, and frequency ω. Let $X = \{u_n(0), \dot{u}_n(0)\}_{n=1}^{N}$ be the solution of the nonlinear equation (4.12) corresponding to the breather with frequency ω. Then the breather energy is

$$E = \sum_{n=1}^{N-1} \left\{ \frac{1}{2}\dot{u}_n^2(0) + \frac{1}{2}g\left[u_{n+1}(0) - u_n(0)\right]^2 + U(u_n(0)) \right\} ,$$

and its centre is

$$m = \sum_{n=1}^{N} np_n ,$$

where the sequence

$$p_n = \frac{E_n}{E} = \frac{1}{ET} \int_0^T \left\{ \frac{1}{2} \dot{u}_n^2(\tau) + \frac{1}{2} g \left[u_{n+1}(\tau) - u_n(\tau) \right]^2 + U \left(u_n(\tau) \right) \right\} d\tau$$

describes the distribution of vibration energy over the chain. $D = 1 + 2R$ is the diameter, and the square of the radius is

$$R^2 = \sum_{n=1}^{N} (n - m)^2 p_n \ .$$

4.4 Properties of Discrete Breathers

The typical form of discrete breathers is shown in Figs. 4.2 and 4.3 and the dependence of their energy E and diameter D on frequency ω is given in Fig. 4.4 (left). The first type is a breather with an integer centre, for which the centre of the vibration energy distribution falls on one oscillator (see Fig. 4.2a, c). The

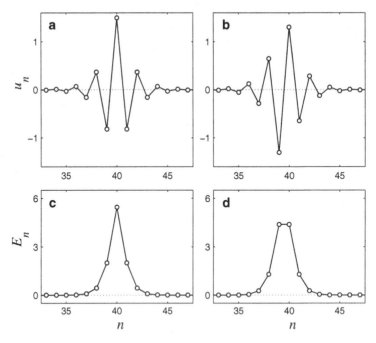

Fig. 4.2 Profile of the discrete breather with an integer (**a**) and a half-integer (**b**) centre at the frequency $\omega = 2.4$ in the chain with parameters $g = 1$, $\alpha = 0$, and $\beta = 1$. Maximum displacements of the chain sites u_n (the displacements at the moment when velocities of all particles are equal to zero) and the corresponding distribution of the energy E_n (**c**) and (**d**) over the chain are shown

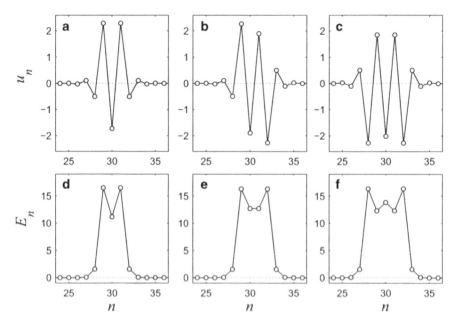

Fig. 4.3 View of the bound state of two discrete breathers located at a distance of two (**a**, **d**) and three (**b**, **e**) sites, and the bound state of three discrete breathers (**c**, **f**) at the frequency $\omega = 2.8$ in the chain with parameters $g = 1$, $\alpha = 0$, and $\beta = 1$. Maximum displacements of the chain sites u_n (the displacements at the moment when velocities of all particles are equal to zero) (**a**–**c**) and the corresponding distribution of the energy E_n (**d**–**f**) in the range of nonlinear vibration localization are shown

second is a breather with a half-integer centre for which the centre of the vibration energy distribution is located in the middle, between the two neighboring oscillators (Fig. 4.2b, d). For $\alpha = 0$, these breathers exist at all frequencies above the phonon frequency spectrum $\omega > \omega_{\rm m}$. With increasing frequency, the energy E grows as ω^4 (see Fig. 4.4a left), and the diameter D tends to 1 for the breather with an integer centre and 2 for the breather with a half-integer centre (Fig. 4.4c left). At high frequencies, the breather with an integer centre can be considered as the excited state of a single oscillator and the breather with a half-integer centre can be treated as the excited state of the two neighboring breathers vibrating in the opposite phase. In the latter case, this breather is essentially the bound state of two breathers with an integer centre. This is confirmed by the fact that the relative difference $\Delta E / E_1$ ($\Delta E = E_{1/2} - E_1$) between the energies E_1 and $E_{1/2}$ of the breathers with the half-integer and integer centres tends monotonically to 1 as $\omega \to \infty$ (Fig. 4.4b left). With decreasing frequency $\omega \to \omega_{\rm m}$, this ratio tends monotonically to zero. The breather energies also tend to zero while their diameters tend to infinity. At low frequencies, the difference between these breathers vanishes almost completely.

There exist other localized vibrations in the chain which are the bound states of the breathers with an integer centre. These states appear when (i) two breathers

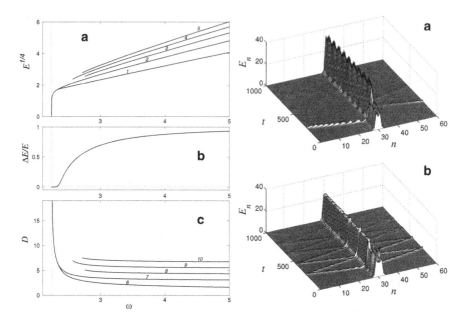

Fig. 4.4 *Left*: Dependencies of $E^{1/4}$ (**a**), the relative difference $\Delta E / E$ in the energies of breathers with half-integer and integer centres of symmetry (**b**), and the diameter D of the discrete breather (**c**) on the frequency ω in the chain with the nonlinear substrate potential ($g = 1$, $\alpha = 0$, and $\beta = 1$). Breathers with integer (*lines 1* and *6*) and half-integer (*lines 2* and *7*) centres, the bound state of breathers located at a distance of two (*lines 3* and *8*) and three (*lines 4* and *9*) sites, and the bound state of three breathers (*lines 5* and *9*). *Right*: Formation of the discrete breather with an integer centre from the bound state of two breathers located at a distance of two sites (**a**) and formation of the discrete breather with a half-integer centre from the bound state of two breathers located at a distance of three sites (**b**) in the chain with parameters $g = 1$, $\alpha = 0$, and $\beta = 1$. Initial frequency of the bound state $\omega = 2.8$. The dependence of the energy distribution E_n over the chain on time t is shown

are separated from each other by two chain links (Fig. 4.3a, d) and by three chain links (Fig. 4.3b, e) and (ii) three breathers are separated from each other by two chain links (Fig. 4.3c, f). They can exist only at significantly high frequencies. The dependencies of their energies and diameters on the frequency are shown in Fig. 4.4 (left). For $\alpha = 0$, the bound state of two breathers exists only at frequencies (i) $\omega \geq 2.76$ for the breathers separated from each other by two chain links and (ii) $\omega \geq 2.57$ for the ones separated by three chain links. The bound state of three breathers exists only at frequencies $\omega \geq 2.72$.

In the absence of the cubic anharmonicity, i.e., when $\alpha = 0$, breathers with integer and half-integer centres are stable for all permissible values of frequencies $\omega > \omega_{\mathrm{m}}$. Bound states of the breathers are stable only for high frequencies. For low frequencies, they lose stability and turn into breathers with integer or half-integer centres, while low-intensity phonons are emitted and the vibration frequency increases (Fig. 4.4 right).

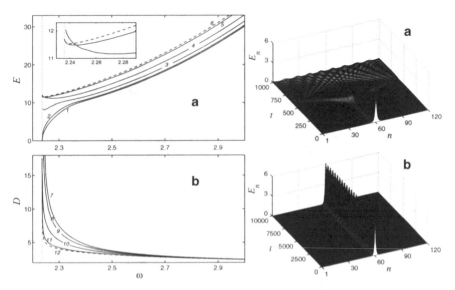

Fig. 4.5 *Left*: Dependencies of energy E (**a**) and diameter D (**b**) of the discrete breather with an integer centre on the frequency ω in the chain with parameters $g = 1$ and $\beta = 1$ for $\alpha = 0$ (*lines 1* and *7*), 0.5 (*lines 2* and *8*), 1.0 (*lines 3* and *9*), 1.5 (*lines 4* and *10*), 2.0 (*lines 5* and *11*), and $3/\sqrt{2}$ (*lines 6* and *12*). *Right*: Discrete breather dynamics in the chain with parameters $g = 1$, $\alpha = 1$, and $\beta = 1$ for frequencies $\omega = 2.24$ (**a**) and $\omega = 2.3$ (**b**). The dependence of the energy distribution E_n over the chain on time t is shown

Let us now consider how the properties of the discrete breathers depend on the cubic anharmonicity parameter α. Low-amplitude, high-frequency breathers can exist only if the inequality (4.11) is satisfied. For $g = 1$ and $\beta = 1$, this inequality reduces to $|\alpha| < \sqrt{57/74} = 0.8777$. Numerical simulation confirms this condition. For $\alpha = 0$ and $\alpha = 0.5$, the discrete breather with integer centre exists for all frequencies $\omega > \omega_m$. In this case, with increasing frequency, the breather energy (and thus also its amplitude) tends monotonically to zero (Fig. 4.5 left). The situation changes for $\alpha > \sqrt{57/74}$. Here, the breather exists for all frequencies $\omega > \omega_m$ as well, but its energy does not tend to zero as the frequency decreases. Moreover, near the upper bound of the phonon spectrum, a decrease in the energy changes to an increase (Fig. 4.5a left). Thus, when the inequality (4.11) is not true near the maximum frequency of the phonon spectrum, high-amplitude breathers exist instead of low-amplitude breathers. Under these conditions, the diameters of these breathers increase steadily with increasing frequency.

At high frequencies, discrete breathers are stable localized vibrations for all values of the cubic anharmonicity parameter. When the inequality (4.11) holds, the breather remains stable for all frequencies $\omega > \omega_m$. If this inequality no longer holds, the breather loses stability at its minimum frequencies. For $\alpha = 1$, the breather is stable at frequency $\omega = 2.3$ and unstable at $\omega = 2.24 > \omega_m = \sqrt{5} = 2.2361$ (Fig. 4.5 right). Numerical simulation of the chain dynamics for

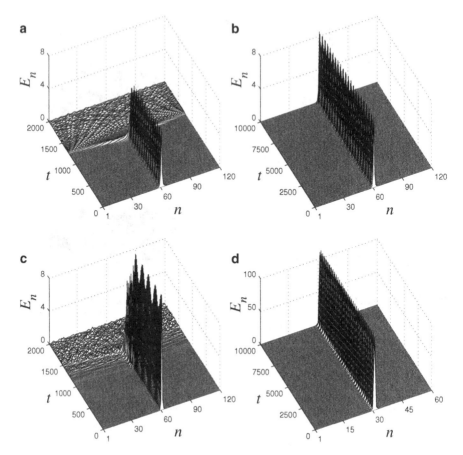

Fig. 4.6 Dynamics of the discrete breather in the chain with parameters $g = 1, \alpha = 2$, and $\beta = 1$ for frequencies $\omega = 2.24$ (**a**), $\omega = 2.3$ (**b**), $\omega = 2.5$ (**c**), and $\omega = 4$ (**d**). The dependence of the energy distribution E_n over the chain on time t is shown

initial conditions corresponding to the discrete breather with an integer centre has shown that, at this frequency, the lifetime of this localized vibration is $t = 460$ (before destruction, the breather performs 164 oscillations) (Fig. 4.5a right). At the frequency $\omega = 2.3$, the breather remains stable during the whole time of numerical integration $t = 10^4$.

For strong cubic anharmonicity $\alpha \geq 2$, the anharmonic oscillator potential $U(u)$ becomes a double well. This causes the breathers to become unstable at higher frequencies, when the oscillation amplitude becomes comparable with the distance between the walls. For $\alpha = 2$, the breather is unstable for frequencies $\omega = 2.24$ and $\omega = 2.5$, and stable for $\omega = 2.3$, as well as for the higher frequency $\omega = 4$ (Fig. 4.6).

The localized vibrations discussed above exist by virtue of the nonlinearity of oscillators compensating for the dispersion. Here, the dispersion and nonlinearity have different origins: the dispersion results from the site–site interaction and the nonlinearity is a substrate potential property. We now consider a case where both nonlinearity and dispersion are due only to the site–site interaction. We will show that localized high-frequency nonlinear vibrations can exist there.

For simplicity, we assume that the substrate potential $U(u)$ is absent. Then the dimensionless Hamiltonian can be written in the form

$$H = \sum_n \left[\frac{1}{2} \dot{u}_n^2 + V(u_{n+1} - u_n)^2 \right] , \tag{4.13}$$

where the potential $V(\rho)$ of the site–site interaction is normalized by the conditions

$$V(0) = 0 , \qquad V'(0) = 0 , \qquad V''(0) = 1 . \tag{4.14}$$

The Hamiltonian (2.13) gives the equations of motion

$$\ddot{u}_n = F(u_{n+1} - u_n) - F(u_n - u_{n-1}) , \quad n = 0, \pm 1, \pm 2, \dots , \tag{4.15}$$

where the function $F(\rho) = V'(\rho)$.

For low-amplitude displacements $|u_n| \ll 1$, by virtue of the normalization conditions (4.14), the equations of motion can be written in the form of the linear system

$$\ddot{u}_n = u_{n+1} - 2u_n + u_{n-1} , \quad n = 0, \pm 1, \pm 2, \dots , \tag{4.16}$$

which gives the dispersion equation

$$\omega(q) = \sqrt{2(1 - \cos q)} = 2 \sin \frac{q}{2} ,$$

where $q \in [0, \pi]$ is the wavenumber. It follows from the dispersion equation that the phonon frequency spectrum consists of the band $[0, 2]$.

For definiteness, we restrict attention to the cubic and quartic anharmonicity of the site–site interaction potential $V(\rho)$. Then this potential has the form

$$V(\rho) = \frac{1}{2}\rho^2 - \frac{1}{3}\alpha\rho^3 + \frac{1}{4}\beta\rho^4 , \tag{4.17}$$

where $\alpha = -d^3 V/d\rho^3|_{\rho=0} \geq 0$ and $\beta = d^4 V/d\rho^4|_{\rho=0} > 0$ are the parameters of the cubic and quartic anharmonicity, respectively.

Vibration self-localization in this one-dimensional anharmonic chain was also investigated analytically by Kosevich and Kovalev [2]. Using an asymptotic method, they have shown that, in the limit of low-amplitude displacements, interaction potential nonlinearity can lead to the existence of self-localized nonlinear vibrations (breathers) (4.17). The frequencies of these vibrations can only lie above the frequency band of the corresponding harmonic chain. Here the magnitude of the splitting of the vibration frequency from the edge of the phonon spectrum is also a small parameter, which means that an asymptotic expansion of the solution to the nonlinear equation can be obtained. In this approach, one can only get the solution of (2.15) in the continuum approximation. Low-amplitude breathers with frequencies $\omega > 1$ can exist only for

$$\alpha^2 < \frac{3\beta}{4} \, . \tag{4.18}$$

These breathers exist for all frequencies $\omega > 2$ [3]. Let us consider their properties.

A typical view of discrete breathers is shown in Fig. 4.7 (left) and the dependencies of their energy E and diameter D on the frequency ω are given in Figs. 4.7

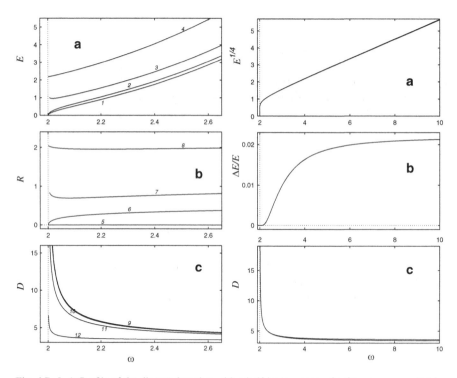

Fig. 4.7 *Left*: Profile of the discrete breather with a half-integer centre for frequency $\omega = 2.2$ in the chain with parameters $\beta = 1$ and $\alpha = 0$ (**a**), 0.5 (**b**), 1 (**c**), and 2 (**d**). Chain site displacements u_n are shown at the time when the velocities of all particles are equal to zero. *Right*: Dependencies of $E^{1/4}$ (**a**), relative difference $\Delta E / E$ in the energies of the breathers with half-integer and integer centres of symmetry (**b**), and diameter of the discrete breather D (**c**) on frequency ω in the chain with the nonlinear site–site potential (2.17) and parameters $\alpha = 0$ and $\beta = 1$. *Upper* and *bottom lines* (**a**) correspond to breathers with an integer centre. *Bottom* and *upper lines* (**c**) correspond to breathers with half-integer centres

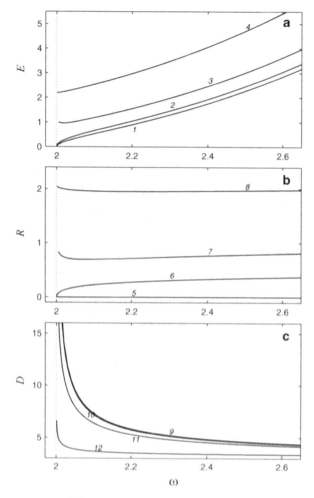

Fig. 4.8 Dependencies of $E^{1/4}$ (**a**), full extension of the chain R (**b**), and diameter D (**c**) of the discrete breather with integer centre on its frequency ω in the chain with the nonlinear site–site potential ($g = 1$, $\beta = 1$) for $\alpha = 0$ (*lines 1, 5, 9*), $\alpha = 0.5$ (*lines 2, 6, 10*), $\alpha = 1$ (*lines 3, 7, 11*), and $\alpha = 2$ (*lines 4, 8, 12*)

(right) and 4.8. There exist only two types of breather, i.e., with integer and half-integer centres. For $\alpha = 0$, these breathers exist for all frequencies lying above the phonon frequency spectrum $\omega > 1$. With increasing frequency, the energy E increases as ω^4 (Fig. 4.7a right) and the diameters of breathers with integer and half-integer centres D tend to 1 and 2, respectively (Fig. 4.7c right). Here, one cannot yet consider the breather with integer centre as the bound state of two breathers with integer centres. The relative difference $\Delta E / E_1$ in the energies of the breathers with half-integer and integer centres does not exceed 0.22 (Fig. 4.7 right). Here, the breather with integer centre has a higher energy $\Delta E = E_1 - E_{1/2} > 0$. When the frequency is decreased $\omega \rightarrow 2$, this ratio tends monotonically to zero, the breather energy also tends to zero, and the diameter tends to infinity. For low frequencies, the difference between these breathers practically vanishes.

In the absence of cubic anharmonicity ($\alpha = 0$), discrete breathers are stable for all frequencies $\omega > 2$. Numerical solution of (4.12) has shown that, when $\beta > 0$, the discrete breather exists for all values of the cubic anharmonicity parameter. The form of the breather depends essentially on the parameter $\alpha \geq 0$. When $\alpha = 0$, the breather is a symmetrical vibration and chain compression does not occur in the region of its localization (Fig. 4.7a left). When $\alpha > 0$, the vibration becomes asymmetrical. The local extension of the chain takes place in the vibration localization region (Fig. 4.7b–d left). The total compression of the chain $R = u_{+\infty} - u_{-\infty}$ steadily increases with increasing cubic anharmonicity.

Let us consider how the properties of the breather depend on the cubic anharmonicity parameter $\alpha \geq 0$. Low-amplitude, high-frequency breathers can exist only if the inequality (4.18) holds. When $\beta = 1$, this inequality reduces to $|\alpha| < \sqrt{3/4} = 0.866$. Numerical simulation confirms this condition. When $\alpha = 0$ and $\alpha = 0.5$, the discrete breather with half-integer centre exists at all frequencies $\omega > 2$. With decreasing frequency, the breather energy and total chain extension R tend monotonically to zero (Fig. 4.8). The picture changes when $\alpha > \sqrt{3/4}$. Here, the breather also exists for all frequencies $\omega > 2$, but its energy and total chain extension do not tend to zero as the frequency increases. Moreover, near the upper bound of the phonon spectrum, a decrease in energy changes to and increase (Fig. 4.8a). Thus, when the inequality (4.18) fails, in the chain with the anharmonic site–site interaction, instead of low-amplitude breathers, there exist high-amplitude breathers at frequencies near the upper bound of the phonon spectrum. In this case, the diameter of these breathers increases steadily with increasing frequency.

When the inequality (4.18) holds, the breather remains stable for all frequencies $\omega > 2$. If this inequality is no longer true, the discrete breather becomes unstable. The instability becomes more pronounced for higher values of the cubic anharmonicity α. Numerical simulation of the chain dynamics with the initial condition corresponding to the discrete breather with half-integer centre has shown that, when $\beta = 1$ and $\omega = 2.2$, the breather remains stable during the whole simulation time if $\alpha = 0, 0.5 < 0.866$ (Fig. 4.9d, c). When $\alpha = 1$, the lifetime of the localized vibration is $t = 3{,}500$ (the breather performs 1.2×10^3 vibrations before destruction), and when $\alpha = 2$, its lifetime is equal to $t = 2{,}000$ (Fig. 4.9b, a).

The investigation discussed above shows that discrete breathers are stable localized periodic vibrations. A necessary condition for their existence is the hard

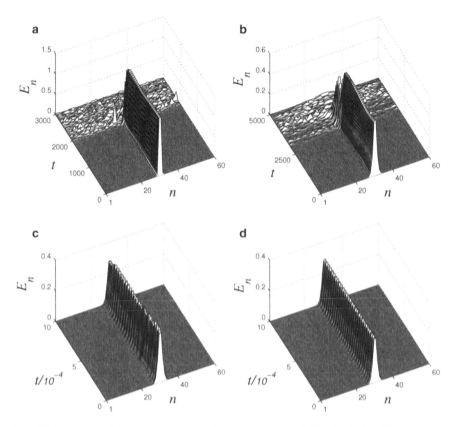

Fig. 4.9 Dynamics of the discrete breather with frequency $\omega = 2.2$ in the chain with parameters $g = 1$ and $\beta = 1$ at $\alpha = 2$ (**a**), 1 (**b**), 0.5 (**c**), and 0 (**d**). The dependence of the energy distribution E_n over the chain on time t is shown

anharmonicity of a chain. In such a chain, breathers can form stable spatial-temporal structures.

References

1. Landau, L.D., Lifshitz, E.M.: Course of Theoretical Physics. 1: Mechanics. Pergamon Press, Oxford/New York/Toronto (1976)
2. Kosevich, A.M., Kovalev, A.S.: Selflocalization of vibrations in a one-dimensional anharmonic chain. Zh. Eksp. Teor. **67**, 1793 (1974)
3. Flach, S., Willis, C.R.: Discrete breathers. Phys. Rep. D **295**, 181 (1998)

Chapter 5
Ratchets

The standard Frenkel–Kontorova model describes a symmetric substrate potential. However, if the system symmetry is broken, new effects related to this phenomenon appear, namely, the so-called saw-tooth or ratchet dynamics. A system under external perturbation exhibits ratchet dynamics if the average particle flux is nonzero, while all average values of external factors (stationary forces, gradients of temperature and concentration, chemical potential, and so on) are equal to zero. Systems displaying ratchet dynamics are currently attracting a lot of attention because they can help to explain the physics of molecular motors and pumps, and also because they open up new possibilities for various applications of nanotechnology.

A comprehensive review of the features of ratchet dynamics can be found in [1, 2]. For a molecular system to exhibit ratchet dynamics, it must satisfy the following conditions:

- The system must be out of thermal equilibrium. This requirement results from the fact that a system in thermal equilibrium cannot display systematic displacement in a preferential direction due to the second law of thermodynamics. Therefore, the system must be undergoing the action of an external perturbation, e.g., an external force with zero average value.
- The spatial-temporal symmetry of the system should be broken as a result of either an asymmetric substrate potential or an asymmetric external force. More stringent rules are formulated in [3,4].

An asymmetry in the potential profile of the substrate (its saw-tooth profile) should lead to the specific dynamics of particles. A view of the asymmetric piecewise linear potential profile is shown in Fig. 5.1. Such a saw-tooth-like potential profile was used for the first time by Feynman in his lectures to illustrate the second law of thermodynamics [5]. He considered the mechanism of a saw-tooth wheel with a ratchet to show that this mechanism could not convert random thermal fluctuations into useful work. But the situation changes dramatically if this mechanism is embedded in a colored thermal bath. In this case, it can convert colored noise

© Springer International Publishing Switzerland 2015
L.N. Lupichev et al., *Synergetics of Molecular Systems*, Springer Series
in Synergetics, DOI 10.1007/978-3-319-08195-3_5

Fig. 5.1 Piecewise linear saw-tooth-like periodic potential profile $V(x)$ (L is the profile period)

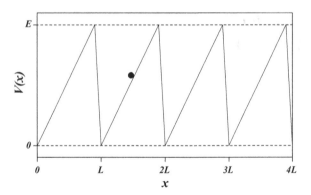

into useful work [6]. Colored noise can induce a continuous one-way flux of particles in the saw-tooth-like periodic potential profile. The simplest models of this phenomenon were investigated in [7], where it was shown that the magnitude and direction of the flux both depend on the form of the profile and the statistical properties of the noise. Such nonequilibrium mechanisms can explain the highly efficient functioning of many molecular biosystems [8–10].

In this chapter we consider ratchet dynamics in detail using the following three models as examples:

- The asymmetric pendulum,
- The FK model with an asymmetrical periodic substrate potential, and
- The ϕ-4 model with an asymmetrical double-well potential.

All these models possess internal asymmetry related to the specific form of the substrate potential. We will show how directed motion can emerge under the action of external forces.

5.1 Asymmetric Pendulum

Particle motion in force fields which are periodic both in space and time serves as a model for many processes in physical systems (classical pendulum, charged particle moving in an electric field, synchronous rotors, Josephson junctions, and others). These systems can possess complex dynamics. For example, studies of pendulums under forced oscillation have revealed complex dynamic processes and chaotic oscillations [11–13].

Symmetric sinusoidal fields are mainly considered when all possible directions of particle motion are completely equivalent. On the other hand, there has recently been great interest in the stochastic dynamics of a particle in the asymmetric periodic ratchet potential profile [14–16]. This concern stems from a number of biological applications [17, 18]. The asymmetry of the potential profile causes asymmetry in the Brownian motion of a particle under the action of colored low-frequency noise, and this in turn means that such noise can produce useful work.

Here, we consider the dynamics of a particle in an asymmetric saw-tooth-like potential profile under the action of an external periodic force. It will be shown that a symmetric periodic external force causes the appearance of a preferred direction in the particle motion with a magnitude and sign that depend on the amplitude and frequency of the force, particle mass, and temperature of the heat bath (a change in any of these parameters can cause a change in the direction of motion). Thus, the action of an external symmetric periodic force with zero average value allows for particles to be selected on the basis of their masses.

5.1.1 Potential Function of an Asymmetric Pendulum

We use the potential function of an asymmetric pendulum given in Fig. 5.2a as the saw-tooth-like periodic potential profile. Let us consider a pendulum composed of

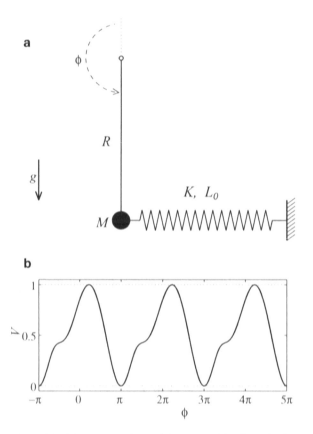

Fig. 5.2 Schematic model of the asymmetric pendulum (**a**) and its dimensionless potential $V(\phi)$ for $l_0 = 2$ and $\kappa = 2$ (**b**)

a point mass M, hanging on a rigid weightless rod of length R and attached to a wall by a weightless spring of length L_0 and rigidity K. This mechanical system provides a simple illustration of asymmetry formation, due here to the presence of the additional spring.

Let ϕ be the angle of rotation of the rod around the point of suspension. Then the potential function of the pendulum has the form

$$P(\phi) = MgR(1 + \cos \phi) + \frac{1}{2}K[L(\phi) - L_0]^2 , \qquad (5.1)$$

where g is the acceleration due to gravity and the current length of the spring is

$$L(\phi) = \sqrt{(L_0 + R \sin \phi)^2 + R^2(1 + \cos \phi)^2}$$

The potential (5.1) is a periodic function of period 2π with minima at the points $\phi = -\pi \pm 2\pi k, k = 0, 1, 2, \ldots$, and maxima at the points $\phi = \phi_0 \pm 2\pi k$, where $\phi_0 \in [-\pi, \pi]$ is the solution of the equation $dP/d\phi = 0$.

We define the dimensionless potential by

$$V(\phi) = \frac{P(\phi)}{P(\phi_0)} = C \left\{ 1 + \cos \phi + \frac{\kappa}{2} \left[\sqrt{(l_0 + \sin \phi)^2 + (1 + \cos \phi)^2} - l_0 \right]^2 \right\} , \qquad (5.2)$$

where $\kappa = KR/Mg$ and $l_0 = L_0/R$ are the dimensionless parameters and the constant C is found from the condition $\max V(\phi) = 1$. The potential (5.2) is an asymmetric periodic function taking values from 0 to 1. For definiteness, we take $l_0 = 2$ and $\kappa = 2$, so the constant $C = 1/3.155110$. A view of the potential $V(\phi)$ is shown in Fig. 5.2b. The potential has the characteristic asymmetrical saw-tooth-like profile with maximum shifted to the right (direction of the 'saw') $\phi_0 = 0.705747$.

5.1.2 Dynamical Equation

Let us consider the dynamics of a particle of mass m in the asymmetrical potential profile $U(x) = \varepsilon V(2\pi x/p)$, where x is the particle coordinate, and p and ε are the period and height of the potential profile, respectively. The Hamiltonian of the particle has the form

$$\mathscr{H} = \frac{1}{2}m \left(\frac{dx}{dt}\right)^2 + U(x) . \qquad (5.3)$$

For convenience, we introduce the dimensionless displacement $\phi = 2\pi x/p$, time $\tau = 2\pi t \sqrt{\varepsilon/m_0 p^2}$, and energy $H = \mathscr{H}/\varepsilon$, where m_0 is the characteristic value of the mass. Then the Hamiltonian (5.3) takes the form

$$H = \frac{1}{2}\mu \left(\frac{d\phi}{d\tau}\right)^2 + V(\phi),$$
(5.4)

where $\mu = m/m_0$ is the dimensionless mass of the particle.

The Langevin dynamics of the particle under the action of an external periodic force is described by the dynamical equation

$$\mu \frac{d^2}{d\tau^2}\phi + F(\phi) + \gamma\mu\frac{d}{d\tau}\phi + \xi + A\cos(\omega\tau + \varphi_0) = 0,$$
(5.5)

where $F(\phi) = dV/d\phi$ is the force function, $\gamma = 1/\tau_r$ is the friction coefficient, and τ_r is the relaxation time. The random force ξ describes the interaction of the particle with the heat bath. A, ω, and φ_0 are the amplitude, angular frequency, and initial phase of the external periodic force, respectively. The random force $\xi(\tau)$ is δ-correlated in time:

$$\langle\xi(\tau_1)\xi(\tau_2)\rangle = 2\gamma\beta\mu\delta(\tau_1 - \tau_2),$$

where β is the dimensionless temperature of the heat bath.

Hereafter we use the initial conditions corresponding to the equilibrium state $\phi(0) = -\pi$, $(d\phi/d\tau)(0) = 0$, and integrate (5.5) by the standard Runge–Kutta method with fourth-order accuracy and constant integration step $d\tau = 0.05$ [19].

5.1.3 Asymmetry of Chaotic Oscillations

We begin by considering the particle dynamics at zero temperature of the heat bath ($\beta = 0$). Forced oscillations of the symmetric pendulum ($\kappa = 0$) have been well studied [11–13]. It is known that these oscillations are chaotic over a wide range of parameter values. For the asymmetric pendulum ($\kappa \neq 0$), the character of the oscillations remains chaotic. Numerical integration of the dynamical equation (5.5) has shown that the chaotic oscillations become asymmetric, i.e., particle drift occurs in the background of the oscillations. This drift does not directly allow the use of the standard Poincaré map because the point set $\{\phi(nT), \phi'(nT)\}_{n=0}^{\infty}$ forms an unbounded set in the phase plane. Here, the prime denotes differentiation with respect to the dimensionless time and $T = 2\pi/\omega$ is the period of the external force.

Let us slightly modify the Poincaré map. Consider the point set

$$\left\{u_n = \phi((n+1)T) - \phi(nT), \; v_n = \phi'((n+1)T)\right\}_{n=1}^{\infty}.$$

This set is bounded for a constant particle drift velocity, which allows us to analyse the nature of the oscillations based on its structure. In Fig. 5.3, the modified Poincaré

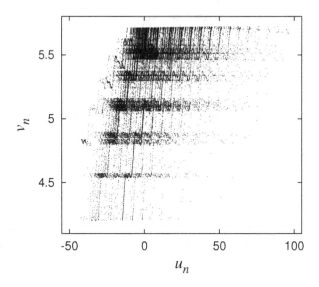

Fig. 5.3 Modified Poincaré map for $\mu = 1$, $\gamma = 0.1$, $\beta = 0$, $A = 0.75$, $T = 100$, and $\varphi_0 = 0$

map consisting of $N = 4 \times 10^4$ points is shown for the parameter values $\mu = 1$, $\gamma = 0.1$, $\beta = 0$, $A = 0.75$, $\varphi_0 = \pi$, and $T = 100$. As can be seen from this figure, the modified Poincaré map has a fine structure which resembles a strange attractor. The characteristic feature of this map is its asymmetry relative to the line $u = 0$. The centre of the point set $\{u_n, v_n\}_{n=1}^N$ turns out to be shifted to the right, indicating that the particle drifts to the right. One can therefore conclude that the particle dynamics in the saw-tooth-like potential profile under the action of the symmetric periodic force has the nature of a bounded chaotic oscillation against a background of uniform motion. The velocity of this uniform motion corresponds to the limit

$$s = \lim_{\tau \to \infty} \frac{\phi(\tau) - \phi(0)}{\tau} \, .$$

Our next aim is to investigate the dependence of s on the parameters A, T, β, and μ.

5.1.4 Asymmetric Particle Drift Velocity

Let us integrate the dynamics equation (5.5) numerically for $0 \le \tau \le 2 \times 10^4$. The drift velocity of a particle can be estimated as

$$s = \left[\phi(2 \times 10^4) - \phi(10^4) \right] \times 10^{-4} \, .$$

The drift velocity is estimated from the second half of the time interval, in which the system is certain to be in thermal equilibrium. The drift velocity s can depend on the initial phase of the force φ_0. Let us average s over φ_0. To do this, we find the average velocity over the set $\{s_n = s(\varphi_0), \ \varphi_0 = 2\pi n/36\}_{n=1}^{36}$, viz.,

$$\bar{s} = \frac{1}{36} \sum_{n=1}^{36} s_n ,$$

and also the standard deviation

$$\sigma = \left[\frac{1}{36} \sum_{n=1}^{36} (s_n - \bar{s})^2 \right]^{1/2} .$$

We thus arrive at the estimate $s = \bar{s} \pm \sigma$.

5.1.5   Frequency Dependence of the Drift Velocity

Let us find the dependence of the drift velocity of a particle s on the period of the external force $T = 2\pi/\omega$. To be specific, we will take $\mu = 1$, $\gamma = 0.1$, and $A = 0.5$. The dependence of s on T for $\beta = 0$ is given in Fig. 5.4a (left). As can be seen from this figure, when $T < 280$, the value of s is very sensitive to a small variation of the period, which can even cause a change in the sign of the velocity. When $T > 280$, the velocity s remains practically unchanged as the period varies.

   Let us introduce a low-amplitude white noise into the external periodic force (thermal fluctuations of the dimensionless temperature $\beta = 0.02$). As can be seen in Fig. 5.4b (left), the addition of the noise leads to smoothing in the dependence $s(T)$. For $15 < T < 95$, the velocity is negative $s < 0$ (the particle drifts against the direction of the saw-tooth-like potential profile), while for $T \geq 95$, the velocity is positive $s > 0$ (the particle drifts in the direction of the saw-tooth-like potential profile). The average velocity $\bar{s}$ increases steadily as the period T increases to a certain limiting value.

   When we add the noise, the disappearance of the singular dependence of s on T in the range of low values of the period proves that the singularity here is due to the nonlinear resonant interaction between the particle and the force. The presence of random forces (thermal fluctuations) leads to the disruption of 'fine' resonances and hence to the disappearance of the singularity (see Fig. 5.4b left).

5.1.6   Temperature Dependence of the Drift Velocity

Comparison of Figs. 5.4a, b (left) shows that the introduction of low-amplitude white noise leads in several cases to a change in the sign and magnitude of the particle drift velocity. We now consider in detail the dependence of this drift velocity s on the dimensionless temperature of the heat bath β.

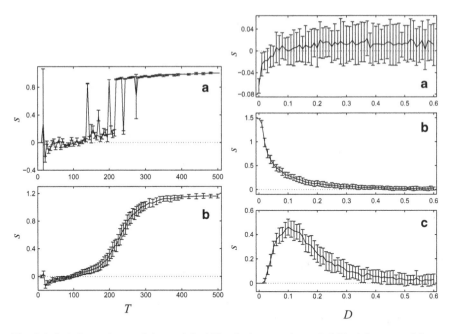

Fig. 5.4 *Left*: Dependence of the particle drift velocity s on the period T of the external force when $\mu = 1$, $\gamma = 0.1$, $A = 0.5$, $\beta = 0$ (**a**), and $\beta = 0.02$ (**b**). *Right*: Dependence of the particle drift velocity s on the temperature of the heat bath β for $A = 0.6$, $T = 100$ (**a**), $A = 0.6$, $T = 500$ (**b**), and $A = 0.3$, $T = 500$ (**c**). Dimensionless mass $\mu = 1$ and $\gamma = 0.1$

Let us choose $\mu = 1$ and $\gamma = 0.1$ and consider first the dependence $s(\beta)$ for the external force amplitude $A = 0.6$ and periods $T = 100$ (Fig. 5.4a right) and $T = 500$ (Fig. 5.4b right). For $T = 100$, temperature growth leads to a much stronger smoothing of the nonlinear resonant interaction between the particle and periodic force and, as a consequence, causes a decrease in the asymmetry of the dynamics. The average value is negative, i.e., $\bar{s} < 0$, only for $\beta < 0.055$. Further growth of β results in a decrease in the average value $\bar{s}$ and increase in the standard deviation σ.

For $T = 500$, the asymmetry in the particle dynamics (positivity of the drift velocity) is a result of higher mobility of a particle moving under the action of a constant force in the direction of the saw-tooth-like potential profile as compared with its movement against this direction. When the force has a long period, the particle velocity at each time will correspond to the current value of the force. The appearance of white noise destroys the synchronization between the particle velocity and the current value of the force, so with an increase in the noise amplitude (temperature of the heat bath), the particle drift velocity tends monotonically to zero.

Let us consider separately the dependence $s(\beta)$ for $A = 0.3$ and $T = 500$ (see Fig. 5.4c right). In the absence of white noise ($\beta = 0$), the external low-amplitude periodic force does not cause particle drift, and the particle only oscillates in the vicinity of its equilibrium position. The introduction of white noise can lead to pinning of the particle and, as a result, to asymmetric drift. For $\beta < 0.02$, the particle remains pinned, while for $\beta > 0.02$, it already drifts in the direction of the saw-tooth-like potential profile. The maximum drift velocity is reached when $\beta = 0.1$ (for this value, the signal/noise ratio turns out to be optimal). A further increase in β already leads to a monotonic decrease in the drift velocity.

Thus, for a low amplitude of the external periodic force, thermalization of the system can be a necessary condition for asymmetric drift, while for a large amplitude, thermalization only causes a decrease in the asymmetric drift of the particle.

5.1.7 Amplitude Dependence of the Drift Velocity

As mentioned above, a small amplitude periodic force does not cause particle drift. For particle drift to occur, the amplitude must exceed a certain threshold value at which depinning happens. Let us consider in detail the dependence of the drift velocity of a particle s on the force amplitude A. We choose $\mu = 1$, $\gamma = 0.1$, and $\beta = 0.02$.

The dependence $s(A)$ for $T = 100$ is shown in Fig. 5.5a (left). For amplitude $A < 0.26$, particle drift does not occur, since the particle remains in the pinned state at all times. For $A = 0.26$, depinning takes place and the particle thus starts

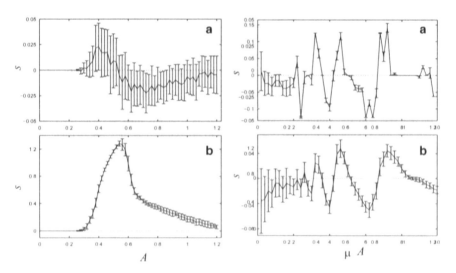

Fig. 5.5 *Left*: Dependence of the particle drift velocity s on the amplitude of the external force A for $\mu = 1$, $\gamma = 0.1$, $\beta = 0.02$, and $T = 100$ (**a**) and $T = 500$ (**b**). *Right*: Dependence of the particle drift velocity s on the dimensionless mass μ when $\gamma = 0.1$, $A = 0.7$, $T = 50$, and $\beta = 0$ (**a**) and $\beta = 0.02$ (**b**)

moving. This motion has the nature of an asymmetric drift: the particle drifts to the right (in the direction of the saw-tooth-like potential profile) for $A < 0.53$ and to the left (against this direction) for $A > 0.53$. In the limit $A \to \infty$, the drift velocity tends to zero. The dependence $s(A)$ for $T = 500$ is shown in Fig. 5.5b (left). Here, for $A < 0.26$, particle drift is also absent. A further increase in the amplitude leads to steady growth in the drift velocity. The velocity reaches a maximum value when $A = 0.54$ and then tends monotonically to zero.

Thus, for asymmetry in the particle dynamics to manifest itself, the amplitude of the external force must be commensurate with the barrier height of the potential profile. For low amplitudes, the particle will be in the pinned state and only oscillate around its equilibrium position with a small amplitude. For very large amplitudes, the potential profile will have little effect on the particle dynamics. For short force periods, a change in its amplitude can cause a change in the direction of the particle drift. For long force periods, a change in its amplitude leads only to a change in the drift velocity, and there is only a single value of the amplitude at which the drift velocity will be at its maximum.

5.1.8 Isotope Dependence of the Particle Drift Velocity

A change in the mass of a particle can lead to a change in its drift direction only when there is a short period external force. Nonlinear resonance manifests itself in the asymmetry of the dynamics. Let us consider in detail the particle dynamics for $\gamma = 0.1$, $A = 0.7$, $T = 50$, and $\beta = 0$ (Fig. 5.5a right) and $\beta = 0.02$ (Fig. 5.5b right).

In the absence of white noise, a severalfold increase in the dimensionless mass μ leads to a change in the sign of the drift velocity s (see Fig. 5.5a right). Furthermore, there even exists an interval of mass values for which particle drift is absent. The introduction of white noise of small amplitude smooths the dependence of s on μ (see Fig. 5.5b right). The drift velocity oscillates when the dimensionless mass is increased, and the range of mass values μ is divided into several intervals in which the drift velocity has constant sign. This dependence of s on μ once again confirms the resonant nature of asymmetry in the particle dynamics under the action of the short-period force. Thus, the action of the periodic force can be used to select particles on the basis of their mass using the asymmetric saw-tooth-like potential profile.

5.1.9 Conclusion

The investigation carried out brings out a number of features of the particle dynamics in an asymmetric saw-tooth-like potential profile under the action of an external periodic force (forced oscillations of an asymmetric pendulum). The particle motion reduces to limited chaotic oscillations against a background of uniform motion, i.e., a symmetric external force leads to the asymmetric drift of a particle. For a low frequency of the force, this dynamics is associated with an asymmetry in the particle mobility, while for a high frequency, it is a manifestation of the nonlinear resonance interaction between the particle and the force. In the second case, a change in the parameters of the system (the amplitude and frequency of the force and particle mass) can lead to a change in both the velocity and the direction of the drift. Low-amplitude white noise (thermal fluctuations of the particle) does not interfere with the asymmetric drift of the particle, and for a small force amplitude, it can even be a necessary condition for drift.

5.2 Ratchet Dynamics of Solitons in the FK Model

The development of modern nonlinear physics has led to the discovery of new nonlinear mechanisms of energy and charge transfer in quasi-one-dimensional molecular systems [20]. One such mechanism is the transport of topological solitons. This mechanism can be implemented in multi-stable systems in which the energy degeneracy of equilibrium states predetermines the possibility of 'state transfer'. A topological soliton (kink, antikink) describes the most effective transition of the system from one equilibrium state to another [20, 21].

The study of topological solitons began with [22]. Here, the nonlinear dynamics of a chain consisting of harmonically coupled particles lying on a sinusoidal substrate were considered for the first time. In all the many subsequent studies, it was always assumed that the potential function of the substrate had a symmetric sinusoidal shape with maxima lying exactly midway between neighboring minima. However, it turns out that, in many biological systems [8–10, 23], the periodic potential of the substrate has an asymmetric saw-tooth-like (ratchet) shape with maxima shifted to one side (Fig. 5.6). Such asymmetry of the substrate potential can significantly influence the nonlinear dynamics of the chain. A study of the isolated particle dynamics in an asymmetric ratchet periodic potential has shown that symmetric colored noise can lead to the directional motion of particles [6, 7]. A similar result holds for topological solitons in the FK chain with a ratchet substrate potential. We will show that, in such an anisotropic system, symmetric low-frequency noise can cause directional motion of solitons.

Fig. 5.6 Schematic model of a linear chain with an asymmetric substrate potential

5.2.1 Asymmetric Chain Model

Here we consider the dynamics of a linear chain with an asymmetric substrate potential. The Hamiltonian of the chain is written in the form

$$\mathcal{H} = \sum_n \left[\frac{1}{2}m\dot{x}_n^2 + \frac{1}{2}\kappa(x_{n+1} - x_n - l)^2 + \varepsilon V(2\pi x_n/l) \right] , \tag{5.6}$$

where m is the mass of a chain link, κ is the force constant of the interparticle interaction, l is the chain period, ε is the barrier height of the substrate potential, x_n is the coordinate of the n th link of the chain, and $V(u)$ is the dimensionless periodic potential of period 2π with zero minima at the points $\pi \pm 2\pi k$, $k = 0, 1, 2, \ldots$, and maxima equal to 1. The chain under consideration has an infinite number of fully equivalent ground states $\{x_n = -l/2 + l(n + k)\}_{n=-\infty}^{n=+\infty}$.

We introduce the anisotropy parameter describing the ratchet potential, viz.,

$$A = \frac{1}{\pi \overline{V}} \int_{-\pi}^{\pi} uV(u)du ,$$

where the average value of the potential is

$$\overline{V} = \frac{1}{\pi} \int_{-\pi}^{\pi} V(u)du .$$

For a symmetric potential, the anisotropy is $A = 0$. The nonzero anisotropy parameter clearly indicates the asymmetry of the substrate potential.

We choose the specific dimensionless substrate potential

$$V(u) = \cos^2\left[\frac{1}{2}\left(u - a\cos^2\frac{u}{2}\right)\right] . \tag{5.7}$$

The parameter a describes the asymmetry of the potential. For $-2 \le a \le 2$, the potential (5.7) has minima and maxima located at the points $\pi \pm 2\pi k$ and $\bar{u} \pm 2\pi k$, $k = 0, 1, 2, \ldots$, respectively, where the point $\bar{u}$ is found from the equation

$$\bar{u} = a\cos^2\frac{\bar{u}}{2} .$$

The potential anisotropy is $A < 0$ for $a < 0$, $A = 0$ for $a = 0$, and $A > 0$ for $a > 0$. When $a = \pm 1.5$, the anisotropy is $A = \pm 0.20478$ and the maximum is at the point $\bar{u} = \pm 1.09412$. The potential is shown in Fig. 5.7.

For convenience, we introduce dimensionless variables such as the displacement $u_n = 2\pi(x_n/l - n)$, time $\tau = t\sqrt{\varepsilon/ml^2}$, and energy $H = \mathcal{H}/\varepsilon$. Then the

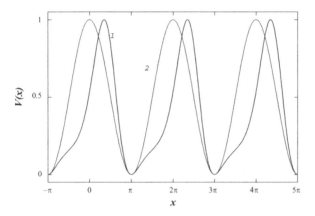

Fig. 5.7 Dimensionless substrate potential $V(x)$ for the anisotropy parameter $a = 1.5$ (*line 1*) and $a = 0$ (*line 2*)

Hamiltonian (5.6) takes the form

$$H = \sum_n \left[\frac{1}{2} u_n'^2 + \frac{1}{2} g (u_{n+1} - u_n)^2 + V(u_n) \right], \tag{5.8}$$

where the prime denotes differentiation with respect to the dimensionless time τ, and the dimensionless cooperativity parameter $g = \kappa l^2/\varepsilon$ describes the magnitude of the interparticle interaction.

5.2.2   Soliton Stationary State

To find the stationary state of a positive (negative) topological soliton, one must solve the minimum problem

$$E = \sum_{n=1}^{N-1} \left[\frac{1}{2} g (u_{n+1} - u_n)^2 + V(u_n) \right] \longrightarrow \min_{u_2,\dots,u_{N-1}} \tag{5.9}$$

with the conditions $u_1 = +\pi$ and $u_N = -\pi$ ($u_1 = -\pi$ and $u_N = +\pi$). It is convenient to characterize the stationary state of the soliton by the position of its centre

$$\bar{n} = \sum_{n=1}^{N-1} \left(n + \frac{1}{2} \right) p_n$$

and the root-mean-square diameter

$$L = 1 + 2 \left[\sum_{n=1}^{N-1} \left(n + \frac{1}{2} - \bar{n} \right)^2 p_n \right]^{1/2} ,$$

where the sequence $p_n = |u_{n+1} - u_n|/2\pi$ describes the distribution of chain deformation.

The minimum problem (5.9) was solved numerically by the method of conjugated gradients [24]. The stationary states of topological solitons were sought in the chain consisting of $N = 200$ links for different values of the cooperativity parameter. To find the Peierls potential profile (the dependence of the energy of the soliton stationary state E on its centre position $\bar{n}$) in the problem (5.9), we fix the displacement $u_{N/2} \in (-\pi, \pi)$. Then, varying the variable $u_{N/2}$ monotonically, we obtain the monotonic change in the soliton centre $\bar{n}$. As a result, the dependence $E(\bar{n})$ can be calculated numerically.

The Peierls potential profile $E(\bar{n})$ is a periodic function with unit period. The anisotropy of the substrate potential (5.7) should lead to anisotropy in the Peierls potential profile. We define the anisotropy of the profile by

$$A = \frac{1}{\overline{\Delta E}} \int_{-1/2}^{1/2} n \left[E \left(n_0 + \frac{1}{2} + n \right) - E(n_0) \right] dn ,$$

where the minimum point of the profile is n_0 and the average value of the profile amplitude is

$$\overline{\Delta E} = \int_{-1/2}^{1/2} \left[E \left(n_0 + \frac{1}{2} + n \right) - E(n_0) \right] dn .$$

A view of the Peierls potential profile is given for $g = 0.1$ and $a = 1.5$ in Fig. 5.8. The profile has amplitude $\Delta E = \max E(n) - \min E(n) = 0.63107$, anisotropy

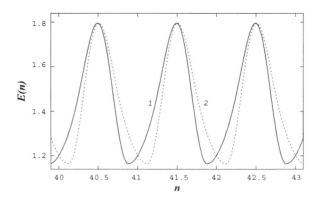

Fig. 5.8 Peierls potential profile $E(\bar{n})$ for the antikink (*line 1*) and kink (*line 2*) in a chain with parameters $g = 0.1$ and $a = 1.5$

Fig. 5.9 Peierls potential profile $E(\bar{n})$ for the antikink (*line 1*) and kink (*line 2*) in a chain with parameters $g = 1$ and $a = 1.5$

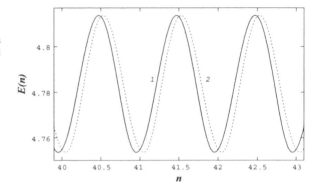

$A = 0.13460$ for the positive soliton (antikink), and $A = -0.13460$ for the negative soliton (kink). The potential profile for the kink is the mirror image of the profile for the antikink relative to the line $\bar{n} = 1/2$. A view of the Peierls potential profile is shown for $g = 1$ and $a = 1.5$ in Fig. 5.9. The Peierls profile has amplitude $\Delta E = 0.06014$ and anisotropy $A = \pm 0.02216$.

When the cooperativity parameter g is increased, both the energy E and the diameter L of the stationary defect increase steadily, and the pinning energy ΔE decreases exponentially (see Fig. 5.10). With decreasing pinning energy, the anisotropy of the potential profile decreases exponentially as well (Fig. 5.11). For $g > 5$, the Peierls profile has an almost symmetrical shape, and for $g > 25$, soliton pinning is almost absent. Therefore, for strong cooperativity, asymmetry of the substrate potential should not affect the motion of a topological soliton. The two directions of motion must be equivalent.

The asymmetry of the substrate potential is manifested in the interaction of opposite-sign solitons. Let us find the dependence of the energy of a pair of opposite-sign solitons on the distance between them. For this purpose, we numerically solve the minimum problem

$$E = \sum_{n=1}^{N-1} \left[\frac{1}{2} g (u_{n+1} - u_n)^2 + V(u_n) \right] \longrightarrow \min_{u_2,\dots,u_{N-1}} \ : \ u_1 = u_N = \pi \ , \qquad (5.10)$$

for the fixed value of the displacement $u_{N/2}$. A solution to the problem (5.10) corresponds to the homogeneous state $u_{N/2} = \pi$ for $u_{N/2} = \pi$, a non-interacting pair of kink–antikink solitons for $u_{N/2} \to 3\pi - 0$, and a non-interacting pair of antikink–kink solitons for $u_{N/2} \to -\pi + 0$.

Let $\{u_n\}_{n=1}^N$ and E be the solution to the problem (5.10) and its energy, respectively. The distance between the solitons is defined as

$$R = \frac{1}{\pi} \sum_{n=1}^{N} (u_n - \pi) \ .$$

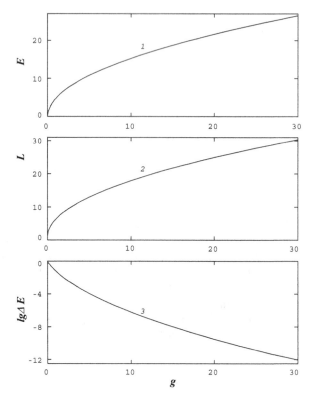

Fig. 5.10 Dependence of the energy E (*line 1*), diameter L (*line 2*), and decimal logarithm of the pinning energy $\lg \Delta E$ (*line 3*) on the cooperativity parameter g for the stationary topological soliton in the chain with anisotropic substrate potential ($a = 1.5$)

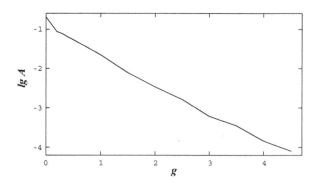

Fig. 5.11 Dependence of the anisotropy of the Peierls potential A on the cooperativity parameter g in a chain with anisotropic substrate ($a = 1.5$)

The distance R depends monotonically on the value of $u_{N/2}$. The distance $R \rightarrow -\infty$ as $u_{N/2} \rightarrow -\pi + 0$, $R = 0$ for $u_{N/2} = \pi$, and $R \rightarrow +\infty$ for $u_{N/2} \rightarrow 3\pi - 0$. The distance between two opposite-sign solitons is negative for the antikink–kink pair and positive for the kink–antikink pair. The dependence $E(R)$ can be obtained from the monotonic dependence of R on $u_{N/2}$.

To find the dependence of the energy of two same-sign solitons, one must solve the minimum problem

$$E = \sum_{n=1}^{N-1} \left[\frac{1}{2} g(u_{n+1} - u_n)^2 + V(u_n) \right] \longrightarrow \min_{u_2,\dots,u_{N-1}} \; : \; u_1 = 3\pi, u_N = -\pi \; ,$$

$$(5.11)$$

for the fixed value of the displacement $3\pi > u_{N/3} \geq \pi \geq u_{2N/3} > -\pi$. Let $\{u_n\}_{n=1}^{N}$ be the solution to the problem (5.11). Then the distance between the solitons is

$$R = \frac{1}{2\pi} \sum_{n=1}^{N} (2\pi - |u_n - \pi|) \; .$$

Varying the displacements $u_{N/3}$ and $u_{2N/3}$, we obtain the dependence of the energy E of a pair same-sign solitons on the distance R between them.

Note that the dependence of the interaction energy of two antikinks on the distance has to coincide exactly with the dependence of the interaction energy of two kinks. Therefore, we restrict ourselves to calculating the interaction potential of two antikinks.

The dependence of the energy E_{+-} of a pair of opposite-sign solitons on the distance R between them is shown in Fig. 5.12. In a chain with symmetric substrate potential $(a = 0)$, the interaction energy is an even function of the distance,

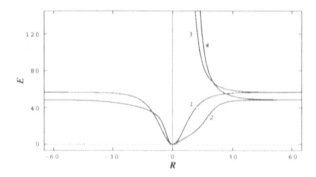

Fig. 5.12 Dependence of the energy E_{+-} of a pair of opposite-sign solitons on the distance R between them for the symmetric (*line 1*) and anisotropic ($a = 1.5$) (*line 2*) substrate potentials, and dependence of the energy E_{++} of a pair of same-sign solitons on the distance R between them for the symmetric (*line 3*) and anisotropic (*line 4*) substrate potentials

i.e., $E_{+-}(-R) = E_{+-}(R)$. The interaction energy of a kink and an antikink coincides exactly with the interaction energy of an antikink and a kink. In the chain with asymmetric substrate potential ($a \neq 0$), the interaction energy of a kink and antikink no longer coincides with the interaction energy of an antikink and kink, i.e., $E_{+-}(-R) \neq -E_{+-}(R)$. As a result, a collision between a kink and an antikink cannot coincide with the results of a collision between an antikink and a kink.

The energy E_{++} of two same-sign solitons increases steadily with decreasing distance between them. The energy $E_{++} \to +\infty$ as $R \to 0$. As can be seen from Fig. 5.12, the anisotropy of the substrate potential does not lead to any qualitative changes in the dependence $E_{++}(R)$. A collision of same-sign solitons can only cause their reflection.

5.2.3 Soliton Dynamics

The Hamiltonian (5.8) gives the equations of motion

$$u_n'' = g(u_{n+1} - 2u_n + u_{n-1}) - \frac{\partial V}{\partial u}(u_n), \quad n = 0, \pm 1, \pm 2, \dots . \tag{5.12}$$

Let us find a soliton solution of (5.12). We search for this in the form of a solitary wave $u_n(\tau) = u(n - s\tau)$, depending smoothly on n, where s is the soliton velocity. Then, one can replace the continuous time derivative by the discrete derivative with respect to n so that $u_n'(\tau) = -s(u_{n+1} - u_n)$. A soliton solution corresponds to the minimum of the discrete Lagrangian

$$\mathscr{L} = \sum_n \left[\frac{1}{2} g(1 - \bar{s}^2)(u_{n+1} - u_n)^2 + V(u_n) \right],$$

where $\bar{s} = s/s_0$ and $s_0 = \sqrt{g}$ is the velocity of long-wave phonons.

To find the shape of a kink (antikink), the following minimum problem was solved numerically:

$$\mathscr{L} = \sum_{n=1}^{N-1} \left[\frac{1}{2} g(1 - \bar{s}^2)(u_{n+1} - u_n)^2 + V(u_n) \right] \tag{5.13}$$

$$\longrightarrow \min_{u_2, \dots, u_{N-1}} : u_1 = -\pi \; (\pi), \; u_N = \pi \; (-\pi) .$$

The method of conjugated gradients [24] was used to solve the problem (5.13) for the chain with $N = 200$ links.

Numerical solution of the problem (5.13) has shown that, for strong cooperativity, the topological soliton has the subsonic velocity spectrum $0 \leq s < s_0$. With

Fig. 5.13 Antikink and kink in the cyclic chain ($g = 25$, $a = 1.5$, $N = 500$, and $s = 2.5$) at the initial time $\tau = 0$ (*line 1*) and time $\tau = 4{,}000$ (*line 2*)

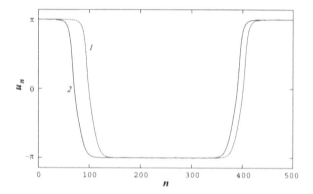

increasing velocity, the soliton diameter decreases monotonically and its energy

$$E = \sum_{n=1}^{N-1} \left[\frac{1}{2} g(1 + \bar{s}^2)(u_{n+1} - u_n)^2 + V(u_n) \right]$$

increases steadily. The diameter $L \to 1$ and energy $E \to +\infty$ as $s \to s_0 - 0$.

Let us numerically model the soliton dynamics. We consider the motion of a kink and an antikink in a cyclic chain. For this purpose, we integrate the equations of motion (5.12) numerically with the periodic boundary conditions $u_{n+N} \equiv u_n$ and $u'_{N+n} \equiv u'_n$ and take $g = 25$, $N = 500$, and $s = 2.5$ ($\bar{s} = 0.5$). The profiles of topological solitons at the initial time $\tau = 0$ are shown in Fig. 5.13. The antikink profile $u = u_n$ is always the mirror image of the kink profile relative to the u-axis. As a result of this symmetry, in a chain with an anisotropic substrate, the kink motion to the right is equivalent to the antikink motion to the left and vice versa.

Numerical solution of the equations of motion has shown the stability of the solitons. The soliton shape at the final time $\tau = 4{,}000$ coincides with the initial shape (see Fig. 5.13). The kink passed 9,989 chain links during this time and the antikink 9,972. The results show that, for strong cooperativity, the substrate anisotropy has almost no effect on the soliton dynamics. The soliton can move freely either to the right or to the left.

5.2.4 Soliton Mobility

Let us introduce relaxation and an external constant force into the equations of motion (5.12) so that they take the form

$$u_n'' = g(u_{n+1} - 2u_n + u_{n-1}) - \frac{\partial V}{\partial u}(u_n) - \gamma u_n' + f , \quad n = 0, \pm 1, \pm 2, \dots ,$$

$$(5.14)$$

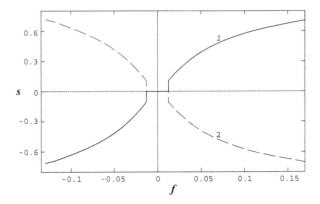

Fig. 5.14 Dependence of the velocity s of a kink (*line 1*) and an antikink (*line 2*) on the value of the constant external force f in the anisotropic chain ($a = 1.5$) for a weak cooperativity $g = 1$ (relaxation time $\tau_0 = 10$)

where $\gamma = 1/\tau_r$ is the friction coefficient, τ_r is the relaxation time of the system, and f is the external constant force. In the system described by (5.14), the topological soliton can move only with a specific velocity, for which the action of the external force is fully compensated by friction.

Let us find the dependence of the soliton velocity on the external force. For this purpose we numerically model the soliton dynamics in a finite chain with free ends for $\tau_r = 10$ ($\gamma = 0.1$) and $a = 1.5$. We consider a stationary soliton at the initial time and analyse its motion under the constant force f. In the absence of pinning, the soliton starts moving. Its velocity increases until it levels out at a value $s(f)$. The dependence $s(f)$ for weak cooperativity $g = 1$ is shown in Fig. 5.14. For this value of the cooperativity, the stationary soliton is pinned. To overcome the pinning, a force $f > 0.0126$ or $f < -0.013$ must be applied. Under the action of such a force, opposite-sign solitons move in opposite directions. If the absolute magnitude of the force is increased, the absolute magnitude of the velocity does so too. The dependence $s(f)$ for strong cooperativity $g = 25$ is shown in Fig. 5.15. In this case, a force of any magnitude causes soliton motion.

As can be seen from the calculations discussed above, the anisotropy of the chain substrate leads to the difference between the velocities of motion to the right and to the left, i.e., $s(f) \neq -s(-f)$. As a result, in a chain with an external force and relaxation, the two directions of soliton motion are no longer equivalent.

Let the external force change sign at regular intervals, i.e.,

$$f(\tau) = f_0 \text{sign} \frac{\sin \pi \tau}{\tau_0} , \tag{5.15}$$

where f_0 is the absolute amplitude of the force and τ_0 is the period of sign change. For sufficiently large values of τ_0, the soliton velocity can be considered to be constant on the interval of constant sign of the force. When the sign of the force

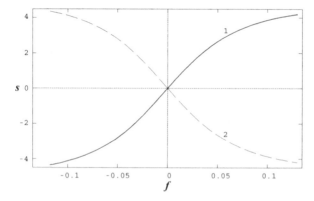

Fig. 5.15 Dependence of the velocity s of the kink (*line 1*) and antikink (*line 2*) on the magnitude of the external force f ($g = 25$, $a = 1.5$, and $\tau_r = 10$)

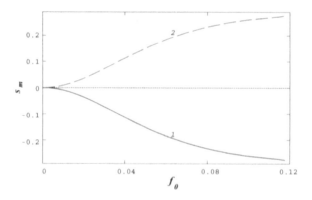

Fig. 5.16 Dependence of the average velocity s_m of the kink (*line 1*) and antikink (*line 2*) on the magnitude of the external force f ($g = 25$, $a = 1.5$, $\tau_r = 10$)

changes, the direction of soliton motion changes. In the limit $\tau_0 \rightarrow \infty$, the average velocity of the soliton is

$$s_m(f_0) = s(f_0) + s(-f_0) \neq 0 .$$

The dependence of $s_m(f_0)$ for $g = 25$ is shown in Fig. 5.16. Therefore, when affected by a low-frequency force (5.15), the kink has to move to the left, while the antikink moves to the right. The velocity depends monotonically on the amplitude of the force. Note that this soliton dynamics can take place only in an anisotropic chain. In a chain with a symmetric substrate potential, the absolute soliton velocity does not change when the sign of the external force changes and the average velocity is therefore $s_m(f_0) = 0$.

5.2.5   Soliton Motion Induced by Low-Frequency Noise

We now examine topological soliton motion in an anisotropic chain ($a = 1.5$) under the action of colored noise with a narrow-band frequency spectrum. For this purpose, we integrate the equations of motion of the finite chain with free ends, viz.,

$$u_1'' = g(u_2 - u_1) - \frac{\partial V}{\partial u}(u_1) - \gamma u_1' + \eta(\tau) \,,$$

$$\vdots$$

$$u_n'' = g(u_{n+1} - 2u_n + u_{n-1}) - \frac{\partial V}{\partial u}(u_n) - \gamma u_n' + \eta(\tau) \,, \qquad (5.16)$$

$$\vdots$$

$$u_N'' = g(u_{N-1} - u_N) - \frac{\partial V}{\partial u}(u_N) - \gamma u_N' + \eta(\tau) \,,$$

where the function $\eta(\tau)$ depends on time according to the Langevin equation

$$\eta'' = -\omega^2 \eta - \Gamma \eta' + \xi \,. \qquad (5.17)$$

In (5.17), the frequency ω defines the average period $T = 2\pi/\omega$ of the colored noise, the parameter Γ describes the width of the frequency spectrum of the colored noise (hereafter, we will take $\Gamma = 1/10T$), and $\xi(\tau)$ is the normal random force with autocorrelation function

$$\langle \xi(\tau_1)\xi(\tau_2) \rangle = 2\Gamma \omega^2 a_\eta^2 \delta(\tau_1 - \tau_2) \,,$$

where a_η is the amplitude of the colored noise.

For the equations of motion (5.16), we take the initial conditions corresponding to the stationary topological soliton centered in the middle of the chain. The soliton dynamics will be considered in a chain consisting of $N = 10^3$ links. To analyse the dynamics, it is convenient to define the soliton centre $\bar{n}$ as the intersection point of the broken line, sequentially connecting the points $\{(n, u_n)\}_{n=1}^N$ with the line $u = \bar{u}$ on the (n, u)-plane, where $\bar{u} \in (-\pi, \pi)$ is the maximum of the anisotropic substrate potential (5.7).

Let $g = 1$, $\gamma = 0.1$, and $a = 1.5$. The dynamics of a kink and an antikink induced by colored noise ($a_\eta = 0.05$ and $T = 10$) is shown in Fig. 5.17 (left). The noise causes opposite-sign solitons to start moving in opposite directions. For short-period noise, the motion occurs as a sequence of random jumps. An increase in the period results in more uniform movement (see Fig. 5.17 (right) for $a_\eta = 0.05$ and $T = 30$). When the noise period increases, the soliton velocity increases as well. The strongest response to the noise is observed at low frequencies (long periods) (see Fig. 5.18 (left) for $\eta = 0.1$ and $T = 200$).

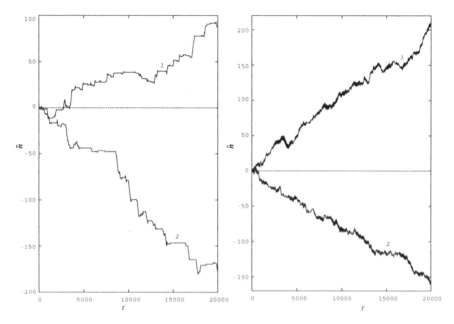

Fig. 5.17 *Left*: Dynamics of a kink (*line 1*) and an antikink (*line 2*) in the anisotropic chain ($a = 1.5$ and $g = 1$) under the action of narrow-band colored noise ($a_\eta = 0.05$ and $T = 10$). *Right*: Dynamics of a kink (*line 1*) and an antikink (*line 2*) in the anisotropic chain ($a = 1.5$ and $g = 1$) under the action of narrow-band colored noise ($a_\eta = 0.05$, $T = 30$)

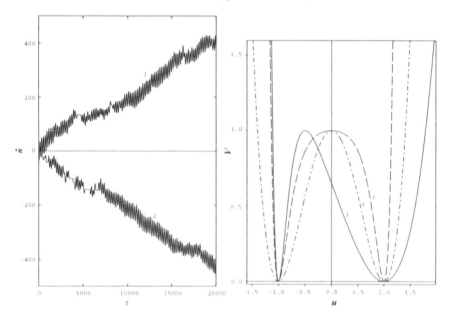

Fig. 5.18 *Left*: Dynamics of the kink (*line 1*) and antikink (*line 2*) in the anisotropic chain ($a = 1.5$, $g = 1$) under the action of low-frequency, narrow-band colored noise ($a_\eta = 0.1$, $T = 200$). *Right*: View of the double-well potential $V(u)$ for $\alpha_1 = 0.5$, $\alpha_2 = 5.0$ (*line 1*), $\alpha_1 = \alpha_2 = 0.5$ (*line 2*), and $\alpha_1 = \alpha_2 = 5.0$ (*line 3*)

5.2.6 Conclusion

The foregoing investigation shows that, in a chain of particles lying on a periodic anisotropic substrate, symmetric low-frequency noise can induce directed motion of topological solitons. Opposite-sign solitons (kink and antikink) start moving in opposite directions. This effect stems from the asymmetry of the substrate potential. The greatest response to the noise is observed in the low-frequency region. Structural anisotropy is a typical feature of many biomolecular systems and it is hoped that this investigation may help to understand the nature of some of their properties.

5.3 Numerical Simulation of Soliton Dynamics in the ϕ-4 Model

In this section we consider the topological soliton dynamics in a chain consisting of asymmetric, energetically degenerate, double-well potentials with weak inter-particle interaction. We will show that in this case the soliton is pinned and the corresponding Peierls potential has a typical asymmetric potential profile. Here, symmetrical external noise, modeling the chain interaction with a heat bath, leads to a nonequilibrium motion of the solitons in a direction that depends on the spectral properties of the noise. The results of this section were first published in [25, 26].

5.3.1 The Model

We will consider the dynamics of a linear chain consisting of bistable monomeric links. Let us assume that the chain monomer can be in two energy-equivalent conformations $\pm l$, and that the corresponding double-well potential has an asymmetric shape (the maximum of the potential barrier is shifted to one of the minima).

The Hamiltonian of the bistable chain is written in the form

$$\mathcal{H} = \sum_n \left[\frac{1}{2} m \dot{x}_n^2 + \frac{1}{2} \kappa (x_{n+1} - x_n)^2 + \varepsilon V(x_n/l) \right], \tag{5.18}$$

where m is the reduced mass of a monomeric link, κ is the force constant of the chain, $V(u)$ is the asymmetric double-well potential with minima ± 1 normalized to zero and a maximum normalized to 1, and ε is the height of the energy barrier between the monomer conformations.

For convenience, we introduce the dimensionless displacement $u_n = x_n/l$, time $\tau = t\sqrt{\varepsilon/ml^2}$, and energy $H = \mathscr{H}/\varepsilon$. Then the Hamiltonian takes the form

$$H = \sum_n \left[\frac{1}{2}u_n'^2 + \frac{1}{2}g(u_{n+1} - u_n)^2 + V(u_n) \right],\qquad (5.19)$$

where the dimensionless parameter $g = \kappa l^2/\varepsilon$ describes the magnitude of the interparticle interaction. It is convenient to specify the asymmetric potential $V(u)$ by the equation

$$V(u) = C\left[(e^{\alpha_1 u} - e^{\alpha_1})(e^{-\alpha_2 u} - e^{\alpha_2})\right]^2 ,\qquad (5.20)$$

where α_1, α_2, and $C > 0$. The potential (5.20) has a double-well shape with minima situated at the points $u = \pm 1$. The constant C normalizes the barrier height to 1:

$$C = \left[(e^{\alpha_1 u_0} - e^{\alpha_1})(e^{-\alpha_2 u_0} - e^{\alpha_2})\right]^{-2} ,$$

where the position u_0 of the potential maximum is obtained by solving $V'(u_0) = 0$.

The potential $V(u)$ is symmetric only if $\alpha_1 = \alpha_2 = \alpha$, when (5.20) takes the particularly simple form

$$V(u) = (e^{\alpha u} - e^\alpha)^2(e^{-\alpha u} - e^\alpha)^2(1 - e^\alpha)^{-4} = \left(\frac{\cosh \alpha x - \cosh \alpha}{1 - \cosh \alpha}\right)^2 .$$

In the general case when $\alpha_1 \neq \alpha_2$, the potential is asymmetric. The maximum of potential barrier is shifted to the right ($u_0 > 0$) for $\alpha_1 > \alpha_2$ and to the left for $\alpha_1 < \alpha_2$. A view of the potential $V(u)$ for the three values of the parameters α_1 and α_2 is shown in Fig. 5.18 (right).

The chain dynamics is described by the dimensionless Langevin equations

$$u_n'' = g(u_{n+1} - 2u_n + u_{n-1}) - F(u_n) - \gamma u_n' + \eta_n , \quad n = 0, \pm 1, \pm 2, \ldots ,\qquad (5.21)$$

where the function $F(u) = dV/du$, $\gamma = 1/\tau_r$ is the friction coefficient, τ_r is the relaxation time, and η_n is the random force describing the interaction of the nth chain link with the heat bath. The random forces η_n are independent and have the correlation functions

$$\langle \eta_{n_1}(\tau_1)\eta_{n_2}(\tau_2)\rangle = 2\gamma\beta\delta_{n_1 n_2}\varphi(\tau_1 - \tau_2) ,$$

where $\beta = kT/\varepsilon$ is the dimensionless temperature and $\varphi(\tau)$ is the dimensionless autocorrelation function normalized to 1, viz., $\int_{-\infty}^{+\infty} \varphi(\tau)d\tau = 1$. The autocorrelation function has the form $\varphi(\tau) = \delta(\tau)$ for white noise and $\varphi(\tau) = (\lambda/2)\exp{-|\lambda \tau|}$ for colored noise, where $\lambda = 1/\tau_c$ and τ_c is the correlation time of the random force.

5.3.2 Asymmetry of Chain Monomer Oscillation

We first consider the dynamics of a single isolated particle in an energetically degenerate, asymmetric potential. Asymmetry of the potential has to lead to asymmetry in the thermal fluctuations of the particle. The probability p of finding a particle in the left well generally does not coincide with the probability $1 - p$ of finding it in the right well. The probability p characterizes the asymmetry of the oscillation. In the case of a symmetric potential, the probabilities of finding the particle in the left and right wells coincide, i.e., $p = 1 - p = 0.5$. Let us examine the dependence of the probability p on the frequency spectrum of colored noise, i.e., its dependence on the correlation time τ_c.

It is convenient to describe the particle dynamics under colored noise by the system of two equations

$$u'' = -F(u) - \gamma u' + \eta \,, \tag{5.22}$$

$$\eta' = \lambda(\xi - \eta) \,, \tag{5.23}$$

where $\xi(\eta)$ is the delta-correlated random force distributed normally, i.e.,

$$\langle \xi(\tau_1)\xi(\tau_2) \rangle = 2\gamma\beta\delta(\tau_1 - \tau_2) \,.$$

Specifically, let us take $\alpha_1 = 0.5$ and $\alpha_2 = 5.0$. Then the asymmetric potential $V(u)$ has a maximum at the point $u_0 = -0.50017$. The curvatures of the left and right wells are $k_- = V''(-1) = 85.17$ and $k_+ = V''(1) = 2.13$, respectively, so the frequency of particle oscillation in the left well, viz., $\omega_- = \sqrt{k_-} = 9.23$, is an order of magnitude higher than that in the right well, viz., $\omega_+ = \sqrt{k_+} = 1.46$.

Let us consider the low-amplitude oscillations. We assume that, in the vicinity of the well at $u = \pm 1$, the potential is $V(u) = k_\pm(u \mp 1)^2/2$. Then the average particle energy is

$$E_\pm = \lim_{t \to \infty} \frac{1}{t} \int_0^t \left[\frac{1}{2}u'^2 + \frac{1}{2}k_\pm(u \mp 1)^2 \right] d\tau$$

$$= 2\gamma\beta \int_0^\infty (\omega^2 + \omega_\pm^2)|H_\pm(\omega)|^2 F(\omega)d\omega \,,$$

where $H_\pm(\omega) = (\omega_\pm^2 - \omega - i\omega\gamma)^{-1}$ is the response function of the Langevin equation and

$$F(\omega) = \frac{1}{2\pi} \int_{-\infty}^{+\infty} \varphi(\tau) \exp(-i\omega\tau)d\tau$$

is the Fourier transformation of the autocorrelation function of the random force. Therefore, the average energy is $E_\pm = K_\pm + P_\pm$, where

$$K_\pm = 2\gamma\beta \int_0^\infty \frac{\omega^2 F(\omega)d\omega}{(\omega_\pm^2 - \omega^2)^2 + \omega^2\gamma^2} \, , \qquad P_\pm = 2\gamma\beta \int_0^\infty \frac{\omega_\pm^2 F(\omega)d\omega}{(\omega_\pm^2 - \omega^2)^2 + \omega^2\gamma^2} \, .$$

For colored noise (an exponentially correlated random force), the Fourier transformation is $F(\omega) = \lambda^2/2\pi(\omega^2 + \lambda^2)$ and the average kinetic and potential energies are $K_\pm = \beta f_K(\omega_\pm, \gamma, \lambda)/2$ and $P_\pm = \beta f_P(\omega_\pm, \gamma, \lambda)/2$, respectively, where the dimensionless functions are

$$f_K(\omega, \gamma, \lambda) = \frac{\lambda^2}{\lambda^2 + \lambda\gamma + \omega^2} \, , \qquad f_P(\omega, \gamma, \lambda) = \frac{\lambda^2(\omega^2 + \lambda^2 - \gamma^2) + \lambda\gamma\omega^2}{(\omega^2 + \lambda^2)^2 - \gamma^2\lambda^2} \, .$$

For infinitely small friction ($\gamma = 0$), the average oscillation energy is

$$E_\pm = \beta\lambda^2/(\lambda^2 + \omega_\pm^2) \, .$$

During one period of oscillation, the particle makes one attempt to overcome the barrier, so the probability of its transition into another well is

$$P_\pm = \frac{\omega_\pm}{2\pi} \exp(-1/E_\pm) \, .$$

The ratio of the probabilities is

$$a = \frac{P_-}{P_+} = \sqrt{\frac{k_-}{k_+}} \exp\left[-\frac{\tau_c^2(k_- - k_+)}{\beta} \right] \, ,$$

and the probability of finding the particle in the left well is $p = 1/(a + 1)$. The dependence of the probability p on the correlation time τ_c is shown in Fig. 5.19. The probability p increases steadily with increasing τ_c. For white noise ($\tau_c = 0$), the ratio $a = \sqrt{k_-/k_+} = 6.321$, the probability $p = 0.137$, and for the low-frequency limit ($\tau_c \to \infty$), the probability is $p = 1$.

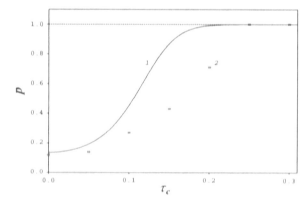

Fig. 5.19 Dependence of the probability p of finding the particle in the left well on the correlation time τ_c of the random force. *Solid line (1)* and *points (2)* show the dependence obtained analytically and numerically, respectively

Table 5.1 Dependence of the probability p on the correlation time of the random force τ_c

τ_c	0.00	0.05	0.10	0.15	0.20	0.25
p	0.12	0.14	0.27	0.53	0.71	1.00

As can be seen from the above analysis, the asymmetric nature of the oscillations depends on both the shape of the double-well potential and the frequency spectrum of the noise. For white noise, the probability of finding the particle in the well is proportional to its width. When the high-frequency components are removed from the noise, this rule is violated, i.e., the particle is more likely to be found in the narrower well on the left. 'Freezing' of high-frequency oscillations occurs and the particle in the well with higher curvature is practically in a non-thermalized state. This leads to a decrease in the probability of the particle leaving the well.

We simulate the particle dynamics numerically. Specifically, we choose $\gamma = 0.01$ and $\tau_r = 100$ and solve the equations of motion (5.22) and (5.23) numerically with the initial conditions $u(0) = u_0$, $u'(0) = 0$, and $\eta(0) = 0$. Numerical integration was performed up to the time $\tau = 10^5$. The dependence of the resulting probability p of finding the particle in the left well on the correlation time of the random force τ_c is given in Table 5.1 and Fig. 5.19. As can be seen from the table and figure, the numerical results confirm our conclusions about the growth of the probability p with increasing correlation time τ_c. The difference between the values p obtained analytically and numerically results from the use of a piecewise-parabolic approximation for the double-well potential.

5.3.3 Stationary States of a Topological Soliton

To find the stationary state of a positive (negative) topological soliton I_+ (I_-), one must solve the minimum problem

$$E = \sum_{n=1}^{N-1} \left[\frac{1}{2} g(u_{n+1} - u_n)^2 + V(u_n) \right] \longrightarrow \min : u_2, \ldots, u_{N-1} , \qquad (5.24)$$

with the conditions $u_1 = +1$ and $u_N = -1$ ($u_1 = -1$ and $u_N = +1$). It is convenient to describe the soliton stationary sate by the position of its centre

$$\overline{n} = \sum_{n=1}^{N-1} n p_n$$

and root mean square diameter

$$L = 1 + \sqrt{\sum_{n=1}^{N-1} (n - \overline{n})^2 p_n} ,$$

Fig. 5.20 Unstable positive (*line 1*) and stable negative (*line 2*) stationary topological solitons ($\alpha_1 = 0.5$, $\alpha_2 = 5.0$, and $g = 1.0$)

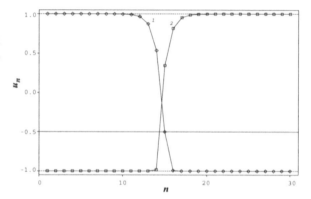

where the sequence

$$p_n = \frac{1}{E} \left\{ V(u_n) + \frac{1}{4} \left[(u_{n+1} - u_n)^2 + (u_{n-1} - u_n)^2 \right] \right\}$$

gives the energy distribution over the chain.

The minimum problem (5.24) was solved by the method of conjugate gradients [24]. The stationary states were sought in the chain consisting of $N = 50$ links and the cooperativity parameter $g = 1$ was used. The results of the calculation showed that the soliton has two stationary states: the stable state ($\overline{n}_1 = 25.2484$, $E_1 = 1.3812$, and $L_1 = 2.196$) and the unstable state ($\overline{n}_2 = 24.6806$, $E_2 = 1.9505$, and $L_2 = 2.147$). The pinning soliton energy is $\Delta E = E_2 - E_1 = 0.5693$. Profiles of these states are shown in Fig. 5.20.

We can also find the energy dependence of the stationary defect on its centre position, i.e., calculate the Peierls potential profile. For this purpose, one must fix the value $u_{N/2} \in (-1, 1)$ in the problem (5.24) and minimize the energy E with respect to the remaining variables. Then, by monotonically varying $u_{N/2}$, we obtain the monotonic change in its centre $\overline{n}$. The centre position is uniquely determined by the coordinate $u_{N/2}$. The dependence of the soliton energy $E(\overline{n})$ on its centre position can thus be obtained numerically. A view of the Peierls potential profile $E(\overline{n})$ is shown in Fig. 5.21 (left). For the value $g = 1$ of the cooperativity parameter used here, the soliton is strongly pinned and its Peierls potential profile has a typical asymmetric ratchet-like profile (the left side of the potential barrier has a steeper shape than the right side). Thus, the topological soliton dynamics in the chain consisting of energetically degenerate, asymmetric double-well potentials can be considered qualitatively as the motion of a particle in the periodic ratchet-like potential profile. Taken together, these results demonstrate that symmetric forces can cause the asymmetric motion of a topological soliton.

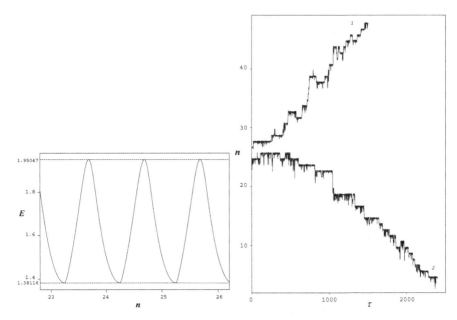

Fig. 5.21 *Left*: Peierls potential profile of the topological soliton ($\alpha_1 = 0.5$, $\alpha_2 = 5.0$, and $g = 1.0$). *Right*: Dynamics of the topological soliton in the thermalized chain with the parameters $\alpha_1 = 0.5$, $\alpha_2 = 5.0$, $g = 1.0$, $\tau_r = 10$, and $\beta = 0.25$ for $\tau_c = 0$ (*line 1*) and $\tau_c = 0.3$ (*line 2*)

5.3.4 Soliton Dynamics in a Thermalized Chain

Let us consider a finite chain with fixed ends. The dynamics of the chain consisting of N links is described by the Langevin equations

$$\left.\begin{aligned} u_n'' &= g(u_{n+1} - 2u_n + u_{n-1}) - F(u_n) - \gamma u_n' + \eta_n \, , \\ \eta_n' &= \lambda(\xi_n - \eta_n) \, , \end{aligned}\right\} \quad n = 2, 3, \ldots, N - 1 \, ,$$

$$(5.25)$$

where $\xi_n(\tau)$ are the normally distributed delta-correlated random forces:

$$\langle \xi_{n_1}(\tau_1)\xi_{n_2}(\tau_2) \rangle = 2\gamma\beta\delta_{n_1 n_2}\delta(\tau_1 - \tau_2) \, .$$

We take the initial conditions of (5.25) corresponding to a stable stationary state of the topological soliton located in the middle of the chain. We fix the coordinate $u_{N/2}$ at the initial time and numerically integrate (5.25) until thermal equilibrium is reached with the thermostat. We then release the coordinate $u_{N/2}$ and analyse the soliton dynamics.

To do this, it is convenient to define the soliton centre $\bar{n}$ as the intersection of the broken line sequentially connecting the points $\{(n, u_n)\}_{n=1}^{N}$ with the line $u = u_0$ on the (n, u)-plane. We choose the chain parameters $\alpha_1 = 0.5$, $\alpha_2 = 5.0$, $g = 1.0$, $\beta = 0.25$, and $\tau_r = 10$. Note that opposite-sign solitons in the model discussed here possess the same properties. Therefore, we restrict attention to the positive soliton dynamics.

The dynamics of the soliton centre in a thermalized chain consisting of $N = 50$ links is shown in Fig. 5.21 (right). We see that, most of the time, the soliton remains in the pinned state, rarely making random jumps. Obviously, if the probability of finding the particle in the left well $p < 0.5$, thermal fluctuations have to cause the soliton drift to the right end or to the left if $p > 0.5$. Indeed, the soliton drifts to the right for $\tau_c = 0$ ($p = 0.12$) and to the left for $\tau_c = 0.3$ ($p = 1.0$).

5.3.5 Conclusion

The numerical analysis performed above shows that, in a thermalized chain consisting of energetically degenerate, asymmetric bistable monomers, the symmetrical thermal fluctuations should lead to the asymmetric motion of a topological soliton. The velocity and direction of the motion depend on both the shape of the asymmetric double-well potential and the frequency spectrum of the noise. This control mechanism of soliton transport may be used in biomolecular systems, since the frequency spectrum of the noise depends significantly on the temperature and conformation state of these systems.

References

1. Braun, O.M., Kivshar, Y.S.: The Frenkel–Kontorova Model. Springer, Berlin/Heidelberg (2004)
2. Braun, O.M., Kivshar, Yu.S.: Model' Frenkelya-Kontrovoi: Kontseptsii, metody, prilozheniya. Fizmatlit, Moscow (2008)
3. Flash, S., Yevtushenko, O., Zolotaryuk, Y.: Directed current due to broken time–space symmetry. Phys. Rev. Lett. **84**, 2358 (2000)
4. Reimann, P.: Supersymmetric ratchets. Phys. Rev. Lett. **86**, 4992 (2001)
5. Feynman, R.P., Leighton, R.B., Sands, M.: The Feynman Lectures on Physics. Addison-Wesley, Reading (1966)
6. Magnasco, M.O.: Forced thermal ratchets. Phys. Rev. Lett. **71**(10), 1477 (1993)
7. Doering, C.R., Horsthemke, W., Riordan, J.: Nonequilibrium fluctuation-induced transport. Phys. Rev. Lett. **72**(19), 2984 (1994)
8. Cordova, N.J., Ermentrout, B., Oster, G.F.: Dynamics of single-motor molecules: the thermal ratchet model. Proc. Natl. Acad. Sci. U. S. A. **89**, 339 (1992)
9. Simon, S.M., Peskin, C.S., Oster, G.F.: What drives the translocation of proteins? Proc. Natl. Acad. Sci. U. S. A. **89**, 3770 (1992)
10. Peskin, C.S., Odell, G.M., Oster, G.F.: Cellular motions and thermal fluctuations: the Browian ratchet. Biophys. J. **65**, 316 (1993)
11. Gwinn, E.G., Westervel, R.M.: Intermittent chaos and low-frequency noise in the driven damped pendulum. Phys. Rev. Lett. **54**, 1613 (1985)

12. Hockett, K., Holmes, P.J.: Josephson junction, Annulus maps, Birkhoff attractors, Horseshoes and rotation sets. Center for Applied Mathematical Report Cornell University (1985)
13. Mun, F.: Chaotic Oscillations. Mir, Moscow (1990)
14. Gorre, L., Ioannidis, E., Silberzan, P.: Rectified motion of a mercury drop in an asymmetric structure. Europhys. Lett. **33**(4), 267 (1996)
15. Gorre-Talini, L., Jeanjean, S., Silberzan, P.: Sorting of Brownian particles by the pulsed application of an asymmetric potential. Phys. Rev. E **56**, 2025 (1997)
16. Sandre, O., Gorre-Talini, L., Ajdari, A., Prost, J., Silberzan, P.: Moving droplets on asymmetrically structured surfaces. Phys. Rev. E **60**, 2964 (1999)
17. Julicher, F., Ajdari, A., Prost, J.: Modeling molecular motors. Rev. Mod. Phys. **69**, 1269 (1997)
18. Bader, J.S., Hammond, R.W., Henck, S.A., Deem, M.W., McDermott, G.A., Bustillo, J.M., Simpson, J.W., Mulhern, G.T., Rothberg, J.M.: DNA transport by a micromachined Brownian ratchet device. Proc. Natl. Acad. Sci. U. S. A. **96**, 13165 (1999)
19. Press, W.H., Teukolsky, S.A., Vetterling, W.T., Flannery, B.P.: Numerical Recipes in Fortran 77: The Art of Scientific Computing. Press Syndicate of the University of Cambridge, Cambridge (1992)
20. Davydov, A.S.: Solitons in Molecular Systems. Naukova Dumka, Kiev (1988)
21. Krumhansl, J.A., Schrieffer, J.R.: Dynamics and statistical mechanics of one-dimensional model Hamiltonian for structural phase transition. Phys. Rev. B **11**, 3535 (1975)
22. Kontorova, T., Frenkel, Y.I.: On the theory of plastic deformation and twinning. II. Zh. Eksp. Teor. Fiz. **8**, 1340 (1938)
23. Satarić, M.V., Tuszyński, J.A., Žakula, R.B.: Kinklike excitations as an energy-transfer mechanism in microtubules. Phys. Rev. E **48**, 589 (1993)
24. Fletcher, R., Reeves, C.M.: Function minimization by conjugate gradients. Comput. J. **7**, 149 (1964)
25. Savin, A.V., Tsironis, G.P., Zolotaryuk, A.V.: Ratchet and switching effects in stochastic kink dynamics. Phys. Lett. A **229**, 279 (1997)
26. Savin, A.V., Tsironis, G.P., Zolotaryuk, A.V.: Reversal effects in stochastic kink dynamics. Phys. Rev. E **56**(3), 2457 (1997)

Chapter 6
Solitons in Polymer Systems

This chapter will focus on the numerical investigation of nonlinear dynamics of localized excitations (acoustic and topological solitons and breathers) in polymer macromolecules. The characteristics of supersonic acoustic solitons in polymer macromolecules will be studied, using the examples of an isolated zigzag macromolecule of polyethylene (PE), a spiral macromolecule of polytetrafluoroethylene (PTFE), and a single-well carbon nanotube. Topological soliton dynamics will be analysed using the crystalline PE and PTFE models. We will discuss the role of topological solitons in the premelting mechanisms of crystals and their structural transitions. Nonlinear localized vibrations, or breathers, will be considered in the case of a *trans* zigzag PE molecule. The quasi-one-dimensional structure of isolated macromolecules, polymer crystals of PE and PTFE, and a single-well carbon nanotube will be shown to lead to the existence of all the basic types of localized nonlinear excitations (acoustic and topological solitons and breathers). The properties of such excitations will be shown to depend significantly on the structure of the polymer macromolecule.

The development of contemporary nonlinear physics has led to the discovery of new fundamental mechanisms which determine, on the molecular level, the progression of many physical processes in crystals and other ordered molecular structures. It is now clear that acoustic solitons may contribute to the most efficient mechanism of energy transfer in such processes as heat conduction and breakdown of solids [1–4]. Topological solitons serve as models of structural defects in polymer crystals, and their mobility ensures the possibility of such processes as plastic deformation [5], relaxation [6], and premelting [7, 8]. Crystal structure defects are described in a natural way using the concept of topological solitons [9, 10], and soliton mobility defines a specific 'soliton' contribution to the thermodynamics and kinetics of polymer crystals. Breathers play a significant role in the mechanisms of energy transfer and relaxation in molecular systems [11–13].

© Springer International Publishing Switzerland 2015
L.N. Lupichev et al., *Synergetics of Molecular Systems*, Springer Series
in Synergetics, DOI 10.1007/978-3-319-08195-3_6

6.1   Acoustic Solitons in Planar Zigzag PE Macromolecules and Spiral PTFE Macromolecules

In molecular systems with pronounced quasi-one-dimensional structure, an acoustic soliton (nonlinear solitary wave) defines a local region of longitudinal compression of intermolecular bonds in an alpha-helix protein molecule, moving along the molecular chain [14–16]. The investigation of nonlinear dynamics of PE macromolecules has shown that the zigzag structure of the chain can lead to the existence of acoustic stretching solitons [17–19].

Today, the PE molecule $(CH_2-)_x$ is the most studied in the class of polymer macromolecules in which the repeating link consists of a single atom. The ground state of the molecule is a planar zigzag conformation of the chain (the helix symbol is 1*2/1). The two-dimensional zigzag structure of the chain leads to peculiar features in its dynamics. It is worth noting that, for most macromolecules in this class, the ground state is a three-dimensional helix rather than a planar zigzag structure. Therefore, it is also of interest to consider the nonlinear dynamics of the PTFE molecule $(CF_2-)_x$, which is analogous to the PE molecule but has the three-dimensional helix structure with symbol 1*13/6 in its ground state.

6.1.1   Stretching Solitons PE Macromolecules

Although the linear dynamics of the polyethylene molecule was studied by Kirkwood [20] over half a century ago, its nonlinear extension has only recently become the subject of theoretical analysis [17–19]. In a study of the low-energy dynamic processes in the polyethylene molecule, the motion of hydrogen atoms with respect to the chain backbone is not important, and the united atom approximation can be used. Let us consider a polyethylene molecule $(CH_2-)_x$ in the *trans* zigzag conformation. At equilibrium, the molecule backbone has the planar zigzag structure characterized by the step $\rho_0 = 1.53\,\text{Å}$ (the equilibrium length of the valence bond H_2C-CH_2), and the zigzag angle $\theta_0 = 113°$ (the equilibrium valence angle $CH_2-CH_2-CH_2$). The *trans* zigzag structure is shown schematically in Fig. 6.1a.

If we let the *trans* zigzag chain be directed along the x-axis and lie in the xy-plane, then the nth site of the chain at equilibrium has coordinates

$$x_n^0 = nl_x, \qquad y_n^0 = (-1)^n l_y/2,$$

where $l_x = \rho_0 \sin(\theta_0/2)$ and $l_y = \rho_0 \cos(\theta_0/2)$ are the longitudinal and transverse steps of the zigzag chain. It is convenient to change from the absolute coordinates x_n and y_n of the nth links to the relative coordinates

$$u_n = x_n - x_n^0, \qquad v_n = (-1)^{n+1}(y_n - y_n^0).$$

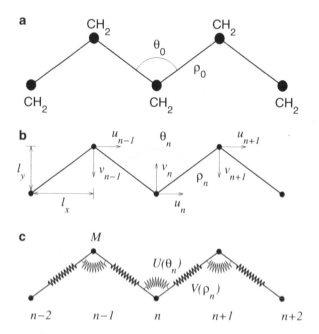

Fig. 6.1 Structure of the polyethylene molecule $(CH_2)_\infty$ (**a**), definition of local coordinate systems (**b**), and the planar mechanical model of the macromolecule (**c**)

Here u_n and v_n define, respectively, the longitudinal and the transverse displacements of the nth link from its equilibrium position, the positive direction of the transverse displacement being taken towards the centre of the zigzag chain (see Fig. 6.1b). The length of the nth valence bond and the cosine of the nth valence angle are, respectively,

$$\rho_n = \sqrt{(l_x - w_n)^2 + (l_y - z_n)^2}\,, \qquad \cos\theta_n = -\frac{a_{n-1}a_n - b_{n-1}b_n}{\rho_{n-1}\rho_n}\,,$$

where $w_n = u_n - u_{n+1}$ and $z_n = v_n + v_{n+1}$ are the longitudinal and transverse compression of the nth chain link, whence $a_n = l_x - w_n$ and $b_n = l_y - z_n$.

The Hamiltonian of the chain can be written as

$$H = \sum_n \left[\frac{1}{2}M(\dot{u}_n^2 + \dot{v}_n^2) + V(\rho_n) + U(\theta_n) \right]. \tag{6.1}$$

Here the mass of the chain link is $M = 14m_p$ (m_p is the proton mass),

$$V(\rho_n) = D_0\left\{1 - \exp\left[-\alpha(\rho_n - \rho_0)\right]\right\}^2 \tag{6.2}$$

is the potential of the nth valence bond and

$$U(\theta_n) = \frac{1}{2} K_\theta (\cos \theta_n - \cos \theta_0)^2 \tag{6.3}$$

is the potential of the nth valence angle. According to [21], the energy of the valence bond $D_0 = 334.72\,\mathrm{kJ\,mol^{-1}}$, the parameter $\alpha = 19.1\,\mathrm{nm^{-1}}$, and the parameter $K_\theta = 130.122\,\mathrm{kJ\,mol^{-1}}$. In [22], the higher energy value $\varepsilon = 529\,\mathrm{kJ\,mol^{-1}}$ was used. The planar mechanical model of the *trans* zigzag chain under consideration is shown in Fig. 6.1c.

The dispersion equation for low-amplitude vibrations of the *trans* zigzag chain was first obtained by Kirkwood [20]. A detailed description of the derivation of this equation is given in [17], so here we will skip most of the intermediate calculations.

The Hamiltonian of the chain (6.1) gives the equations of motion

$$M\ddot{u}_n = -\frac{\partial H}{\partial u_n}\,, \qquad M\ddot{v}_n = -\frac{\partial H}{\partial v_n}\,, \qquad n = 0, \pm 1, \pm 2, \dots\,. \tag{6.4}$$

We linearize these equations and find their solution in the form of a harmonic wave

$$u_n(t) = A \exp\mathrm{i}(\Omega t + kn)\,, \qquad v_n(t) = B \exp\mathrm{i}\big[\Omega t + k(n + 1/2)\big]\,,$$

where Ω is the circular frequency and $-\pi \le k \le \pi$ is the dimensionless wave vector. Then the dispersion equation has the form

$$\Omega_{\pm}^2(k) = \omega_0^2(k) \pm \sqrt{\omega_0^4(k) - \omega_1^4(k)}\,, \tag{6.5}$$

where

$$\omega_0^2(k) = C_1(1 + \cos\theta_0 \cos k) + 2C_2(1 + \cos k)(1 - \cos\theta_0 \cos k)\,,$$
$$\omega_1^4(k) = 8C_1 C_2(1 + \cos k)\sin^2 k\,.$$

Here, $C_1 = K_1/M$ and $C_2 = K_2/M\rho_0^2$ are the rigidity parameters, where

$$K_1 = \frac{\mathrm{d}^2}{\mathrm{d}\rho^2} V(\rho)\bigg|_{\rho=\rho_0} = 2D_0\alpha^2 = 405.53\,\mathrm{N/m}$$

and

$$K_2 = \frac{\mathrm{d}^2}{\mathrm{d}\theta^2} U(\theta)\bigg|_{\theta=\theta_0} = K_\theta \sin^2\theta_0 = 18.308 \times 10^{-20}\,\mathrm{J}$$

are the rigidities of the valence bond and valence angle, respectively, The shape of the dispersion curve (6.5) is shown in Fig. 6.2. The upper branch $\Omega = \Omega_+(k)$

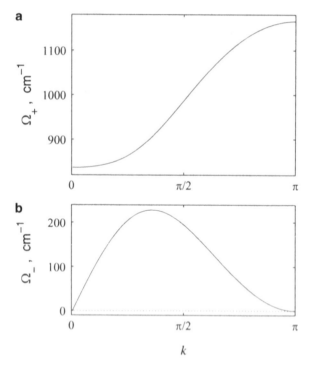

Fig. 6.2 Optical $\Omega_+(k)$ (**a**) and acoustic $\Omega_-(k)$ (**b**) branches of the dispersion curve of the PE *trans* zigzag

corresponds to the high-frequency optical phonons of the *trans* zigzag chain, and the lower branch $\Omega = \Omega_-(k)$ to the low-frequency optical phonons. The frequency of acoustic phonons tends to zero as $k \to 0$ and $k \to \pi$. The long-wave (smooth) longitudinal and bending waves of the zigzag molecular chain correspond to these limiting values of the wave number.

The velocity of the long-wave longitudinal acoustic phonons (speed of sound) is given by the relation

$$c_0 = l_x \lim_{k \to 0} \frac{\Omega_-(k)}{k} = \frac{2\sqrt{K_2/M}\,\tan(\theta_0/2)}{\sqrt{1 + 4\delta\tan(\theta_0/2)}},$$

where the dimensionless parameter

$$\delta = \frac{C_2}{C_1} = \frac{K_2}{K_1\rho_0^2},$$

defines the ratio of the rigidity of the valence angle to that of the valence bond ($\delta = 0.01929$ when $K_\theta = 130.122\,\mathrm{kJ\,mol^{-1}}$ and $\delta = 0.07841$ when

$K_\theta = 529$). As it turns out, the rigidity of the valence bond is two orders of magnitude greater than that of the valence angle. It would seem that it is quite possible to use the approximation of an infinitely rigid valence bond $\delta = 0$ ($K_1 = \infty$), but in this approximation, even for $\delta = 0.01929$, the sound speed $\bar{c}_0 = 2\sqrt{K_2/M}\tan(\theta_0/2) = 8{,}449\,\text{ms}^{-1}$ differs considerably from the exact value of $c_0 = 7{,}790\,\text{ms}^{-1} = 0.92210\bar{c}_0$. Such a substantial shift in the sound speed means that one must take account of the deformation of valence bonds.

The complexity of the equations of motion (6.4) defies analytical investigation (unlike the case of infinitely rigid valence bonds). To obtain the soliton solution, we use a numerical method of soliton analysis [23]. According to this, for every value of the velocity c, the soliton solution $u_n(t) = u(nl_x - ct)$, $v_n(t) = v(nl_x - ct)$, $n = 0, \pm 1, \pm 2, \ldots$, is sought as an extreme point of a certain functional, which, in the continuous approximation, corresponds to the equations of motion of the system.

We now seek a solution of the equations of motion (6.4) in the form of a traveling smooth wave with constant profile. For this purpose, we set $u_n(t) = u(\xi)$ and $v_n(t) = v(\xi)$, where $\xi = nl_x - ct$ is the wave variable, c is the wave velocity, and u and v are smooth functions of ξ. Then the Lagrangian corresponding to the equations of motion (6.4) is

$$L = \sum_n \left[\frac{1}{2} M(\dot{u}_n^2 + \dot{v}_n^2) - V(\rho_n) - U(\theta_n) \right] , \tag{6.6}$$

which can be written in the form

$$\bar{L} = \sum_n \left\{ \frac{c^2 M}{24 l_x^2} \left[16 w_n^2 - (w_n + w_{n+1})^2 + 16(v_{n+1} - v_n)^2 - (v_{n+2} - v_n)^2 \right] \right.$$
$$\left. - V(\rho_n) - U(\theta_n) \right\} . \tag{6.7}$$

The supersonic soliton state of the chain corresponds to the saddle point of the Lagrangian, so it can be sought as the minimum of the functional

$$F = \frac{1}{2} \sum_n \left(\bar{L}_{w_n}^2 + \bar{L}_{v_n}^2 \right) .$$

Thus, in order to find the soliton solution (solitary wave) $\{w_n, v_n\}_{n=1}^N$, one should solve the minimum problem

$$F = \frac{1}{2} \sum_{n=2}^{N-1} \left(\bar{L}_{w_n}^2 + \bar{L}_{v_n}^2 \right) \longrightarrow \min : \ w_1 = w_N = v_1 = 0, \ v_N = 0 \ (l_y) . \tag{6.8}$$

The solution to this problem allows us to find all the soliton solutions of the nonlinear system (6.4), i.e., smooth solitary waves with constant profile. The absence of such solutions for any value c of the velocity implies the impossibility of soliton motion at this velocity. The problem (6.8) was solved numerically by the method of conjugate gradients [24]. A solution was sought in a chain of $N = 400$ links. The initial point of descent was taken in the form of two symmetrical bell-shaped (or kink) profiles $w(n)$ and $v(n)$ centred at the middle of the chain.

The key idea of the method consists in replacing the continuous time derivatives in the Lagrangian (6.6) by their discrete approximations (the transition from the Lagrangian (6.6) to the discrete functional (6.7)). Therefore, this method can only be used to find 'broad' soliton solutions whose shape exhibits a smooth dependence on the number n of a chain site (the length of the chain N must be ten times the width of the soliton solution).

Let us assume that $\{w_n^0, v_n^0\}_{n=1}^N$ is the required soliton solution with symmetry centre located at the site $n = N/2$. Then the relevant soliton is characterized by the energy

$$E = \sum_{n=2}^{N-1} \left\{ \frac{c^2 M}{24 l_x^2} [16 w_n^2 - (w_n + w_{n+1})^2 + 16(v_{n+1} - v_n)^2 - (v_{n+2} - v_n)^2] \right.$$
$$\left. + V(\rho_n) + U(\theta_n) \right\} ,$$

the total compression of the chain is

$$R = \sum_{n=1}^N w_n ,$$

the root-mean-square width given in the chain periods is

$$D = 2 \left[\sum_{n=1}^N (n - m)^2 \frac{w_n}{R} \right]^{1/2} ,$$

and the point

$$m = \frac{1}{2} + \sum_{n=1}^N \frac{n w_n}{R}$$

defines the position of the soliton centre. The other characteristics of the soliton are the maximum value of the valence angle $A_\theta = \max_n (\theta_n)$ and the maximum length of the valence bond $A_\rho = \max_n (\rho_n)$.

Numerical analysis has shown that the form of the soliton solution depends on the dimensionless parameter δ which characterizes the ratio between physical and geometrical anharmonicity. The physical anharmonicity is due to the potential of the valence bond, whereas the geometrical anharmonicity is due to the potential of the valence angle. Geometrical anharmonicity prevails for $\delta < 0.0356$ and physical anharmonicity for $\delta > 0.03556$.

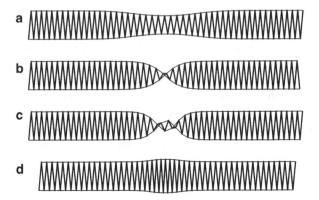

Fig. 6.3 Deformation of the *trans* zigzag chain corresponding to the first type of stretching soliton $\delta = 0.01929$, $s = 1.02$ (**a**), the second type of stretching soliton $\delta = 0.01929$, $s = 1.05$ (**b**), the third type of stretching soliton $\delta = 0.01929$, $s = 1.0738$ (**c**), and the compression soliton $\delta = 0.07841$, $s = 1.035$ (**d**)

When $\delta = 0.01929$, the equations of motion (6.4) have three types of soliton solution. The first corresponds to a solitary wave of longitudinal stretching of the *trans* zigzag chain (see Fig. 6.3a) with amplitude $A_v = \max_n v_n < l_y/2$ (the maximum value of the valence angle in the range of soliton localization is $A_\theta < \pi$) and asymptotic behavior $w_n, v_n \to 0$ as $n \to \infty$. The second type soliton is a solitary wave of large-amplitude longitudinal stretching of the *trans* zigzag chain (see Fig. 6.3b) with asymptotic behavior $w_n, v_n \to 0$ as $n \to -\infty$ and $w_n \to 0$, $v_n \to l_y$ as $n \to +\infty$. This solitary wave describes the sequential unfolding of valence angles from one equilibrium value $\theta_n = \theta_0$ to the other $\theta_n = 2\pi - \theta_0$. As a result, the chain transforms from one ground state $\{w_n \equiv 0, v_n \equiv 0\}$ to another one $\{w_n \equiv 0, v_n \equiv l_y\}$. The third type of soliton solution corresponds to a solitary stretching wave of the *trans* zigzag chain (see Fig. 6.3c) with amplitude $l_y/2 < A_v < l_y$ ($\pi < A_\theta < 2\pi - \theta_0$) and asymptotic behavior $w_n, v_n \to 0$ as $n \to \infty$. This soliton is essentially a bound state of two opposite-sign solitons of the second type.

The dependence of the energy E, root-mean-square width D, and total longitudinal compression of the chain R on the dimensionless soliton velocity $s = c/c_0$ are shown in Fig. 6.4 (left). Solitons of the first type have the supersonic spectrum of velocities $1 < s < s_1 = 1.020$. When the soliton velocity increases, the energy E and total compression R of the chain increase steadily, while the root-mean-square width D decreases. Solitons of the second type have the supersonic band of admissible velocities $s_2 = 1.023 < s < s_3 = 1.062$. In this case, E, D, and R decrease monotonically. Solitons of the third type exist only for velocity $s = s_4 = 1.074$.

A typical shape of the first type of soliton is shown in Fig. 6.4 (right). The soliton has the characteristic bell-shaped profile of a solitary wave with respect to its components w_n, v_n, and θ_n. In the region of soliton localization, the molecular chain exhibits longitudinal stretching ($w_n > 0$) and transverse compression ($v_n > 0$): the

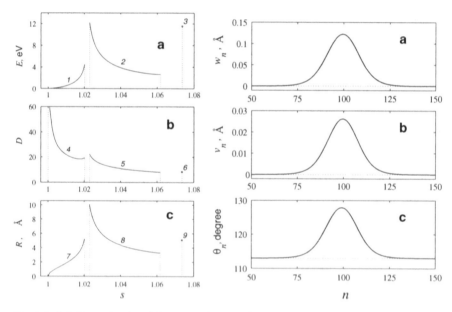

Fig. 6.4 *Left*: Dependencies of the energy E, width D, and total stretching of the chain R on the velocity s of the compression soliton of the *trans* zigzag chain for the first (*lines 1, 4,* and *7*), second (*lines 2, 5,* and *8*), and third (*lines 3, 6,* and *9*) types. *Right*: Profiles of the stretching soliton of the first type. Components w_n (**a**), v_n (**b**), and θ_n (**c**) are shown at the initial time $t = 0$ and the final time $t = 1{,}613.8$ after passing 99,999.69 chain links ($s = 1.015$)

valence angles increase and the valence bonds stretch out. The existence of these solitons in the zigzag chain is due to the geometrical nonlinearity of the chain rather than the inherent (physical) anharmonicity of intermolecular potentials. This is the fundamental distinction between the *trans* zigzag model and the model of the two-dimensional alpha-helix.

As the velocity increases, the energy and amplitude of the soliton increase steadily and attain their maximum values $E_{\mathrm{m}} = 4.6\,\mathrm{eV}$ and $R_{\mathrm{m}} = 5.3\,\text{Å}$, respectively, at $s = s_1$. The soliton width decreases, but always exceeds 18 chain links, i.e., the stretching soliton always complies with the assumption we made when we found its shape, i.e., that its profile depends smoothly on the number of chain links. The specific values of energy E, width D, amplitude R, increments in the angle $\Delta\theta = A_\theta - \theta_0$, and valence bond $\Delta\rho = A_\rho - \rho_0$ are given in Table 6.1. As can be seen from this table, with increasing velocity s, in the region of soliton localization, the deformations of angles and bonds increase steadily but always satisfy $\Delta\theta < 27°$ for the valence angle and $\Delta\rho < 0.05\,\text{Å}$ for the valence bond.

With the second value of the parameter $\delta = 0.078419$, the equations of motion (6.4) have a soliton solution corresponding to a solitary wave of longitudinal compression of the *trans* zigzag chain (see Fig. 6.3d). In the region of soliton localization, compression of valence angles and bonds occurs. The soliton has a finite range of velocities $1 < s \leq 1.035$.

Table 6.1 Energy E, width D, total compression R, torsion Ψ, amplitudes A_r, A_ρ, A_θ, and A_δ of the helix stretching soliton for different values of the dimensionless velocity s

s	E (kJ mol^{-1})	D	R (Å)	Ψ (deg)	A_r (Å)	A_ρ (Å)	A_θ (deg)	A_δ (deg)
1.005	8.8	20.3	−0.44	7.8	0.006	−0.010	−1.6	−0.3
1.010	27.8	15.9	−0.66	11.4	0.013	−0.019	−3.3	−0.5
1.015	55.5	13.6	−0.88	14.5	0.020	−0.028	−5.1	−0.7
1.020	97.2	12.2	−1.10	17.4	0.028	−0.038	−7.2	−0.9
1.025	158.3	11.3	−1.35	20.5	0.038	−0.048	−9.5	−1.1
1.030	252.4	10.8	−1.65	24.3	0.049	−0.059	−12.2	−1.3
1.035	418.6	10.6	−2.11	29.9	0.062	−0.071	−15.8	−1.5
1.040	941.0	11.5	−3.34	44.9	0.085	−0.088	−21.8	−1.7

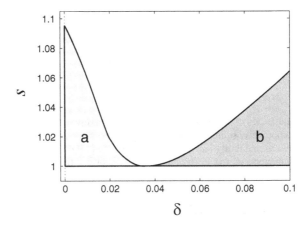

Fig. 6.5 Existence regions of the acoustic solitons of stretching (**a**) and compression (**b**) of the *trans* zigzag chain in the space of dimensionless parameters δ and s

The dependence of the velocity range of the acoustic soliton on the dimensionless parameter δ is shown in Fig. 6.5. The soliton is a solitary wave of stretching for $\delta < \delta_0 = 0.0356$ and a solitary wave of compression of the *trans* zigzag chain for $\delta > \delta_0$. In the chain with $\delta = 0$ (the approximation of an infinitely rigid valence bond), the soliton velocity range is $1 < s < s_1 = 1.095$. As δ increases, the upper limit of the velocities s_1 decreases steadily. At a threshold value $\delta = \delta_0$, the geometrical and physical anharmonicity cancel out, and the velocity range vanishes ($s_1 = 1$). A further increase in δ leads to a monotonic growth of the velocity range.

Numerical simulation of the dynamics has shown that the first type soliton is dynamically stable at all values of the velocity $1 < s < s_1$. It moves with a constant velocity and completely retains its original shape. For example, at the initial dimensionless velocity $s = 1.015$ ($c = 7{,}906.85$ ms^{-1}, $\delta = 0.01929$), the soliton passed 99,999.694 chain links in 1,613.8 ps and had the final velocity $s = 1.014995$ ($c = 7{,}906.81$ ms^{-1}). As shown in Fig. 6.4 (right), the final shape of the soliton is exactly the same as the initial one. The solitons interact as elastic particles.

Fig. 6.6 Elastic collision between stretching solitons of the first type in the cyclic chain consisting of $N = 500$ links. The time dependencies of the longitudinal (**a**) and transverse (**b**) displacements of the chain links are shown

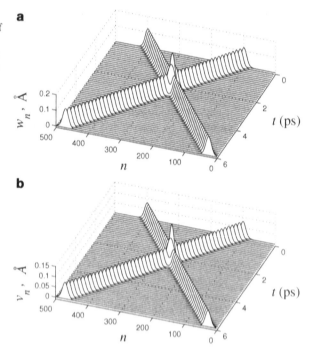

Their collisions result in elastic reflection without emission of phonons or change in shape (see Fig. 6.6). Only near the limiting velocity s_1 does the soliton interaction become inelastic. Collision is then accompanied by phonon emission. Thus, near the sound speed, the stretching solitons belonging to the first type exhibit clear particle-like properties. Solitons of the second type are unstable. When moving, they emit phonons and quickly decay. The third type of soliton is stable at the velocity $s = s_4$. It moves along the chain at a constant velocity and retains its shape. The interaction between solitons of the third type is not elastic. Indeed, collisions lead to their destruction.

The above analysis of the *trans* zigzag model has shown that, in an isolated planar polyethylene macromolecule, dynamically stable stretching solitons can exist, with a relatively narrow spectrum of supersonic velocities. The existence of the solitons is due to the geometrical anharmonicity of the zigzag chain rather than the physical anharmonicity of the intermolecular interaction potentials. A more detailed description of the results is given in [17, 18, 25].

6.1.2 Stretching Solitons in a PTFE Macromolecule

A molecule of PTFE in the crystal state has helix conformation $1*13/6$ with lattice periods $a = b = 5.59\,\text{Å}$ and $c = 16.88\,\text{Å}$ [26]. Let us use the united atom approximation in which a segment of the PTFE macromolecule CF_2 is treated as a unified particle of mass $M = 50\,m_p$, where m_p is the proton mass. Then the nth chain link in the equilibrium state is defined by the radius vector

$$\mathbf{R} = \left(R_0\cos(n\Delta\phi),\ R_0\sin(n\Delta\phi),\ n\Delta\phi\right),\qquad(6.9)$$

where R_0 is the helix radius, and $\Delta\phi = 12\pi/13 = 166.15°$ and $\Delta z = c/13 = 1.298\,\text{Å}$ are the angular and longitudinal helix step, respectively.

The helix radius can be determined from the length of the valence bond C–C, viz., $\rho_0 = 1.533\,\text{Å}$. Indeed, it follows from the helix equation (6.9) that the square of the valence bond length is

$$\rho_0^2 = \left|\mathbf{R}_{n+1} - \mathbf{R}_n\right| = 2R_0^2(1 - \cos\Delta\phi) + \Delta z^2,$$

whence the helix radius is

$$R_0 = \sqrt{\frac{\rho_0^2 - \Delta z^2}{2(1 - \cos\Delta\phi)}} = 0.410\,\text{Å}.$$

The valence angle in the equilibrium state is $\theta_0 = \arccos\left[-(\mathbf{e}_{n-1},\mathbf{e}_n)/\rho_0^2\right]$, where the vector $\mathbf{e}_n = \mathbf{R}_{n+1} - \mathbf{R}_n$ determines the direction of the nth valence bond. After elementary transformations, we find that

$$\theta_0 = \pi - \arccos\frac{4R_0^2\sin^2(\Delta\phi/2)\cos\Delta\phi + \Delta z^2}{\rho_0^2} = 116.30°.$$

The dihedral (torsional) angle in the equilibrium state is

$$\eta_n = \arccos\frac{(\mathbf{v}_{n-1},\mathbf{v}_n)}{|\mathbf{v}_{n-1}||\mathbf{v}_n|} = \arccos\frac{h^2\cos\Delta\phi + \sin^2\Delta\phi}{h^2 + \sin^2\Delta\phi},$$

where $\mathbf{v}_n = (\mathbf{e}_n,\mathbf{e}_{n+1})$ is the vector product of the vectors $\mathbf{e}_n$ and $\mathbf{e}_{n+1}$ and $h = \Delta z/R_0$ is the dimensionless longitudinal helix step. We also introduce the rotation angle around the nth bond $\delta_n = \pi - \eta_n$, where η_n is the nth dihedral angle. In the equilibrium state, the rotation angle is $\delta_0 = \pi - \eta_0 = 16.32°$.

Let x_n, y_n, and z_n be the coordinates of the nth chain site. We change from Cartesian to cylindrical coordinates:

$$x_n = (R_0 + r_n)\cos(n\Delta\phi + \varphi_n),$$
$$y_n = (R_0 + r_n)\sin(n\Delta\phi + \varphi_n),$$
$$z_n = n\Delta z + h_n,$$

where r_n, φ_n, and h_n are the radial, angular, and longitudinal displacements of the nth chain link from its equilibrium state, respectively. Then the Hamiltonian of the chain has the form

$$H = \sum_n \left\{ \frac{M}{2} \left[\dot{r}_n^2 + \dot{\varphi}_n^2 (R_0 + r_n)^2 + \dot{h}_n^2 \right] + V(\rho_n) + U(\theta_n) + W(\delta_n) \right\} ,$$

(6.10)

where the dot denotes differentiation with respect to time t and ρ_n, θ_n, and δ_n are the length of the nth valence bond, the valence angle, and the rotation angle, respectively.

The potential of the valence bond $V(\rho_n)$ is given by (6.2), in which the length of the nth bond is

$$\rho_n = (a_{n,1} + b_n^2)^{1/2} ,$$

with

$$a_{n,1} = d_n^2 + d_{n+1}^2 - 2 d_n d_{n+1} c_{n,1} , \qquad b_n = \Delta z + h_{n+1} - h_n ,$$

and $d_n = R_0 + r_n$, $c_{n,1} = \cos(\Delta\phi + \varphi_{n+1} - \varphi_n)$. The energy of the valence bond is $D_0 = 334.72 \, \text{kJ mol}^{-1}$ and the anharmonicity parameter is $\alpha = 1.91 \, \text{Å}$ [21].

The deformation energy of the valence angle is described by the potential (6.3), in which the nth valence angle is

$$\theta_n = \arccos\left(-\frac{a_{n,2} + b_{n-1}b_n}{\rho_{n-1}\rho_n} \right) ,$$

(6.11)

with

$$a_{n,2} = d_{n-1}d_n c_{n-1,1} + d_n d_{n+1} c_{n,1} - d_n^2 - d_{n-1}d_{n+1} c_{n,2} ,$$

$$c_{n,2} = \cos(2\Delta\phi + \varphi_{n+1} - \varphi_{n-1}) .$$

The energy $K_\theta = 529 \, \text{kJ mol}^{-1}$ [22]. The potential of internal rotation $W(\delta_n)$ characterizes the slowdown of chain link rotation around the nth valence bond. The nth rotation angle is

$$\delta_n = \arccos\left(-\frac{b_n b_{n+1} a_{n,2} + b_{n-1} b_n a_{n+1,2} - b_n^2 a_{n,4} - b_{n-1} b_{n+1} a_{n,1} + a_{n,3} a_{n+1,3}}{\sqrt{\beta_n \beta_{n+1}}} \right).$$

where

$$a_{n,3} = d_{n-1}d_n s_{n-1,1} + d_n d_{n+1} s_{n,1} - d_{n-1}d_{n+1} s_{n,2} ,$$

$$a_{n,4} = d_n d_{n+2} c_{n+1,2} - d_n d_{n+1} c_{n,1} - d_{n-1}d_{n+2} c_{n,3} + d_{n-1}d_{n+1} c_{n,2} ,$$

$$s_{n,1} = \sin(\Delta\phi + \varphi_{n+1} - \varphi_n) ,$$

$$s_{n,2} = \sin(2\Delta\phi + \varphi_{n+1} - \varphi_{n-1}) ,$$

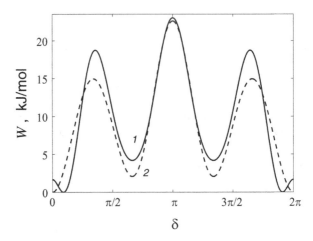

Fig. 6.7 Potential $W(\delta)$ of rotation around the valence bond C–C for macromolecules of PTFE (*line 1*) and PE (*line 2*)

$$c_{n,3} = \cos(3\Delta\phi + \varphi_{n+2} - \varphi_{n-1}) \, ,$$
$$\beta_n = a_{n-1,1}b_n^2 + a_{n,1}b_{n-1}^2 - 2a_{n,2}b_{n-1}b_n + a_{n,3}^2 \, .$$

A view of the potential for the PF macromolecule is presented in Fig. 6.7. The absolute minimum $\delta_0 = 0$ of the potential corresponds to the *trans* conformation, while the other two minima $\delta_1 \approx 2\pi/3$ and $\delta_2 \approx 4\pi/3$ refer to the *gauche* conformation. The potential is characterized by three values:

- The height $\epsilon_1 = U(\pi/3)$ of the potential barrier between *trans* and *gauche* conformations,
- The second minimum $\epsilon_2 = U(2\pi/3)$ of the potential, corresponding to the *gauche* conformation energy, and
- The maximum value $\epsilon_3 = U(\pi)$, corresponding to the energy level of the masked conformation.

According to [26], the values are $\epsilon_1 = 14.94 \, \text{kJ mol}^{-1}$, $\epsilon_2 = 2.0768 \, \text{kJ mol}^{-1}$, and $\epsilon_3 = 22.6 \, \text{kJ mol}^{-1}$.

A view of the potential for the PTFE macromolecule is shown in Fig. 6.7. The difference between these two potentials relates to the polarity of the C–F bond and the fact that the van der Waals radius of fluoride is greater than that of hydrogen. All these effects also contribute to the total potential energy of rotation. In contrast to PE, which has three rotary isomers on each C–C bond, the one with minimum energy (*trans*) and two others with higher energy, viz., *gauche* (+) and *gauche* (−), PTFE has four rotary isomers. Two of them, viz., *trans* (+) and *trans* (−), have identical minimum energy, i.e., $\delta_1 = \delta_0$, $\delta_2 = 2\pi - \delta_0$, $W(\delta_1) = W(\delta_2) = 0$, while the two others, *gauche* (+) and *gauche* (−), have higher energies, i.e., $\delta_3 \approx 2\pi/3$ and $\delta_4 \approx 4\pi/3$, $W(\delta_3) = W(\delta_4) > 0$.

The rotational potential is characterized by four values:

- The height $\epsilon_0 = W(0)$ of the potential barrier between the two *trans* conformations,
- The height $\epsilon_1 = W(\pi/3)$ of the potential barrier between the *trans* and *gauche* conformations,
- The *gauche* conformation energy $\epsilon_2 = W(2\pi/3)$, and
- The height $\epsilon_3 = W(\pi)$ of the barrier between the *gauche* conformations.

According to [26], $\epsilon_0 = 1.674 \, \text{kJ mol}^{-1}$, $\epsilon_1 = 18.42 \, \text{kJ mol}^{-1}$, $\epsilon_2 = 4.186 \, \text{kJ mol}^{-1}$, and $\epsilon_3 = 23.02 \, \text{kJ mol}^{-1}$.

For the numerical modeling of the dynamics, it is convenient to represent the rotational potential by the equation

$$W(\delta) = \left[C_1 Z_\alpha(\delta) + C_2 Z_\beta(\delta) - C_3 \right]^2 , \tag{6.12}$$

where the one-parameter functions are

$$Z_\alpha(\delta) = \frac{(1 + \alpha) \sin^2(\delta/2)}{1 + \alpha \sin^2(\delta/2)} , \qquad Z_\beta(\delta) = \left[\frac{(1 + \beta) \sin(3\delta/2)}{1 - \beta \sin(3\delta/2)} \right]^2 .$$

The parameter values $C_1 = 3.411 \, \text{kJ}^{1/2} \, \text{mol}^{-1/2}$, $C_2 = 2.681 \, \text{kJ}^{1/2} \, \text{mol}^{-1/2}$, and $C_3 = 1.294 \, \text{kJ}^{1/2} \, \text{mol}^{-1/2}$, together with $\alpha = 14.6125$ and $\beta = 4.0028 \times 10^{-3}$, are uniquely determined by the following equations:

$$W(0) = C_3^2 = \epsilon_0 ,$$

$$W(\delta_0) = \left[C_1 Z_\alpha(\delta_0) + C_2 Z_\beta(\delta_0) - C_3 \right]^2 = 0 ,$$

$$W(\pi/3) = \left[C_1 \frac{1 + \alpha}{4 + \alpha} + C_2 \left(\frac{1 + \beta}{1 - \beta} \right)^2 - C_3 \right]^2 = \epsilon_1 ,$$

$$W(2\pi/3) = \left(3C_1 \frac{1 + \alpha}{4 + 3\alpha} - C_3 \right)^2 = \epsilon_2 ,$$

$$W(\pi) = (C_1 + C_2 - C_3)^2 = \epsilon_3 .$$

A view of the potential (6.12) for the given values of the parameters is shown in Fig. 6.7. The potential has absolute minimum at $\delta = \delta_0$ and $2\pi - \delta_0$. Note that the rotational potential $W(\delta)$ of PE is symmetric relative to the point of absolute minimum $\delta_0 = 0$. The cubic anharmonicity of the potential at this point is equal to zero. In the case of PTFE, the minimum δ_0 is no longer the symmetry point. The cubic anharmonicity at this point is not equal to zero. As a result, one can expect the existence of torsion solitons, caused by the cubic anharmonicity of the rotational potential.

To analyse the small-amplitude vibrations of the chain, it is convenient to change from the cylindrical coordinates r_n, φ_n, and h_n to the local ones:

$$\mathbf{u}_n = \begin{pmatrix} u_{n,1} \\ u_{n,2} \\ u_{n,3} \end{pmatrix}$$

$$= \begin{pmatrix} \cos(n\Delta\phi) & \sin(n\Delta\phi) & 0 \\ -\sin(n\Delta\phi) & \cos(n\Delta\phi) & 0 \\ 0 & 0 & 1 \end{pmatrix} \begin{pmatrix} (R_0 + r_n)\cos(n\Delta\phi + \varphi_n) - R_0\cos(n\Delta\phi) \\ (R_0 + r_n)\sin(n\Delta\phi + \varphi_n) - R_0\sin(n\Delta\phi) \\ h_n \end{pmatrix}.$$

In this coordinate system, the Hamiltonian (6.10) has the form

$$H = \sum_n \left[\frac{1}{2}M(\dot{\mathbf{u}}_n, \dot{\mathbf{u}}_n) + V(\mathbf{u}_n, \mathbf{u}_{n+1}) + U(\mathbf{u}_{n-1}, \mathbf{u}_n, \mathbf{u}_{n+1}) \right.$$
$$\left. + W(\mathbf{u}_{n-1}, \mathbf{u}_n, \mathbf{u}_{n+1}, \mathbf{u}_{n+2}) \right]. \quad (6.13)$$

The following equations of motion correspond to the Hamiltonian (6.13):

$$-M\ddot{\mathbf{u}}_n = \mathbf{V}_1(\mathbf{u}_n, \mathbf{u}_{n+1}) + \mathbf{V}_2(\mathbf{u}_{n-1}, \mathbf{u}_n) + \mathbf{U}_1(\mathbf{u}_n, \mathbf{u}_{n+1}, \mathbf{u}_{n+2})$$
$$+\mathbf{U}_2(\mathbf{u}_{n-1}, \mathbf{u}_n, \mathbf{u}_{n+1}) + \mathbf{U}_3(\mathbf{u}_{n-2}, \mathbf{u}_{n-1}, \mathbf{u}_n)$$
$$+\mathbf{W}_1(\mathbf{u}_n, \mathbf{u}_{n+1}, \mathbf{u}_{n+2}, \mathbf{u}_{n+3}) + \mathbf{W}_2(\mathbf{u}_{n-1}, \mathbf{u}_n, \mathbf{u}_{n+1}, \mathbf{u}_{n+2})$$
$$+\mathbf{W}_3(\mathbf{u}_{n-2}, \mathbf{u}_{n-1}, \mathbf{u}_n, \mathbf{u}_{n+1}) + \mathbf{W}_4(\mathbf{u}_{n-3}, \mathbf{u}_{n-2}, \mathbf{u}_{n-1}, \mathbf{u}_n), \quad (6.14)$$

where the vectors are

$$\mathbf{V}_i(\mathbf{u}_1, \mathbf{u}_2) = \frac{\partial}{\partial \mathbf{u}_i}V, \quad i = 1, 2,$$

$$\mathbf{U}_i(\mathbf{u}_1, \mathbf{u}_2, \mathbf{u}_3) = \frac{\partial}{\partial \mathbf{u}_i}U, \quad i = 1, 2, 3,$$

$$\mathbf{W}_i(\mathbf{u}_1, \mathbf{u}_2, \mathbf{u}_3, \mathbf{u}_4) = \frac{\partial}{\partial \mathbf{u}_i}W, \quad i = 1, 2, 3, 4.$$

The linear approximation for the nonlinear equations (6.14) takes the form

$$-M\ddot{\mathbf{u}}_n = B_1\mathbf{u}_n + B_2(\mathbf{u}_{n-1} + \mathbf{u}_{n+1}) + B_3(\mathbf{u}_{n-2} + \mathbf{u}_{n+2}) + B_4(\mathbf{u}_{n-3} + \mathbf{u}_{n+3}), \quad (6.15)$$

where the constant matrices are

$$B_1 = V_{11} + V_{22} + U_{11} + U_{22} + U_{33} + W_{11} + W_{22} + W_{33} + W_{44},$$
$$B_2 = V_{12} + U_{12} + U_{23} + W_{12} + W_{23} + W_{34},$$

$$B_3 = U_{13} + W_{13} + W_{24} ,$$

$$B_4 = W_{14} .$$

Here

$$V_{ij} = \frac{\partial^2 V}{\partial u_i \, \partial u_j}(0,0) , \quad i, j = 1, 2 ,$$

$$U_{ij} = \frac{\partial^2 U}{\partial u_i \, \partial u_j}(0,0,0) , \quad i, j = 1, 2, 3 ,$$

$$W_{ij} = \frac{\partial^2 W}{\partial u_i \, \partial u_j}(0,0,0,0) , \quad i, j = 1, 2, 3, 4 .$$

Let us search for a solution of the linear system (6.8) in the form of a harmonic wave

$$\mathbf{u}_n = \mathbf{A} \exp \left[i(qn - \omega t) \right] . \tag{6.16}$$

Substituting (6.16) into the linear equation (6.15), we obtain the dispersion equation

$$\left| B_1 + 2B_2 \cos q + 2B_3 \cos(2q) + 2B_4 \cos(3q) - \omega^2 E \right| = 0 , \tag{6.17}$$

where E is the unit matrix. The dispersion equation (6.17) is an algebraic equation of third order in the variable ω^2. The corresponding algebraic curve has three branches: two acoustic $\omega = \omega_t(q)$, $\omega = \omega_l(q)$, and one optical $\omega = \omega_o(q)$, with $\omega_t(q) \leq \omega_l(q) \leq \omega_o(q)$.

A view of the dispersion curves is presented in Fig. 6.8. The lower curve $\omega = \omega_t(q)$ gives the dispersion law for acoustic phonons corresponding to torsional oscillations, while the middle curve $\omega = \omega_l(q)$ gives the dispersion law for acoustic phonons corresponding to longitudinal oscillations of the helix. The upper curve $\omega = \omega_o(q)$ corresponds to the high-frequency optical phonons in the helix.

The velocity $v_l = \Delta z \lim_{q \to 0} \omega_l(q)/q = 6{,}978.6$ m/s of the long-wavelength longitudinal phonons exceeds the velocity $v_t = \Delta z \lim_{q \to 0} \omega_t(q)/q = 5{,}585.3$ m/s of the torsional phonons. The ratio of these velocities is $s_t = v_t/v_l = 0.80035$.

The Hamiltonian of the chain (6.10) gives the equations of motion

$$M \ddot{r}_n - M(R_0 + r_n)\dot{\varphi}_n^2 + \frac{\partial}{\partial r_n} P = 0 ,$$

$$M(R_0 + r_n)^2 \ddot{\varphi}_n + 2M(R_0 + r_n)\dot{\varphi}_n \dot{r}_n + \frac{\partial}{\partial \varphi_n} P = 0 , \tag{6.18}$$

$$M \ddot{h}_n^2 + \frac{\partial}{\partial h_n} P = 0 ,$$

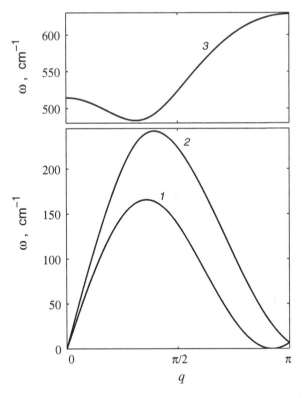

Fig. 6.8 Dispersion curves $\omega = \omega_t(q)$ (*line 1*), $\omega = \omega_1(q)$ (*line 2*), and $\omega = \omega_0(q)$ (*line 3*) for an isolated PTFE molecule

where the potential energy is

$$P = \sum_n \left[V(\rho_n) + U(\theta_n) + W(\delta_n) \right] .$$

The complexity of the equations of motion precludes analytical investigation, so to analyse the soliton solutions, we turn once again to the numerical methods of [14]. We search for a solution to the system of equations (6.18) in the form of a smooth traveling solitary wave with constant profile. For this purpose, we set $r_n(t) = r(\xi)$, $\varphi_n(t) = \varphi(\xi)$, and $h_n(t) = h(\xi)$, where $\xi = n\Delta z - vt$ is the wave variable, v is the wave velocity, and r, φ, and h are smooth functions of ξ.

We use the discrete approximation to the time derivatives:

$$\dot{r}_n = -v(r_{n+1} - r_{n-1})/2\Delta z ,$$
$$\dot{\varphi}_n = v(\psi_{n+1} - 5\psi_n - 2\psi_{n-1})/6\Delta z ,$$
$$\ddot{r}_n = v^2(r_{n+1} - 2r_n + r_{n-1})/\Delta z^2 ,$$

$$\ddot{\varphi}_n = -v^2(\psi_{n+1} - 15\psi_n + 15\psi_{n-1} - \psi_{n-2})/12\Delta z^2 ,$$

$$\ddot{h}_n = -v^2(w_{n+1} - 15w_n + 15w_{n-1} - w_{n-2})/12\Delta z^2 ,$$

where $\psi_n = \varphi_{n+1} - \varphi_n$ is the relative rotation and $w_n = h_{n+1} - h_n$ is the relative displacement. Then the equations of motion (6.18) can be written in the form of discrete equations in the variables r_n, ψ_n, and w_n:

$$F_{1,n} = -c_1(r_{n+1} - 2r_n + r_{n-1})$$

$$+ c_2(\psi_{n+1} - 5\psi_n - 2\psi_{n-1})^2(R_0 + r_n)$$

$$+ F_1(r_{n-3}, \ldots, r_{n+3}; \psi_{n-3}, \ldots, \psi_{n-2}; w_{n-3}, \ldots, w_{n+2}) = 0 ,$$

$$F_{2,n} = c_3(\psi_{n+1} - 15\psi_n + 15\psi_{n-1} - \psi_{n-2})$$

$$+ c_4(r_{n+1} - r_{n-1})(\psi_{n+1} - 5\psi_n - 2\psi_{n-1})/(R_0 + r_n) \qquad (6.19)$$

$$+ F_2(r_{n-3}, \ldots, r_{n+3}; \psi_{n-3}, \ldots, \psi_{n-2}; w_{n-3}, \ldots, w_{n+2}) = 0 ,$$

$$F_{3,n} = c_3(w_{n+1} - 15w_n + 15w_{n-1} - w_{n-2})$$

$$+ F_3(r_{n-3}, \ldots, r_{n+3}; \psi_{n-3}, \ldots, \psi_{n-2}; w_{n-3}, \ldots, w_{n+2}) = 0 ,$$

where the coefficients are $c_1 = v^2/\Delta z^2$, $c_2 = c_1/36$, $c_3 = c_1/12$, and $c_4 = c_1/6$.

We search numerically for a soliton solution $\{r_n, \psi_n, w_n\}_{n=1}^N$ to the discrete equations (6.13) as a solution of the constrained minimum problem

$$F = \frac{1}{2} \sum_{n=2}^{N-1} (F_{1,n}^2 + F_{2,n}^2 + F_{3,n}^2) \longrightarrow \min, \quad r_1 = r_N = \psi_1 = \psi_N = w_1 = w_N = 0.$$

$$(6.20)$$

The solution to this problem can be used to find numerically all the soliton solutions (solitary waves with constant profile) of the equations of motion (6.18). The absence of such solutions at any value of the velocity v implies the impossibility of soliton motion at that velocity.

The problem (6.20) was solved numerically by the method of conjugate gradients [24]. A solution was sought on a chain consisting of $N = 400$ links (this value of N ensures that the solution shape is independent of the zero boundary conditions). The initial point of descent was taken in the form of three symmetric bell-shaped profiles $r(n)$, $\psi(n)$, and $w(n)$, centered on the middle of the chain.

Every soliton solution $\{r_n, \psi_n, w_n\}_{n=1}^N$ is characterized by the energy

$$E = \sum_{n=2}^{N-1} \left\{ \frac{Mv^2}{8\Delta z^2} \left[(r_{n+1} - r_{n-1})^2 + (R_0 + r_n)^2(\psi_n + \psi_{n-1})^2 + (w_n + w_{n-1})^2 \right] \right.$$

$$+ V(\rho_n) + U(\theta_n) + W(\delta_n) \Big\} ,$$

the total torsion $\Psi = \sum_{n=1}^{N} \psi_n$, and the chain compression $R = \sum_{n=1}^{N} w_n$. The root-mean-square width of the solution, given in periods of the chain, is

$$D = 2 \left[\sum_{n=1}^{N} \frac{(n - \bar{n})w_n}{R} \right]^{1/2},$$

where $\bar{n} = \sum_{n=1}^{N} n w_n / R$ determines the position of the soliton centre. The solution is also characterized by the amplitude of the transverse displacement of chain links $A_r = \max_n r_n$, the amplitudes of valence bond deformations $A_\rho = \min_n (\rho_n - \rho_0)$, the valence angles $A_\theta = \min_n (\theta_n - \theta_0)$, and the rotational angles $A_\delta = \max_n (\delta_n - \delta_0)$.

Numerical solution of the problem (6.20) has shown the existence of soliton solutions of two types. The first type solution describes the propagation of a torsional solitary wave along the chain. A typical view of this solution is shown in Fig. 6.9 (left). In the region of soliton localization, the helix is expanded relative to the transverse component r_n, monotonically twisted relative to the angular variable φ_n, and monotonically squeezed relative to the longitudinal variable h_n. As can be

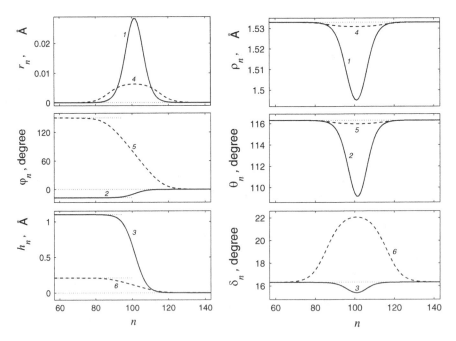

Fig. 6.9 *Left*: Profiles of the longitudinal compression soliton with components r_n, φ_n, and h_n (*lines 1–3*) at velocity $s = 1.02$ and the torsional soliton (*lines 4–6*) at $s = 0.82$. *Right*: Length change of the valence bonds ρ_n, valence angle θ_n, and rotational angles δ_n in the localization region of the compression soliton (*lines 1–3*) at velocity $s = 1.02$ and the torsional soliton (*lines 4–6*) at $s = 0.82$

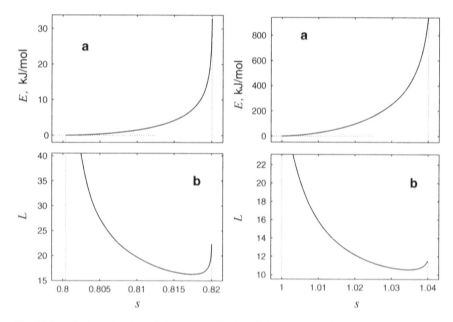

Fig. 6.10 *Left*: dependencies of the energy E (**a**) and depth D (**b**) of the torsional soliton on the dimensionless velocity s. *Right*: dependencies of the energy E (**a**) and depth D (**b**) of the compression soliton on the dimensionless velocity s

seen from Fig. 6.9 (right), all these deformations of the helix result basically from a local increase in the rotational angles δ_n, i.e., squeezing dihedral angles of the chain.

The second solution describes the propagation of a solitary wave of longitudinal compression along the chain. The form of the solution is shown in Fig. 6.9 (left). In the localization region of the soliton, the helix is expanded relative to the transverse component r_n, slightly untwisted relative to the angular variable φ_n, and squeezed relative to the longitudinal variable h_n. As can be seen in Fig. 6.9 (right), these deformations are due mainly to squeezing the valence angles and valence bonds. The torsional angles are thus weakly deformed.

The torsional solitons occupy the finite interval $s_t < s < 0.820034$ of permissible values of the dimensionless velocity $s = v/v_1$. The dependencies of the energy E and soliton width D on s are shown in Fig. 6.10 (left). With increasing velocity, the soliton energy steadily increases. The soliton width decreases monotonically up to the minimum value $D = 16.3$ at $s = 0.818$, and then begins to grow monotonically. Table 6.2 gives concrete values of the energy E, width D, total compression R, torsion of the helix Ψ, amplitude of transverse expansion of the chain A_r, and extreme values of the strains of valence bonds A_ρ, valence angles A_δ, and rotational angles A_δ. It follows from this table that the amplitude of longitudinal compression of the chain grows monotonically with increasing velocity and reaches the maximum value $-0.23\,\text{Å}$ at the maximum value of the velocity. The transverse expansion of the helix does not reach noticeable values. At all values of the velocity, the valence bonds and angles are almost undeformed, in contrast to the rotational

Table 6.2 Dependence of the energy E, depth D, total compression R, torsion Ψ, amplitudes A_r, A_ρ, A_θ, and A_δ of the torsion soliton on the dimensionless velocity s

s	E (kJ mol^{-1})	D	R (Å)	Ψ (deg)	A_r (Å)	A_ρ (Å)	A_θ (deg)	A_δ (deg)
0.8025	0.11	39.9	−0.012	−11.8	0.0003	−0.0001	−0.006	0.32
0.8050	0.37	27.5	−0.019	−18.5	0.0006	−0.0002	−0.017	0.72
0.8075	0.81	22.5	−0.026	−24.6	0.0010	−0.0003	−0.033	1.16
0.8100	1.48	19.7	−0.033	−31.1	0.0014	−0.0005	−0.052	1.65
0.8125	2.51	17.9	−0.043	−38.6	0.0020	−0.0007	−0.080	2.21
0.8150	4.19	16.8	−0.056	−48.4	0.0028	−0.0010	−0.120	2.88
0.8175	7.47	16.3	−0.079	−64.0	0.0038	−0.0013	−0.179	3.77
0.8200	28.61	21.0	−0.201	−144.8	0.0063	−0.0021	−0.325	5.70

Table 6.3 Dependence of the energy E, depth D, total compression R, torsion Ψ, amplitudes A_r, A_ρ, A_θ, and A_δ of the soliton of helix expansion on the dimensionless velocity s

s	E (kJ mol^{-1})	D	R (Å)	Ψ (deg)	A_r (Å)	A_ρ (Å)	A_θ (deg)	A_δ (deg)
1.005	8.8	20.3	−0.44	7.8	0.006	−0.010	−1.6	−0.3
1.010	27.8	15.9	−0.66	11.4	0.013	−0.019	−3.3	−0.5
1.015	55.5	13.6	−0.88	14.5	0.020	−0.028	−5.1	−0.7
1.020	97.2	12.2	−1.10	17.4	0.028	−0.038	−7.2	−0.9
1.025	158.3	11.3	−1.35	20.5	0.038	−0.048	−9.5	−1.1
1.030	252.4	10.8	−1.65	24.3	0.049	−0.059	−12.2	−1.3
1.035	418.6	10.6	−2.11	29.9	0.062	−0.071	−15.8	−1.5
1.040	941.0	11.5	−3.34	44.9	0.085	−0.088	−21.8	−1.7

angles. The compression amplitude of the torsion angles increases monotonically with increasing velocity. The maximum compression is at the right end of the velocity interval where the total torsion of the helix $\Psi = -163.5°$.

The solitons of helix compression occupy the interval of permissible values $1 < s < 1.04$ of the dimensionless velocity. The dependencies of the energy E and width D of the soliton on s are presented in Fig. 6.10 (right). The energy of the soliton grows monotonically with growth in the velocity, and the width decreases monotonically down to the minimum value $D = 10.19$ at $s = 1.035$, and then begins to grow steadily. Concrete values of the energy E, width D, total compression R, torsion of the helix Ψ, amplitude of transverse expansion of the chain A_r, extremal strain of the valence bonds A_ρ, valence angles A_θ, and rotational angles A_δ are presented in Table 6.3. The amplitude of longitudinal compression of the chain increases steadily with increasing velocity and reaches the maximum value of 3.34 Å on the right end of the velocity interval. The amplitude of transverse expansion can reach a value of 0.085 Å, which exceeds by more than one order of magnitude the maximum value of helix expansion for the torsion soliton. The strains of the valence bonds and angles can also reach large values. In this case, the dihedral angles are deformed weakly. Extension of the torsion angle at the maximum value of the velocity does not exceed 1.7°.

Numerical simulation of the soliton dynamics has demonstrated the dynamical stability of the solitons at all permissible values of velocity. The solitons move along the chain with a constant velocity and maintain their shape. Modeling of the soliton collision has shown that the solitons interact with each other like elastic particles at velocities close to the left edges of the soliton velocity intervals: when they collide, they reflect from each other without changing their shape and without phonon emission. At velocities close to the left edges of the soliton velocity intervals, the soliton interaction is no longer elastic, and the soliton collision is already accompanied by low-amplitude phonon emission (see Fig. 6.11).

Our investigation of the nonlinear dynamics of PTFE allows us to conclude that, in an isolated polymer macromolecule having the structure of a three-dimensional helix, two types of supersonic solitons can exist simultaneously: solitons of torsion

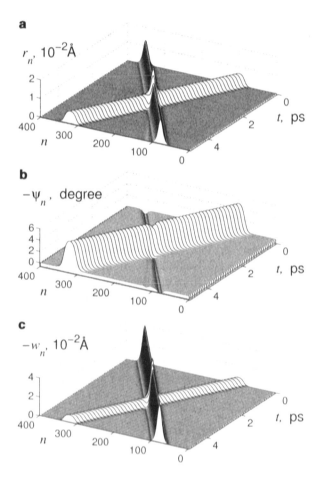

Fig. 6.11 Collision of the torsion soliton ($s = 0.82$) with the solitons of longitudinal compression ($s = -1.01$) in the helix. The dependencies of the transverse displacement of chain links r_n (**a**), relative rotation ψ_n (**b**), and relative longitudinal displacement w_n (**c**) on n and t are shown

and solitons of longitudinal compression of the helix. The first type of soliton corresponds to solitary waves of rotary displacements, and the second to solitary waves of longitudinal displacements of chain links. The helix intertwisting is mainly realized by compression of the dihedral (torsion) angles, while longitudinal compression is due to compression of the valence angles and bonds. The solitons have finite supersonic velocity intervals: the torsional soliton has a higher velocity interval than the velocities of long-wave torsional phonons, and the compression soliton has a higher velocity interval than the velocities of long-wave longitudinal phonons. The solitons are dynamically stable for all permissible values of the velocity and have particle-like properties (inelasticity of their interaction is exhibited only at the maximum velocity values). A more detailed description of the results is given in [27, 28].

6.2 Planar Solitary Waves in Graphite Layers and Soliton-Like Excitations in Carbon Nanotubes

Carbon nanotubes (CNTs) are cylindrical macromolecules of diameter exceeding half a nanometer and a length of up to several microns. Similar structures were obtained more than 50 years ago through thermal decomposition of carbon oxide on an iron contact [29]. However, nanotubes themselves were obtained only about 20 years ago as by-products of the synthesis of fullerene C_{60} [30]. Carbon nanotubes now attract attention due to their unique properties [31]. A nanotube is a quasi-one-dimensional molecular structure with pronounced nonlinear properties. It was shown in [32] that, in the continuum approximation, the nonlinear dynamics of such structures can be described by the Korteweg–de Vries equation, i.e., supersonic longitudinal compression solitons can exist in nanotubes.

The aim of this section is a numerical investigation of the motion of supersonic acoustic solitons in ideal single-wall CNTs. If the nanotube radius increases infinitely, the nanotube takes the form of a planar graphite layer and its acoustic waves become planar waves in the layer. It will be shown that stable solitary waves (solitons) exist only in the flat layer. Supersonic motion of these excitations in CNTs is always accompanied by continuous phonon emission. This emission is due to the variation in nanotube diameter in the region of excitation localization, and its intensity decreases with increasing CNT radius. Phonon emission results in the slowdown of supersonic solitons, implying that they have a finite lifetime.

6.2.1 Structure of Graphite Layers

A graphite layer is a plane covered by regular hexagons with carbon atoms at their vertices (see Fig. 6.12a). In this two-dimensional lattice, each carbon atom is bonded to the three neighboring atoms by valence bonds, thereby forming regular hexagons.

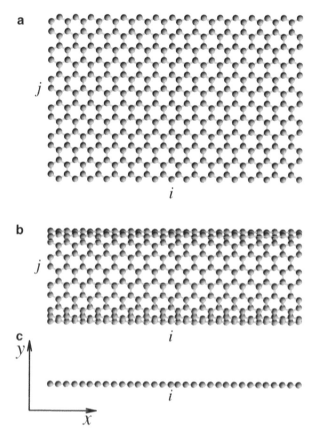

Fig. 6.12 Structure of a flat graphite layer (**a**), carbon nanotube (**b**) with chirality (10,10), and reduced one-dimensional chain (**c**) used to model the propagation of symmetric plane waves

It is convenient to denote lattice sites by two indices i and j. Let $x_{i,j}$, $y_{i,j}$, and $z_{i,j}$ be the coordinates of the lattice site with the corresponding indices. In equilibrium, we have

$$x_{i,j}^{0} = ia_x , \qquad y_{i,j}^{0} = (-1)^{i+j}\rho_0/4 + ja_y , \qquad z_{i,j}^{0} = 0 , \qquad (6.21)$$

where $a_x = \rho_0\sqrt{3}/2$, $a_y = 3\rho_0/2$, and ρ_0 is the equilibrium valence bond length.

   Here and in what follows, we will take into account only the interaction between the nearest neighbor carbon atoms bound by valence bonds and describe the interaction itself by the many-particle Brenner potential [33]. To specify the electrostatic interaction energy between two carbon atoms with two-dimensional indices **i** and **j**, one must also know the positions of the other carbon atoms bonded to these two atoms. Let r_{ij}, $\theta_{i,1}$, $\theta_{i,2}$, $\theta_{j,1}$, and $\theta_{j,2}$ be the bond length and the angles formed by the valence bonds, respectively (see Fig. 6.13).

Fig. 6.13 Configuration of a
carbon structure in the
vicinity of the valence bond
between carbon atoms with
indices **i** and **j**

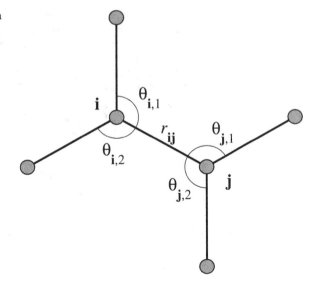

Then the interaction energy is

$$U_{ij} = V_R(r_{ij}) - \frac{1}{2}(B_{ij} + B_{ji})V_A(r_{ij}) , \qquad (6.22)$$

where the repulsive part of the potential is

$$V_R(r) = \frac{D}{S-1} \exp\left[-\sqrt{2S}\beta(r-r_0)\right] ,$$

its attractive part is

$$V_A(r) = \frac{DS}{S-1} \exp\left[-\sqrt{\frac{2}{S}}\beta(r-r_0)\right] ,$$

and the parameters are $r_0 = 1.315\,\text{Å}$, $D = 6.325\,\text{eV}$, $\beta = 1.5\,\text{Å}^{-1}$, and $S = 1.29$.
The coefficients are

$$B_{ij} = \left[1 + G(\theta_{i,1}) + G(\theta_{i,2})\right]^{-\delta} , \qquad B_{ji} = \left[1 + G(\theta_{j,1}) + G(\theta_{j,2})\right]^{-\delta} ,$$

where

$$G(\theta) = a_0\left[1 + \frac{c_0^2}{d_0^2} - \frac{c_0^2}{d_0^2 + (1 + \cos\theta)^2}\right] ,$$

and the parameters are $a_0 = 0.011304$, $c_0 = 19$, $d_0 = 2.5$, and $\delta = 0.80469$.

To find the equilibrium valence bond length in the graphite lattice described by (6.21), one must solve the minimum problem

$$U_{ij} \longrightarrow \min_{r_{ij}} , \qquad (6.23)$$

with the constraints $\theta_{i,1} = \theta_{i,2} = \theta_{j,1} = \theta_{j,2} = 2\pi/3$. Numerical solution of the problem (6.23) yields the value $R_0 = 1.419\,\text{Å}$, which is in good agreement with the experimental valence bond length $|CC| = 1.42\,\text{Å}$. The valence bond energy is $E_0 = -U_{ij}(R_0) = 4.92\,\text{eV}$.

6.2.2   Dispersion Equation of a Planar Wave in a Graphite Layer

Here we consider a symmetric planar wave propagating along the x-axis in a graphite layer. For such motion, all sites with the same value of the index i always have equal displacements, i.e., $x_{i,j} \equiv x_i$, $y_{i,j} \equiv y_i$, and $z_{i,j} \equiv z_i$. Therefore, instead of the dynamics of a two-dimensional lattice (Fig. 6.12a), it suffices to consider the dynamics of a reduced one-dimensional chain (see Fig. 6.12c) described by the Hamiltonian

$$H = \sum_i \left[\frac{1}{2} M(\dot{\mathbf{r}}_i, \dot{\mathbf{r}}_i) + V(\mathbf{r}_{i-1}, \mathbf{r}_i, \mathbf{r}_{i+1}, \mathbf{r}_{i+2}) \right] , \qquad (6.24)$$

where $M = 12m_p$ is the mass of a site in the reduced chain (doubled carbon atom mass, $m_p = 1.67261 \times 10^{-27}$ kg is the proton mass). The vector $\mathbf{r}_i = (x_i, y_i, z_i)$ specifies the position of the ith site, and the site–site interaction potential is

$$V(\mathbf{r}_{i-1}, \mathbf{r}_i, \mathbf{r}_{i+1}, \mathbf{r}_{i+2}) = U_{i,1} + U_{i,2} + U_{i,3} . \qquad (6.25)$$

Here, the potential $U_{i,1}$ is given by (6.4) and describes the valence bond between the sites $(i-1, j)$ and (i, j). The potentials $U_{i,2}$ and $U_{i,3}$ describe the bonds between the sites (i, j) and $(i, j+1)$, and (i, j) and $(i+1, j)$, respectively.

The Hamiltonian (6.24) gives the equations of motion

$$-M\ddot{\mathbf{r}}_i = F_1(\mathbf{r}_i, \mathbf{r}_{i+1}, \mathbf{r}_{i+2}, \mathbf{r}_{i+3}) + F_2(\mathbf{r}_{i-1}, \mathbf{r}_i, \mathbf{r}_{i+1}, \mathbf{r}_{i+2})$$

$$+ F_3(\mathbf{r}_{i-2}, \mathbf{r}_{i-1}, \mathbf{r}_i, \mathbf{r}_{i+1}) + F_4(\mathbf{r}_{i-3}, \mathbf{r}_{i-2}, \mathbf{r}_{i-1}, \mathbf{r}_i) , \qquad (6.26)$$

where

$$F_k = \frac{\partial}{\partial \mathbf{r}_k} U(\mathbf{r}_1, \mathbf{r}_2, \mathbf{r}_3, \mathbf{r}_4) , \quad k = 1, 2, 3, 4 .$$

For convenience, we change from the absolute coordinates $\mathbf{r}_i(t)$ to the relative displacements $\mathbf{u}_i(t) = \mathbf{r}_i(t) - \mathbf{r}_i^0$, where $\mathbf{r}_i^0$ is the equilibrium position of the i th site. To analyse small-amplitude oscillations, we use a linear approximation. For displacements $|\mathbf{u}_i| \ll \rho_0$, we change from the nonlinear equations of motion (6.26) to the linear system

$$-M\ddot{\mathbf{u}}_i = B_1\mathbf{u}_i + B_2(\mathbf{u}_{i-1} + \mathbf{u}_{i+1}) + B_3(\mathbf{u}_{i-2} + \mathbf{u}_{i+2}) + B_4(\mathbf{u}_{i-3} + \mathbf{u}_{i+3}) , \qquad (6.27)$$

where the matrices $B_1 = F_{11} + F_{22} + F_{33} + F_{44}$, $B_2 = F_{12} + F_{23} + F_{34}$, $B_3 = F_{13} + F_{24}$, $B_4 = F_{14}$, and the matrix

$$F_{kl} = \frac{\partial^2 V}{\partial \mathbf{u}_k \partial \mathbf{u}_l}(0,0,0,0) , \quad k,l = 1,2,3,4 .$$

We search for the solution of (6.9) in the form of a wave

$$\mathbf{u}_i(t) = \mathbf{A} \exp[i(qi - \omega t)] , \qquad (6.28)$$

where ω, $\mathbf{A}$, and $q \in [0, \pi]$ are the wave frequency, amplitude vector, and dimensionless wave number. Substituting (6.28) into the linear system (6.27), we obtain the dispersion equation

$$|B_1 + 2B_2 \cos q + 2B_3 \cos(2q) + 2B_4 \cos 3q - \omega^2 E| = 0 , \qquad (6.29)$$

where E is the unit matrix.

The dispersion equation (6.29) is a third-order polynomial with respect to the square of the frequency ω^2. The corresponding algebraic curve has three branches: $0 \leq \omega_z(q) \leq \omega_y(q) \leq \omega_x(q)$ (see Fig. 6.14a left). The first branch $\omega = \omega_z(q)$ describes the dispersion of the transverse planar waves in the graphite layer when the lattice sites are displaced from the layer plane and move along the z-axis. The second branch $\omega = \omega_y(q)$ corresponds to the dispersion of the transverse planar waves when the lattice sites move along the y-axis, but remaining in the lattice plane. The third branch $\omega = \omega_x(q)$ relates to the dispersion of the plane longitudinal waves when the lattice sites move along the x-axis in the plane of the two-dimensional lattice.

The dispersion curves are shown in Fig. 6.14a (left). All three frequencies are equal to zero for the wave number $q = 0$. The frequencies increase steadily with increasing q. Their maximal values are reached at $q = \pi$, where $\omega_x(\pi) = 121 \, \text{cm}^{-1}$, $\omega_y(\pi) = 97 \, \text{cm}^{-1}$, and $\omega_z(\pi) = 54 \, \text{cm}^{-1}$. Let us determine the velocities of the corresponding long-wavelength phonons. The velocity of planar longitudinal phonons is $v_x = a_x \lim_{q \to 0} \omega_x(q)/q = 15{,}878.5 \, \text{m/s}$, while the velocity of planar transverse phonons is $v_y = a_x \lim_{q \to 0} \omega_y(q)/q = 10{,}670.7 \, \text{m/s}$. The ratio of the velocities is $v_y/v_x = 0.672$. Non-planar phonons have zero velocity $v_z = a_x \lim_{q \to 0} \omega_z(q)/q = 0$.

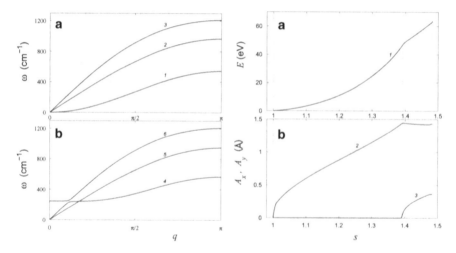

Fig. 6.14 *Left*: Dispersion curves for a flat graphite layer (**a**) and a (5,5) carbon nanotube (**b**). $\omega = \omega_z(q)$ (*line 1*), $\omega = \omega_y(q)$ (*line 2*), $\omega = \omega_x(q)$ (*line 3*), $\omega = \omega_o(q)$ (*line 4*), $\omega = \omega_t(q)$ (*line 5*) and $\omega = \omega_t(q)$ (*line 6*). *Right*: Dependencies of the energy E (**a**) (*line 1*), amplitudes (**b**) A_x (*line 2*) and A_y (*line 3*) on the dimensionless velocity $s = v/v_x$ for the planar solitary wave in a graphite layer

6.2.3 Plane Solitary Waves in a Graphite Layer

To find a planar solitary wave (soliton), we use the pseudospectral method [34, 35]. A detailed description of this method can be found in [36], where the dynamics of planar solitary waves were modeled numerically in a two-dimensional hexagonal lattice. This method is applicable almost without any changes to the problem of finding planar solitary waves in graphite layers. Therefore, we omit the description of the method and directly state the results.

The pseudospectral method can give an unambiguous answer to the problem of the existence of broad solitons as well as narrow ones whose widths are comparable to the chain period. In terms of the relative displacements $\mathbf{u}_i(t)$, the equations of motion (6.8) have the form

$$-M\ddot{\mathbf{u}}_i = F_1(\mathbf{u}_i, \mathbf{u}_{i+1}, \mathbf{u}_{i+2}, \mathbf{u}_{i+3}) + F_2(\mathbf{u}_{i-1}, \mathbf{u}_i, \mathbf{u}_{i+1}, \mathbf{u}_{i+2})$$
$$+ F_3(\mathbf{u}_{i-2}, \mathbf{u}_{i-1}, \mathbf{u}_i, \mathbf{u}_{i+1}) + F_4(\mathbf{u}_{i-3}, \mathbf{u}_{i-2}, \mathbf{u}_{i-1}, \mathbf{u}_i),$$
$$i = 0, \pm 1, \pm 2, \ldots . \tag{6.30}$$

The solution of the nonlinear equations (6.30) is sought in the form of a solitary wave of stationary profile $\mathbf{u}_i(t) = \mathbf{u}(\xi) = (u_x(\xi), u_y(\xi), u_z(\xi))$, where $\xi = a_x i - vt$ is the wave variable. The vector function $\mathbf{u}(\xi)$ determines the shape of the wave propagating along the x-axis with constant velocity v. As $\xi \to \pm\infty$, the derivatives of all three components of this function must tend to zero $\mathbf{u}' \to 0$ (solitariness condition).

Numerical analysis of the nonlinear equations (6.30) has shown that only solitary waves of longitudinal compression can exist in a planar graphite layer, and that their supersonic dimensionless velocities lie in the finite interval $1 < s = v/v_x < s_2 = 1.48$. The solitary wave $\mathbf{u}$ is characterized by the energy

$$
E = \sum_i \left[\frac{1}{2} M v^2 \left(\mathbf{u}'(i a_x), \mathbf{u}'(i a_x) \right) \right.
$$
$$
\left. + V \left(\mathbf{u}((i-1)a_x), \mathbf{u}(i a_x), \mathbf{u}((i+1)a_x), \mathbf{u}((i+2)a_x) \right) \right],
$$

and by the vector amplitude $\mathbf{A} = (A_x, A_y, A_z)$, where

$$
A_x = \max_\xi \left[u_x(\xi + 1) - u_x(\xi) \right], \qquad A_y = \max_\xi \left[u_y(\xi + 1) - u_y(\xi) \right],
$$
$$
A_z = \max_\xi \left[u_z(\xi + 1) - u_x(\xi) \right].
$$

For all permissible values of the solitary wave velocity, the lattice site displacements occur in the lattice plane, i.e., the amplitude $A_z = 0$. The wave energy E increases monotonically with increasing propagation velocity (see Fig. 6.14a right). The amplitude of the lattice compression A_x along the x-axis also grows monotonically for $1 < s < s_1 = 1.39$ (Fig. 6.14b right). In this case, the displacement amplitude $A_y = 0$, i.e., the lattice sites are only displaced along the x-axis. This component of the displacement has the form of a solitary wave (Fig. 6.15a left). For velocities in the interval $s_1 < s < s_2$, the second component of the site displacement is also nonzero. Here, with increasing velocity, the amplitude A_y increases steadily and the displacements have the form of a solitary wave in the two components x and y (see Fig. 6.15b left).

Numerical integration of (6.30) shows that the planar solitary waves are stable for all values of the dimensionless velocity in the range $1 < s < s_2$. They propagate along the chain with constant velocity and completely conserve their initial shapes. As shown in [36], the planar solitary wave in a two-dimensional hexagonal lattice can propagate in any direction (only the maximum velocity depends on the direction of propagation).

6.2.4 Structure of a Carbon Nanotube

An ideal single-wall carbon nanotube (Fig. 6.12b) is a cylindrically convoluted planar graphite layer ribbon. The direction of convolution determines the chirality of the CNT, denoted by the set of symbols (m, n). For convenience of calculation, we

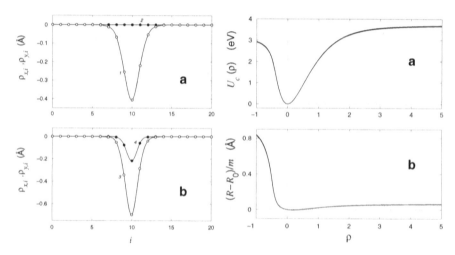

Fig. 6.15 *Left*: profile of a solitary wave in a graphite layer for the dimensionless velocities $s = 1.25$ (**a**) and $s = 1.47$ (**b**). The relative displacements along the x-axis, viz., $\rho_{x,i} = u_{x,i+1} - u_{x,i}$ (*lines 1* and *3*) and along the y-axis, viz., $\rho_{y,i} = u_{y,i+1} - u_{y,i}$ (*lines 2* and *4*) are shown. *Right*: changes in the deformation energy U_c (**a**) and radius R of an (m, m)-nanotube ($m = 5, 10$, and 20) (**b**) as a function of the compression of the nanotube step ρ

consider CNTs with equal chirality symbols (m, m). Note, that all results obtained are also valid for any CNT, regardless of the symbols (m, n).

CNTs with chirality (10,10) are most stable. This nanotube is shown in Fig. 6.12b. In the equilibrium position, carbon atoms in the infinite (m, m) nanotube have coordinates

$$x^0_{i,j} = i\,\Delta x \,, \qquad y^0_{i,j} = R\cos\varphi_{i,j} \,, \qquad z^0_{i,j} = R\sin\varphi_{i,j} \,, \tag{6.31}$$

where the indices $i = 0, \pm 1, \pm 2, \ldots$ and $j = 1, 2, \ldots, m+m$ define the numbering of nanotube sites in the longitudinal and transverse directions, respectively. Δx and R are the longitudinal step and nanotube radius, respectively. The angles are

$$\varphi_{i,j} = (j - 1)\frac{\pi}{m} - \frac{1}{2}\big[1 - (-1)^{i+j}\big]\Delta\varphi \,,$$

where $\Delta\varphi$ is the angle of the transverse displacement. Stationary values of the parameters Δx_0, R_0, and $\Delta\varphi_0$ can be found as minima of the function

$$E(\Delta x, R, \Delta\varphi) = \frac{1}{3}(E_1 + E_2 + E_3) \,,$$

where E_1 is the energy of the valence bond between the $(1, 1)$ and $(1, 2)$ sites, found from (6.4), E_2 is the bonding energy between sites $(1, 1)$ and $(2, 1)$, and E_3 is the bonding energy between sites $(1, 1)$ and $(0, 1)$. The dependencies of Δx_0, R_0, $\Delta\varphi_0$,

Table 6.4 Dependencies of the longitudinal step Δx_0, radius R_0, angle of transverse displacement $\Delta \varphi_0$, and energy of one valence bond E_0 on the chirality symbols (m, m) of the nanotube

m	Δx_0 (Å)	R_0 (Å)	$\Delta \varphi_0$ (grad)	E_0 (eV)
5	1.23387	3.42	11.89	4.8655
7	1.23153	4.76	8.53	4.8912
10	1.23036	6.79	5.99	4.9048
15	1.22975	10.18	4.00	4.9120
20	1.22954	13.56	3.00	4.9146
30	1.22939	20.34	2.00	4.9164
50	1.22932	33.89	1.20	4.9173
100	1.22929	67.77	0.60	4.9177
∞	1.22928	∞	0	4.9178

and the energy of the valence bond E_0 are given in Table 6.4. As $m \rightarrow \infty$, the radius of the CNT tends to infinity and the nanotube becomes a two-dimensional lattice (graphite layer).

6.2.5 Dispersion Equation of Longitudinal Waves in Nanotube

Let $x_{i,j}$, $y_{i,j}$, and $z_{i,j}$ be the coordinates of the site (i, j) in the nanotube. We change from Cartesian coordinates to cylindrical $h_{i,j}$, $r_{i,j}$, and $\phi_{i,j}$:

$$x_{i,j} = i\,\Delta x + h_{i,j} ,$$
$$y_{i,j} = (R + r_{i,j}) \cos(\varphi_{i,j} + \phi_{i,j}) , \qquad (6.32)$$
$$z_{i,j} = (R + r_{i,j}) \sin(\varphi_{i,j} + \phi_{i,j}) ,$$

noting that, in equilibrium $h_{i,j} = 0$, $\phi_{i,j} = 0$, and $r_{i,j} = 0$. Next, we assume that the displacements of atoms with the same second indices j are equal, i.e.,

$$h_{i,j} \equiv h_i , \qquad r_{i,j} \equiv r_i , \qquad \phi_{i,j} \equiv \phi_i .$$

Then the Hamiltonian of the system takes the form

$$H = \sum_i \left\{ \frac{1}{2} M \left[\dot{r}_i^2 + (R + r_i)^2 \dot{\phi}_i^2 + \dot{h}_i^2 \right] + V_i \right\} , \qquad (6.33)$$

where

$$V_i = V(r_{i-1}, \phi_{i-1}, h_{i-1}; r_i, \phi_i, h_i; r_{i+1}, \phi_{i+1}, h_{i+1}; r_{i+2}, \phi_{i+2}, h_{i+2})$$
$$= U_{i,1} + U_{i,2} + U_{i,3} .$$

Here, the potential $U_{i,1}$ is given by (6.4) and describes the valence bond between the sites $(i - 1, j)$ and (i, j). The potentials $U_{i,2}$ and $U_{i,3}$ describe the bonds between the sites (i, j) and $(i, j + 1)$, and (i, j) and $(i + 1, j)$, respectively.

It is convenient to analyze low-amplitude vibrations by introducing local orthogonal coordinates $u_{i,1} = (R + r_i) \cos \phi_i - R$, $u_{i,2} = (R + r_i) \sin \phi_i$, and $u_{i,3} = h_i$, in which the Hamiltonian of the system has the form

$$H = \sum_i \left[\frac{1}{2} M(\dot{\mathbf{u}}_i, \dot{\mathbf{u}}_i) + V(\mathbf{u}_{i-1}, \mathbf{u}_i, \mathbf{u}_{i+1}, \mathbf{u}_{i+2}) \right], \tag{6.34}$$

where $\mathbf{u}_i = (u_{i,1}, u_{i,2}, u_{i,3})$. In the linear approximation, the equations of motion (6.34) correspond to the Hamiltonian (6.27).

We will search for a solution of the linear system in the form of a linear wave (6.28). Then the dispersion equation with respect to the frequency squared ω^2 is a cubic equation. The corresponding algebraic curve has three branches: $0 \le \omega_o(q) \le \omega_t(q) \le \omega_l(q)$ (see Fig. 6.14b left). The first branch $\omega = \omega_o(q)$ describes the optical (transverse) nanotube vibrations, for which the deformation is mostly due to the displacement of the local radius r_i. The second branch $\omega = \omega_t(q)$ describes the torsional nanotube vibrations, for which the deformation is due to the displacement of the local angles ϕ_i, and the third branch corresponds to the longitudinal vibrations, for which deformation is due to the local longitudinal displacements h_i. Note that, for small values of the dimensionless wave number q, the frequencies $\omega_o(q)$ and $\omega_l(q)$ correspond to longitudinal and transverse vibrations of the nanotube, respectively.

The dispersion curves are shown in Fig. 6.14b (left). For $q = 0$, the frequencies $\omega_o(q)$ and $\omega_t(q)$ are equal to zero and $\omega_l(q) > 0$. The frequency $\omega_l(q) \to 0$ as the nanotube radius increases. The frequencies increase monotonically with q and reach their maximum values at $q = \pi$. We define the velocity of long-wavelength torsional phonons as $v_t = \Delta x \lim_{q \to 0} \omega_t(q)/q$ and the velocity of longitudinal phonons as $v_l = \Delta x \lim_{q \to 0} \omega_o(q)/q$. The dependencies of the limiting values of frequencies and velocities of long-wavelength phonons v_t and v_l on the chirality symbols of CNT (m, m) are given in Table 6.5.

With increasing index m, i.e., with increasing radius R, the CNT becomes a planar layer and the dispersion curves $\omega_o(q)$, $\omega_t(q)$, and $\omega_l(q)$ approach the corresponding curves $\omega_o(q)$, $\omega_t(q)$, and $\omega_l(q)$ for a planar layer: as $m \to \infty$, the frequencies $\omega_o(q) \to \omega_z(q)$, $\omega_t(q) \to \omega_y(q)$, $\omega_l(q) \to \omega_x(q)$, and the velocities $v_t \to v_y$, $v_l \to v_x$.

6.2.6 Soliton-Like Excitations of Nanotubes

To analyze the nonlinear nanotube dynamics, it is useful to consider the nanotube effective potential of the longitudinal compression:

$$U_c(\rho) = \min_{R, \Delta \varphi} \left[E(\Delta x_0 + \rho, R, \Delta \varphi) - E(\Delta x_0, R_0, \Delta \varphi_0) \right],$$

Table 6.5 Dependencies of
the limiting values of
frequencies $\omega_l(0)$, $\omega_o(\pi)$,
$\omega_t(0)$, $\omega_l(\pi)$, and velocities
of long-wavelength phonons
v_t and v_l on the chirality
symbols (m, m) of the
nanotube. The frequencies are
given in reciprocal
centimeters

m	$\omega_l(0)$	$\omega_o(\pi)$	$\omega_t(\pi)$	$\omega_l(\pi)$	v_t (m/s)	v_l (m/s)
5	24.5	56.7	94.7	120.6	10,442.3	15,766.1
7	17.6	55.5	95.6	120.8	10,551.5	15,782.8
10	12.4	54.9	96.1	120.9	10,611.4	15,792.9
15	8.3	54.5	96.4	120.9	10,644.2	15,798.5
20	6.2	54.3	96.5	121.0	10,655.7	15,808.2
30	4.1	54.2	96.6	121.0	10,664.0	15,820.7
50	2.5	54.2	96.6	121.0	10,668.3	15,848.5
100	1.2	54.2	96.6	121.0	10,670.1	15,877.1
∞	0	54.2	96.6	121.0	10,670.7	15,878.5

which describes a change in the energy of one valence bond in the nanotube
when each longitudinal bond is stretched by an amount ρ. When the nanotube is
stretched, both its energy and radius are changed. The effective potential $U_c(\rho)$
and the nanotube radius R are shown as functions of the longitudinal compression
ρ in Fig. 6.15 (right). As can be seen from this figure, $\rho = 0$, the radius of the
spiral $R = R_0$, and the deformation energy $U_c(0) = 0$ in the ground state. Both
compression and stretching of the nanotube lead to an increase in its radius. The
shape of the potential $U_c(\rho)$ shows that the nanotube has a pronounced negative
anharmonicity. The deformation energy increases much faster for compression than
for stretching.

The negative anharmonicity of the nanotube suggests that supersonic solitons
of longitudinal compression can exist in it. This was noted for the first time in
[32], where it was shown that, in the continuum approximation (without taking into
account variations in the nanotube radius), the nanotube nonlinear dynamics can be
described by the Korteweg–de Vries equation with soliton solutions.

Let us consider the one-dimensional effective torsional potential of the nanotube,
viz.,

$$U_t(\phi) = \min_{\Delta x, R} \left[E(\Delta x, R, \Delta \varphi_0 + \phi) - E(\Delta x_0, R_0, \Delta \varphi_0) \right],$$

which describes the nanotube energy variation when each nanotube link is twisted
through an angle ϕ. The torsional potential is shown in Fig. 6.16 (left). As can be
seen from this figure, the nanotube has fourth order negative anharmonicity with
respect to the angular variable. This anharmonicity does not ensure the existence of
acoustic solitons (soliton solutions can only exist for systems with positive quartic
anharmonicity). Therefore, one should not expect the existence of torsional solitary
waves in nanotubes.

The following equations of motion correspond to the Hamiltonian (6.33):

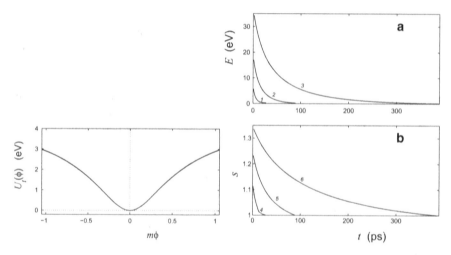

Fig. 6.16 *Left*: Dependence of the deformation energy U_t of the (m, m) nanotube ($m = 5, 10$, and 20) on the torsion angle ϕ. *Right*: Drop in the energy E (**a**) and the dimensionless velocity $s = v/v_1$ (**b**) of a soliton-like excitation propagating along the (m, m) nanotube for $m = 10$ (*lines 1 and 4*), $m = 20$ (*lines 2 and 5*), and $m = 40$ (*lines 3 and 6*)

$$\left.\begin{array}{l} M\ddot{r}_i - M(R + r_i)\dot{\phi}_i^2 + \dfrac{\partial}{\partial r_i}P = 0\,, \\[2mm] M(R + r_i)^2\ddot{\phi}_i + 2M(R + r_i)\dot{\phi}_i\dot{r}_i + \dfrac{\partial}{\partial\phi_i}P = 0, \\[2mm] M\ddot{h}_i + \dfrac{\partial}{\partial h_i}P = 0\,, \end{array}\right\} \quad i = 0, \pm 1, \pm 2, \ldots\,,$$

$$(6.35)$$

where the potential energy is

$$P = \sum_i V\left(r_{i-1}, \phi_{i-1}, h_{i-1}; r_i, \phi_i, h_i; r_{i+1}, \phi_{i+1}, h_{i+1}; r_{i+2}, \phi_{i+2}, h_{i+2}\right)\,.$$

Numerical analysis of the discrete system (6.35) has shown that, in contrast to the case of a graphite layer, the equations of motion for a nanotube have no exact solutions describing the propagation of solitary waves (solitons) of longitudinal compression along the chain.

The fundamental difference between a plane layer and a nanotube is that compression of a nanotube modifies the surface curvature, whereas a graphite layer always remains planar. If the transverse displacements in the equations of motion for a nanotube (6.35) are forbidden, i.e., all $r_i = 0$, just as for a graphite layer, the exact soliton solutions appear, describing longitudinal compression solitons in a finite supersonic velocity range. However, as soon as transverse displacements of the nanotube sites become possible, these solutions are no longer exact. Therefore, only soliton-like excitations can be discussed. These soliton-like excitations move with supersonic velocities $v > v_1$, but while moving, they always emit longitudinal acoustic phonons. The emission intensity depends on the nanotube radius. Indeed, the smaller the radius, the greater the emission rate. When the excitation is propagating, its energy and velocity decrease in time (see Fig. 6.16 right). Therefore, the soliton excitations in CNTs have a finite lifetime which increases steadily with increasing nanotube radius. The equations of motion (6.35) have no solutions describing the propagation of solitary longitudinal torsional waves along the nanotube.

Thus, in CNTs there can exist only soliton-like solutions describing the propagation of the localized region of longitudinal compression, and this is always accompanied by phonon emission. To check this claim, we numerically model the excitation dynamics in a finite fragment of the nanotube with fixed ends. We consider the equations of motion (6.35) with $i = 1, 2, \ldots, 1,000$ and use the condition of fixed ends by setting $\dot{r}_i$, $\dot{h}_i$, and $\dot{\phi}_i \equiv 0$ for $i = 1$ and $i = 1,000$. Then we integrate the equations of motion with the initial conditions corresponding to the ground state ($r_i = 0$, $\dot{r}_i = 0$, $h_i = 0$, $\dot{h}_i = 0$, $\phi_i = 0$, and $\dot{\phi}_i = 0$, $i = 2, 3, \ldots, 1,000$) with the first link displaced, namely, $r_1(0) = A_r$, $h_1(0) = A_x$, and $\phi_1(0) = A_\phi$. The results of numerical integration are shown in Fig. 6.17.

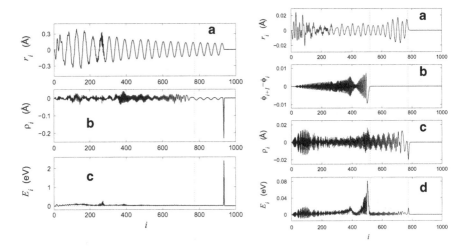

Fig. 6.17 *Left*: Distributions of the transverse displacements r_i (**a**), longitudinal compression $\rho_i = h_{i+1} - h_i$ (**b**), and E_i energy (**c**) for the (10,10) nanotube at time $t = 6\,\text{ps}$. The initial deformations are $A_r = 0$, $A_h = 1\,\text{Å}$, and $A_\phi = 0$. *Vertical dashed lines* indicate the region through which the long-wavelength longitudinal phonons pass during the numerical integration. *Right*: Formation of two oscillating wave packets caused by the initial torsion of a finite fragment of the (10,10) nanotube (the amplitudes are $A_r = 0$, $A_h = 0$, and $A_\phi = \Delta\varphi/3$). Distributions of the transverse displacements r_i (**a**), relative rotations $\phi_{i+1} - \phi_i$ (**b**), longitudinal compression $\rho_i = h_{i+1} - h_i$ (**c**), and energy E_i (**d**) along the chain at time $t = 6\,\text{ps}$. *Vertical dashed lines* indicate the region through which the long-wavelength torsional phonons and longitudinal deformation phonons pass during the numerical integration

The system of equations (6.35) was integrated numerically using the standard fourth-order Runge–Kutta method with a constant integration step [37]. Integration accuracy was estimated using the conservation of the total energy integral (6.33). For the integration step $\Delta t = 0.5 \times 10^{-15}$ s, the system energy was conserved up to six significant figures for the entire time of numerical integration.

To be specific, let us consider a (10,10) CNT. For the longitudinal compression of the first link by 1 Å ($A_r = 0$, $A_x = 1$ Å, and $A_\phi = 0$), a localized region of longitudinal compression is formed in the chain and propagates with supersonic velocity $v > v_l$ (Fig. 6.17 left). The propagation of this soliton-like excitation is accompanied by the emission of longitudinal phonons. In addition, an oscillating wave packet is formed which propagates with subsonic velocity $v < v_l$. The main part of the initial deformation energy is concentrated in the soliton-like excitation, but it is then completely spent on phonon emission in further motion.

In the initial twisting of a nanotube ($A_r = 0$, $A_x = 0$, and $A_\phi = \Delta\varphi/3$), only two oscillating wave packets appear (see Fig. 6.17 right). The first is formed by the torsional phonons and moves with velocity $v = v_t$, while the second is formed by the longitudinal phonons and moves with the higher velocity $v = v_l$.

Thus, the numerical modeling confirms the conclusion that acoustic solitons do not exist in carbon nanotubes. There exist only soliton-like excitations whose propagation is always accompanied by the emission of longitudinal phonons. The emission amplitude decreases with increasing CNT radius, but it is never equal to zero. A more detailed description of the results is given in [38].

6.3 Topological Solitons in a Quasi-One-Dimensional Polymer Crystal

The acoustic solitons discussed above can exist only in isolated linear molecules of PE and PTFE. In crystals, each molecule is in a dense environment of neighboring macromolecules which form its substrate. In this case, only topological solitons can exist. The displacements of chain atoms associated with the soliton must be consistent with the substrate structure. In this section we consider the features of topological soliton dynamics in a quasi-one-dimensional crystal formed by linear polymers in a parallel arrangement. The defect structure of polymer crystals is naturally described in terms of the static topological solitons [9, 10], while their mobility determines the specific 'soliton contribution' to the thermodynamics and kinetics of polymer crystals. Topological solitons will be exemplified by the simplest quasi-one-dimensional polymer crystal of PE and PTFE.

6.3.1 Topological Solitons in Crystalline PE

Each molecule of PE $(CH_2)_\infty$ in a crystal is in the *trans* zigzag conformation, i.e., its backbone has a planar zigzag structure which is described by the spacing

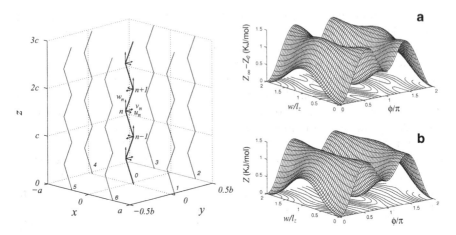

Fig. 6.18 *Left*: Schematic representation of crystalline PE. The central *trans* zigzag backbone (*line 0*) and the six neighbor chains (*lines 1–6*) are shown. The local coordinates for the central chain are shown. *Right*: Two-dimensional surface $z = Z_\infty(u(\phi), v(\phi), w) - Z_0$ (**a**) and its approximation $z = Z(u(\phi), v(\phi), w)$ (**b**)

$\rho_0 = 1.53\,\text{Å}$ (equilibrium length of the valence bond $H_2C{-}CH_2$) and by the zigzag angle $\theta_0 = 113°$ (equilibrium valence angle $CH_2{-}CH_2{-}CH_2$).

In a study of the low-energy nonlinear dynamical processes, the motion of hydrogen atoms with respect to the main chain is not essential, so the united atom approximation can be used. Therefore, we will consider each CH_2 group as a single particle. In this approach, it is impossible to obtain an orthorhombic lattice, and only a monoclinic lattice can be stable [39]. Therefore, as a model system, we consider this lattice with the periods

$$a' = \frac{1}{2}\sqrt{a^2 + b^2}\,, \qquad b' = a\,, \qquad c' = c\,,$$

where the following parameters (Fig. 6.18 left) give a realistic density for the system: $a = 4.51\,\text{Å}$, $b = 7.031\,\text{Å}$, $c = 2\rho_0 \sin(\theta_0/2) = 2.552\,\text{Å}$.

The corresponding structure of the crystal is shown schematically in Fig. 6.18 (left). All macromolecules are situated in the parallel planes. The structure of the crystal is completely determined by the zigzag angle and three parameters a, b, and c. Let us consider an isolated PE molecule. We choose the coordinate system in such a manner that the sites of the *trans* zigzag chain in the equilibrium position have coordinates

$$x_n^{(0)} = (-1)^n \frac{l_x}{2}\,, \qquad y_n^{(0)} = 0\,, \qquad z_n^{(0)} = nl_z\,,$$

where $n = 0, \pm 1, \pm 2, \ldots$ is the number of a chain site and $l_x = \rho_0 \cos(\theta_0/2)$ and $l_z = c/2 = \rho_0 \sin(\theta_0/2)$ are the transverse and longitudinal constants of the zigzag chain. Let us label the nearest six chains of the crystal by the numbers from 1 to 6, as shown in Fig. 6.18 (left). Then, in the equilibrium position, the nth site of the k chain has coordinates

$$x_n^{(k)} = x_n^{(0)} + e_x^{(k)}, \qquad y_n^{(k)} = y_n^{(0)} + e_y^{(k)}, \qquad z_n^{(k)} = z_n^{(0)},$$

where $n = 0, \pm 1, \pm 2, \ldots, k = 1, 2, \ldots, 6$, and

$$e_x^{(1)} = a, \ e_x^{(2)} = \frac{a}{2}, \ e_x^{(3)} = -\frac{a}{2}, \ e_x^{(4)} = -a, \ e_x^{(5)} = -\frac{a}{2}, \ e_x^{(6)} = \frac{a}{2},$$

$$e_y^{(1)} = 0, \ e_y^{(2)} = \frac{b}{2}, \ e_y^{(3)} = \frac{b}{2}, \ e_y^{(4)} = 0, \ e_y^{(5)} = -\frac{b}{2}, \ e_y^{(6)} = -\frac{b}{2}.$$

Below, we consider only the dynamics of the central chain ($k = 0$) and take into account its interaction with the six neighboring chains ($k = 1, 2, \ldots, 6$), which are assumed to be immobile (the fixed neighbor approximation). It is convenient to change from the absolute coordinates of the nth site of the zeroth *trans* zigzag chain x_n, y_n, and z_n to the relative coordinates

$$u_n = (-1)^{n+1}(x_n - x_n^{(0)}), \quad v_n = (-1)^{n+1} y_n, \quad w_n = z_n - z_n^{(0)}, \quad n = 0, \pm 1, \pm 2, \ldots.$$

The local systems of coordinates are shown schematically in Fig. 6.18 (left).

The Hamiltonian of the chain can be written as follows

$$H = \sum_n \left[\frac{1}{2} M(\dot{\mathbf{u}}_n, \dot{\mathbf{u}}_n) + V(\rho_n) + U(\theta_n) + W(\delta_n) + Z(u_n, v_n, w_n) \right], \qquad (6.36)$$

where the first term describes the kinetic energy of the nth site, the second is the deformation energy of the nth valence bond, the third is the deformation energy of the nth valence angle, the fourth is the deformation energy of the nth torsional angle, and the last term is the energy of interaction of the nth site with the six neighboring chains (the substrate potential). $M = 14 m_p$ is the mass of the united atom (m_p is the proton mass) and the vector $\mathbf{u}_n = (u_n, v_n, w_n)$. The length of the nth valence bond is $\rho_n = (a_{n,1}^2 + a_{n,2}^2 + a_{n,3}^2)^{1/2}$, where $a_{n,1} = u_n + u_{n+1} - l_x, a_{n,2} = v_n + v_{n+1}$, and $a_{n,3} = w_{n+1} - w_n + l_z$. The cosine of the nth valence angle is

$$\cos \theta_n = \frac{a_{n-1,1} a_{n,1} + a_{n-1,2} a_{n,2} - a_{n-1,3} a_{n,3}}{\rho_{n-1} \rho_n},$$

and the cosine of the nth torsional angle is

$$\cos \delta_n = \frac{-b_{n,1} b_{n+1,1} - b_{n,2} b_{n+1,2} + b_{n,3} b_{n+1,3}}{\beta_n \beta_{n+1}}.$$

where $\beta_n = (b_{n,1}^2 + b_{n,2}^2 + b_{n,3}^2)^{1/2}$ and

$$b_{n,1} = a_{n-1,2}a_{n,3} + a_{n,2}a_{n-1,3} , \qquad b_{n,2} = a_{n-1,1}a_{n,3} + a_{n,1}a_{n-1,3} ,$$

$$b_{n,3} = a_{n-1,2}a_{n,1} - a_{n,2}a_{n-1,1} .$$

Following [21,40], we take the potential of the valence bond to be (6.2), that of the valence angle to be (6.3), and that of the torsional angle in the form

$$W(\delta_n) = C_1 + C_2 \cos \delta_n + C_3 \cos 3\delta_n , \tag{6.37}$$

where the parameters $D_0 = 334.72\,\text{kJ}\,\text{mol}^{-1}$, $\alpha = 1.91\,\text{Å}^{-1}$, $K_\theta = 130.122\,\text{kJ}\,\text{mol}^{-1}$, $C_1 = 8.37\,\text{kJ}\,\text{mol}^{-1}$, $C_2 = 1.675\,\text{kJ}\,\text{mol}^{-1}$, and $C_3 = 6.695\,\text{kJ}\,\text{mol}^{-1}$.

The substrate potential $Z(u, v, w)$ in the Hamiltonian (6.36) is given by the infinite series

$$Z_\infty(u, v, w) = \sum_{n=-\infty}^{+\infty} \sum_{k=1}^{6} V_{LJ}(r_{k,n}) , \tag{6.38}$$

where the Lennard-Jones potential

$$V_{LJ}(r_{k,n}) = 4\epsilon \left[(\sigma/r_{k,n})^{12} - (\sigma/r_{k,n})^{6} \right]$$

describes the interaction of the zeroth site of the central chain with the nth site of the kth chain. The distance between these sites is given by

$$r_{k,n} = \left\{ \left[u + x_n^{(k)} - l_x/2 \right]^2 + \left[v + y_n^{(k)} \right]^2 + \left[w - z_n^{(k)} \right]^2 \right\}^{1/2} .$$

In accordance with [40], the parameters appearing in the Lennard-Jones potential are $\sigma = 3.8\,\text{Å}$ and $\epsilon = 0.4937\,\text{kJ}\,\text{mol}^{-1}$. The potential (6.38) is a periodic function of the variable w with period $2l_z$.

The use of the infinite series for numerical calculation (6.38) is not reasonable because the power decrease in the terms requires summation of a large number of terms ($-50 \le n \le 50$), and therefore long computation times. For this reason, we use a simpler analytical potential as substrate potential, viz.,

$$Z(u, v, w) = \varepsilon_w \sin^2(\pi w/l_z) + \frac{1}{2} K_v \left[1 + \varepsilon_v \sin^2(\pi w/l_z) \right] v^2$$

$$+ \frac{1}{2} K_u \left[1 + \varepsilon_u \sin^2(\pi w/l_z) \right] \left\{ u - \frac{1}{2} l_x \left[1 - \cos(\pi w/l_z) \right] \right\}^2 . \tag{6.39}$$

The two-dimensional shape of this potential was first used in the generalization of the Frenkel–Kontorova model for zigzag-like molecular chains [41].

The potential (6.39) is a periodic function in the variable w with period $2l_z$. The minima of the potential coincide with the site positions in the *trans* zigzag chains of the undeformed crystal.

The large number of free parameters in the potential (6.39) (ε_u, ε_v, ε_w, K_u, and K_v) allows accurate approximation of the potential (6.38). Let us consider a longitudinal displacement and rotation of the *trans* zigzag chain around the $x = 0$ and $y = 0$ axes. The local coordinates of the chain sites depend on the angle ϕ according to $u(\phi) = (l_x/2)\cos\phi$ and $v(\phi) = (l_y/2)\sin\phi$. To find appropriate values of the free parameters of the potential (6.39), we solve the minimum problem

$$\left[\int_0^{2\pi} \int_0^{2l_z} Z_{er}^2\big(u(\phi), v(\phi), w\big)\,d\phi\,dw \right]^{1/2} \longrightarrow \min_{\varepsilon_u,\varepsilon_v,\varepsilon_w,K_u,K_v} , \qquad (6.40)$$

where $Z_{er}\big(u(\phi), v(\phi), w\big) = Z\big(u(\phi), v(\phi), w\big) - Z_\infty\big(u(\phi), v(\phi), w\big) + Z_\infty(0,0,0)$.

The numerical study of the problem (6.40) has shown that the error in approximating the infinite series (6.38) by the analytic potential (6.39) is minimum when the parameters take the values

$$\varepsilon_u = 0.0674265 \text{ kJ mol}^{-1}, \quad \varepsilon_v = 0.0418353 \text{ kJ mol}^{-1}, \quad \varepsilon_w = 0.1490124 \text{ kJ mol}^{-1},$$

$$K_u = 2.169513 \text{ kJ/Å mol}^2 , \quad K_v = 13.683865 \text{ kJ/Å mol}^2 .$$

The approximation accuracy is illustrated in Fig. 6.18 (right), which shows that the surface $z = Z\big(u(\phi), v(\phi), w\big)$ practically coincides with $z = Z_\infty\big(u(\phi), v(\phi), w\big) - Z_0$. The approximation error is

$$\frac{1}{4\pi l_z} \int_0^{2\pi} \int_0^{2l_z} \big|Z_{er}\big(u(\phi), v(\phi), w\big)\big|\,d\phi\,dw = 0.02833 \text{ KJ mol}^{-1} .$$

Let us consider the possible local topological defects of the *trans* zigzag chain. As can be seen in Fig. 6.18 (left), the *trans* zigzag chain can be transformed into itself by three isometric transformations: a shift along the z-axis by one zigzag period $w = 2l_z$, a shift by the half period $w = l_z$, and a rotation of the whole chain through the angle $\phi = \pi$ around the z-axis. The transition from the point $(0,0)$ to the point $(2l_z,0)$ in the space (w, ϕ) corresponds to the first transformation, from $(0,0)$ to (l_z, π) corresponds to the second, and from $(0,0)$ to $(0, 2\pi)$ to the third. All other isometric transformations which preserve the form of the *trans* zigzag chain are compositions of these three and the three inverse transformations. Consequently, only three basic types of local topological defect are possible: pure stretching (compression), stretching (compression) accompanied by twist, and pure twist of the *trans* zigzag chain.

The equations of motion corresponding to the Hamiltonian (6.36) have the form

$$M\ddot{u}_n = -\frac{\partial H}{\partial u_n} , \quad M\ddot{v}_n = -\frac{\partial H}{\partial v_n} , \quad M\ddot{w}_n = -\frac{\partial H}{\partial w_n} , \quad n = 0, \pm 1, \pm 2, \dots .$$

The complexity of this system precludes analytical study, so we use the variational techniques proposed in [23, 42].

We search for solutions of the equations (6.41) in the form of a traveling wave with smooth constant profile. For this purpose, we put $u_n(t) = u(\xi)$, $v_n(t) = v(\xi)$, and $w_n(t) = w(\xi)$, where $\xi = nl_z - vt$ and v are the wave variable and the wave velocity, respectively, and the functions u, v, and w are assumed to depend smoothly on the variable ξ. Then the Lagrangian corresponding to the equations of motion (6.41), viz.,

$$L = \sum_n \left[\frac{1}{2} M \left(\dot{u}_n^2 + \dot{v}_n^2 + \dot{w}_n^2 \right) - V(\rho_n) - U(\theta_n) - W(\delta_n) - Z(u_n, v_n, w_n) \right] ,$$

can be written in the form

$$\overline{L} = \sum_n \left\{ \frac{v^2 M}{24 l_z^2} \left[16(u_{n+1} - u_n)^2 - (u_{n+2} - u_n)^2 + 16(v_{n+1} - v_n)^2 - (v_{n+2} - v_n)^2 \right. \right.$$
$$\left. + 16(w_{n+1} - w_n)^2 - (w_{n+2} - w_n)^2 \right]$$
$$\left. - V(\rho_n) - U(\theta_n) - W(\delta_n) - Z(u_n, v_n, w_n) \right\} .$$

A soliton solution can correspond to a saddle point of the Lagrangian $\overline{L}$, so it is convenient to search for it as a minimum of the functional

$$F = \frac{1}{2} \sum_n \left(\overline{L}_{u_n}^2 + \overline{L}_{v_n}^2 + \overline{L}_{w_n}^2 \right) .$$

Therefore, to find soliton solutions (solitary waves) $\{u_n, v_n, w_n\}_{n=1}^N$, one must numerically solve the minimum problem

$$F = \frac{1}{2} \sum_{n=2}^{N-1} \left(\overline{L}_{u_n}^2 + \overline{L}_{v_n}^2 + \overline{L}_{w_n}^2 \right) \longrightarrow \min_{u_2, v_2, w_2, \ldots, u_{N-1}, v_{N-1}, w_{N-1}} , \tag{6.41}$$

$$u_1 = u_{-\infty} , \quad v_1 = v_{-\infty} , \quad w_1 = w_{-\infty} , \tag{6.42}$$

$$u_N = u_{+\infty} , \quad v_N = v_{+\infty} , \quad w_N = w_{+\infty} . \tag{6.43}$$

The boundary conditions (6.42) and (6.43) for the problem (6.41) determine the type of soliton solution, and the number N should be chosen sufficiently large to prevent any dependence of the soliton shape on N. The solution of the problem given by (6.41), (6.42), and (6.43) allows us to find numerically all soliton solutions, i.e., smooth solitary waves with constant profile solving the nonlinear system (6.41).

The absence of such solutions for a certain value v implies the impossibility of soliton motion at this velocity.

We consider the possible local topological defects of the *trans* zigzag chain. As can be seen in Fig. 6.18 (left), the *trans* zigzag chain can be transformed into itself by the following three isometric transformations: a shift along the z-axis by one zigzag period $w = 2l_z$, a shift by the half period $w = l_z$, and a rotation of the whole chain through the angle $\phi = \pi$ around the z-axis. The transition from the point $(0,0)$ to the point $(2l_z, 0)$ in the space (w, ϕ) corresponds to the first transformation, from $(0,0)$ to (l_z, π) corresponds to the second, and from $(0,0)$ to $(0, 2\pi)$ the third. Each of these transformations involves overcoming the potential barrier (see Fig. 6.18 right). All other isometric transformations which preserve the form of the *trans* zigzag chain are compositions of these three and the three inverse transformations.

Consequently, only three basic types of local topological defect are possible: pure stretching (compression), stretching (compression) accompanied by twist, and pure twist of the *trans* zigzag chain. Let us determine the rotational angle of the nth unit of the chain, viz.,

$$\phi_n = \arg \left[l_x - u_n - u_{n+1} + i(v_n + v_{n+1}) \right] ,$$

where i is the imaginary unit. The topological charge of the defect can defined as the two-dimensional vector $\mathbf{q} = (q_1, q_2)$, where

$$q_1 = (w_{+\infty} - w_{-\infty})/2l_z , \qquad q_2 = (\phi_{+\infty} - \phi_{-\infty})/2\pi .$$

Then the first type of defect has the topological charge $\mathbf{q} = (1, 0)$ $[\mathbf{q} = (-1, 0)]$, the second $\mathbf{q} = (0.5, \pm0.5)$ $[\mathbf{q} = (-0.5, \pm0.5)]$, and the third $\mathbf{q} = (0, \pm1)$.

To find a soliton solution with topological charge $\mathbf{q} = (q_1, q_2)$, one must solve the minimum problem (6.41) with the boundary conditions

$$u_{-\infty} = 0 , \quad v_{-\infty} = 0 , \quad w_{-\infty} = 0 ,$$

$$u_{+\infty} = \frac{1}{2} l_x (1 - \cos 2q_2 \pi) , \quad v_{+\infty} = 0 , \quad w_{+\infty} = 2l_z q_1 .$$

The solution of this problem was sought numerically by the method of conjugate gradients [24] with $N = 500$. The starting point was chosen to be

$$u_n = \frac{1}{2} l_x \left[1 - \cos(2q_2 \pi r_n) \right] , \quad v_n = \frac{1}{2} l_x \sin(2q_2 \pi r_n) , \quad w_n = 2l_z q_1 r_n ,$$

where

$$r_n = \frac{1}{2} \left\{ 1 + \tanh \left[\left(n - \frac{N-1}{2} \right) \mu \right] \right\} ,$$

and μ is an adjustable parameter.

Each soliton solution $\{u_n^0, v_n^0, w_n^0\}_{n=1}^N$ corresponds to a topological soliton with energy

$$
\begin{aligned}
E = \sum_n \Bigg\{ & \frac{v^2 M}{24 l_z^2} \Big[16(u_{n+1}^0 - u_n^0)^2 - (u_{n+2}^0 - u_n^0)^2 + 16(v_{n+1}^0 - v_n^0)^2 - (v_{n+2}^0 - v_n^0)^2 \\
& + 16(w_{n+1}^0 - w_n^0)^2 - (w_{n+2}^0 - w_n^0)^2 \Big] \\
& + V(\rho_n) + U(\theta_n) + W(\delta_n) + Z(u_n^0, v_n^0, w_n^0) \Bigg\} \; ,
\end{aligned}
$$

and root-mean-square width, given in chain periods,

$$
D = 2 \Bigg[\sum_{n=1}^N (n - n_c)^2 p_n \Bigg]^{1/2} \; ,
$$

where the point

$$
n_c = \frac{1}{2} + \sum_{n=1}^N n p_n
$$

determines the position of the soliton centre. The sequence p_n is given by

$$
p_n = \begin{cases} (w_{n+1} - w_n)/C & \text{for } q_2 = 0 \; , \\ (\phi_{n+1} - \phi_n)/C & \text{for } q_2 \neq 0 \; , \end{cases}
$$

where the constant C is found from the normalization condition $\sum_{n=1}^N p_n = 1$.

Let us define the parameters which describe the amplitudes of chain deformation in the region of soliton localization, i.e., the maximum deformations of the valence bonds

$$
\Delta \rho = \max_{n=1,2,\ldots,N} |\rho_n - \rho_0| \; ,
$$

the valence angle

$$
\Delta \theta = \max_{n=1,2,\ldots,N} |\theta_n - \theta_0| \; ,
$$

and the torsional angle

$$
\Delta \delta = \max_{n=1,2,\ldots,N} |\delta_n - \pi| \; .
$$

The rest mass of the soliton can also be found as the limit

$$M_0 = \lim_{v \to 0} 2 \frac{E(v) - E(0)}{v^2}.$$

Numerical solution of the minimum problem given by (6.41)–(6.43) shows that the equations of motion (6.41) have soliton solutions with the topological charge $\mathbf{q} = (1, 0)$ only for the dimensionless velocities $s = v/v_a$ in the subsonic interval $0 \leq s \leq 0.79$, whereas solutions with topological charge $\mathbf{q} = (-1, 0)$ exist in the velocity interval $0 \leq s \leq 0.66$. The topological solitons with charge $\mathbf{q} = (\pm 0.5, \pm 0.5)$ exist only for $0 \leq s \leq 0.4$, while solitons with $\mathbf{q} = (0, \pm 1)$ exist only for $0 \leq s \leq 0.35$. Here, $v_a = 7,810$ m/s is the velocity of planar longitudinal sound (velocity of long-wave planar acoustic phonons in an isolated *trans* zigzag chain). The torsional sound velocity (velocity of the long-wave torsional acoustic phonons in an isolated *trans* zigzag chain) is $v_t = 7,613$ m/s [8].

The typical profile of the topological solitons with charge $\mathbf{q} = (1, 0)$ [$\mathbf{q} = (-1, 0)$] is shown in Fig. 6.19 (left). The soliton has the form of a kink (antikink) with respect to the w_n component, and the typical bell-shaped profile of a solitary wave with respect to the u_n component. All the chain particles remain in the *trans* zigzag plane ($v_n \equiv 0$). When $q_1 = 1$, there is stretching of the zigzag backbone in the localization region (valence bonds and angles are stretched), and when $q_1 = -1$,

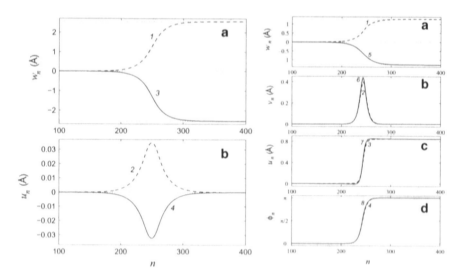

Fig. 6.19 *Left*: Profiles of the topological soliton with charge $\mathbf{q} = (1, 0)$ (*lines 1* and *2* show components w_n and u_n, respectively) after the passage of 10^4 chain sites ($\tau = 40,010.2$, $s = 0.2499$) and the soliton with charge $\mathbf{q} = (-1, 0)$ ($\tau = 40,011$, $s = 0.2499$) (*lines 3* and *4*). Initial soliton velocity $s = 0.25$. *Right*: Profiles of the topological soliton with charge $\mathbf{q} = (0.5, 0.5)$ (*lines 1–4* show components w_n, v_n, u_n, and ϕ_n, respectively) after the passage of 10^4 chain sites ($\tau = 40,127.2$, $s = 0.2492$) and the soliton with charge $\mathbf{q} = (-0.5, 0.5)$ ($\tau = 40,174$, $s = 0.2488$) (*lines 5–8*). Initial soliton velocity $s = 0.25$

there is compression of the zigzag chain (valence bonds and angles are compressed). In the soliton localization region, the deformations of the valence bonds do not exceed 0.01 Å, while the maximum deformations of the valence angles can reach 8°. Therefore, the stretching (compression) of the *trans* zigzag chain is realised mainly due to stretching (compression) of the planar valence angles.

The profile of the soliton with topological charge $\mathbf{q} = (0.5, 0.5)$ [$\mathbf{q} = (-0.5, 0.5)$] is shown in Fig. 6.19 (right). The soliton has the form of a kink (antikink) with respect to the components w_n and u_n and a bell-shaped profile with respect to the component v_n. If $q_1 = 0.5$, in the soliton localization region, there is stretching and torsion of the zigzag backbone (the valence bonds and angles are stretched, and the torsion angles are compressed), and if $q_1 = -0.5$, there is compression and stretching (the valence bonds and angles as well as the torsion angles are compressed). The rotations of the chain sites ϕ_n also have the form of a kink. The deformations of the valence bonds do not exceed 0.007 Å. In this case, the stretching (compression) and torsion of the *trans* zigzag chain are realized mainly through deformation of the torsional and valence angles.

The soliton solution with topological charge $\mathbf{q} = (0, 1)$ is shown in Fig. 6.20 (left). All three soliton components w_n, v_n, and u_n have the form of a solitary wave, and in combination they describe a localized rotation of the chain through an angle 2π (the soliton has the form of a kink with respect to the rotation). In the region of

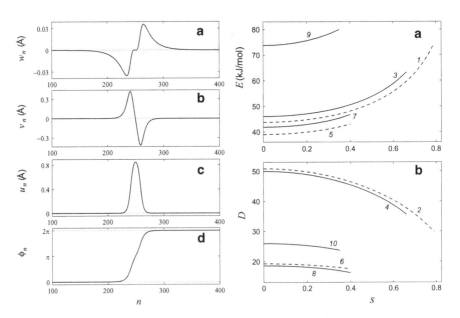

Fig. 6.20 *Left*: Profiles of the topological soliton with charge $\mathbf{q} = (0, 1)$ (components w_n, v_n, u_n, and ϕ_n) after the passage of 10^4 chain sites ($\tau = 40{,}219.4$, $s = 0.2486$). Initial soliton velocity $s = 0.25$. *Right*: Dependence on the dimensionless velocity s of the energy E (**a**) and width D (**b**) of the soliton with topological charge $\mathbf{q} = (1, 0)$ (*lines 1* and *2*), $\mathbf{q} = (-1, 0)$ (*lines 3* and *4*), $\mathbf{q} = (0.5, 0.5)$ (*lines 5* and *6*), and $\mathbf{q} = (-0.5, 0.5)$ (*lines 7* and *8*), $\mathbf{q} = (0, 1)$ (*lines 9* and *10*)

soliton localization, a very small stretching of the valence bonds, accompanied by a compression of the valence angles, occurs, i.e., the torsion of the zigzag backbone is accomplished mainly through deformation of the torsional angles.

The dependencies of the energy E and width D of the solitons on the dimensionless velocity s are shown in Fig. 6.20 (right). When the velocity grows, the soliton energies increase steadily and the widths decrease monotonically. The soliton with topological charge $\mathbf{q} = (0, \pm 1)$ has maximum energy while the soliton with charge $\mathbf{q} = (0.5, \pm 0.5)$ has minimum energy. The stretching solitons ($q_2 = 0$) are approximately two times broader than the torsional solitons ($q_2 \neq 0$).

Topological excitations in a single chain can be formed only as pairs of topological solitons with opposite signs (the sum of the topological charges has to be equal to zero). Therefore, the formation energy of the soliton pair with topological charges $\mathbf{q}$ and $-\mathbf{q}$ must not be less than the sum of their energies: $E_c(\mathbf{q}) = E(\mathbf{q}) + E(-\mathbf{q})$.

Numerical modeling has shown that the topological solitons are dynamically stable for all permissible values of the velocity. They propagate with constant velocity and retain their initial forms. For the initial value $s = 0.25$ of the dimensionless velocity, the soliton with topological charge $\mathbf{q} = (1, 0)$ passes 10^4 chain sites during the dimensionless time $\tau = 40{,}010.2$ (the time unit of the dimensionless time, $\tau = v_a t / l_z$, corresponds to the time of one chain site passage by the longitudinal sound $t_0 = l_z / v_a = 1.633 \times 10^{-14}$ s). At the final time, this soliton has velocity $s = 0.2499$ and its profile coincides totally with the initial one (see Fig. 6.19 left). The soliton with topological charge $\mathbf{q} = (-1, 0)$, after the passage of 10^4 chain sites, has velocity $s = 0.2499$ at the final time $\tau = 40{,}011$. Its profile also remains unchanged, as shown in Fig. 6.19 (left). Similar results were obtained for the solitons with topological charges $\mathbf{q} = (0.5, 0.5)$ and $(-0.5, 0.5)$ (see Fig. 6.19 right). After the passage of 10^4 chain sites, the solitons with $\mathbf{q} = (0.5, 0.5)$ and $\mathbf{q} = (-0.5, 0.5)$ have velocities $s = 0.2492$ and $s = 0.2488$, respectively. The topological soliton with $\mathbf{q} = (0, 1)$ passes 10^4 chain sites in the time $\tau = 40{,}219.4$ ($s = 0.2486$). The profile of this soliton at the final time is also the same as the initial one (see Fig. 6.20 left).

Let us model the head-on collision of topological solitons. We consider the collision of solitons with topological charges $\mathbf{q}_1$ and $\mathbf{q}_2$. Solition collision will be examined in a finite chain of $N_1 = 10^3$ sites. Viscous friction is introduced at the chain ends to ensure absorption of phonon radiation after the collision. Initially, the solitons move towards each other with the same velocity $s = 0.25$. The dependence of the results of the soliton collision on their topological charges $\mathbf{q}_1$ and $\mathbf{q}_2$ is presented in Table 6.6. It is obvious that the collision cannot lead to a change in the total topological charge of the chain $\mathbf{q}_1 + \mathbf{q}_2$. Therefore, complete recombination is only possible for solitons with different signs on their charges ($\mathbf{q}_1 = -\mathbf{q}_2$), when the total charge of the chain equals zero. If this total charge does not equal zero, the soliton collision can lead to their reflection, passage through one another, a change in their type (change in the topological charges), or to partial recombination (change in the number of solitons).

Table 6.6 Dependence of the result of a soliton collision on their topological charges q_1 and q_2

Chain state before collision		Chain state after collision
q_1	q_2	
$(0, \pm 1)$	$(0, \pm 1)$	$(0, \pm 1) + (0, \pm 1)$
		(elastic reflection)
$(0, \pm 1)$	$(0, \mp 1)$	Breather + phonon radiation
		(recombination of solitons)
$(0, 1)$	$(\pm 0.5, -0.5)$	$(\pm 0.5, 0.5)$ + breather + phonon radiation
		(partial recombination of solitons)
$(0, 1)$	$(\pm 0.5, 0.5)$	$(\pm 0.5, 0.5) + (0, 1)$
		(passage of solitons through each other)
$(0, -1)$	$(\pm 0.5, 0.5)$	$(\pm 0.5, -0.5)$ + breather + phonon radiation
		(partial recombination of solitons)
$(0, -1)$	$(\pm 0.5, -0.5)$	$(\pm 0.5, -0.5) + (0, -1)$
		(passage of solitons through each other)
$(0, \pm 1)$	$(1, 0)$	$(0.5, \pm 0.5) + (0.5, \pm 0.5)$ + phonon radiation
		(change of the soliton type)
$(0, \pm 1)$	$(-1, 0)$	$(-0.5, \pm 0.5) + (-0.5, \pm 0.5)$ + phonon radiation
		(change of the type of solitons)
$(\pm 0.5, \pm 0.5)$	$(\mp 0.5, \mp 0.5)$	Breather + phonon radiation
		(soliton recombination, Fig. 6.21 left)
$(\pm 0.5, \mp 0.5)$	$(\mp 0.5, \pm 0.5)$	Breather + phonon radiation
		(soliton recombination)
$(\pm 0.5, \pm 0.5)$	$(\pm 0.5, \pm 0.5)$	$(\pm 0.5, \pm 0.5) + (\pm 0.5, \pm 0.5)$
		(elastic reflection of solitons)
$(\pm 0.5, \pm 0.5)$	$(\mp 0.5, \pm 0.5)$	$(\mp 0.5, \pm 0.5) + (\pm 0.5, \pm 0.5)$
		(passage of solitons through each other)
$(\pm 1, 0)$	$(\mp 0.5, 0.5)$	$(\pm 0.5, 0.5)$ + phonon radiation
		(partial recombination of solitons, Fig. 6.21 right)
$(\pm 1, 0)$	$(\mp 0.5, -0.5)$	$(\pm 0.5, -0.5)$ + phonon radiation
		(partial recombination of solitons)
$(\pm 1, 0)$	$(\pm 0.5, \pm 0.5)$	$(\pm 1, 0) + (\pm 0.5, \pm 0.5)$
		(elastic reflection of solitons)
$(\pm 1, 0)$	$(\mp 1, 0)$	$(\mp 1, 0) + (\pm 1, 0)$ + phonon radiation
		(passage of solitons through each other)
$(\pm 1, 0)$	$(\pm 1, 0)$	$(\pm 1, 0) + (\pm 1, 0)$
		(elastic reflection of solitons)

Solitons with the same topological charges ($q_1 = q_2$) manifest elastic reflection (which occurs without phonon emission or excitation). The collision of solitons with opposite signs $q_1 = (0, \pm 1)$ and $q_2 = (0, \mp 1)$ results in their recombination. As this takes place, breather-like localized torsional vibrations form in the chain. The recombination is accompanied by intensive phonon emission. The collision of

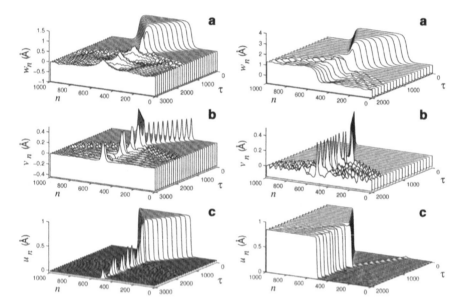

Fig. 6.21 *Left*: Recombination of solitons with topological charges $q_1 = (0.5, 0.5)$ and $q_2 = (-0.5, -0.5)$. The collision is accompanied by intensive phonon emission. The formation of breather-like localized vibrations occurs in the chain. *Right*: Formation of the soliton with topological charge $q = (0.5, 0.5)$ as a result of the collision of solitons with charges $q_1 = (1, 0)$ and $q_2 = (-0.5, 0.5)$. The collision is accompanied by intensive phonon emission

solitons with charges $q_1 = (\pm 0.5, \pm 0.5)$ $[q_1 = (\pm 0.5, \mp 0.5)]$ and $q_2 = -q_1$ gives similar results (see Fig. 6.21 left). The collision of solitons with charges $q_1 = (\pm 1, 0)$ and $q_2 = -q_1$ leads to the solitons passing through one another, accompanied by insignificant phonon emission.

The collision of the soliton with charge $q_1 = (q_{1,1}, q_{1,2}) = (0, \pm 1)$ and the soliton with charge $q_2 = (q_{2,1}, q_{2,2})$ leads to the solitons passing through one another if $q_{1,2} q_{2,2} > 0$ (soliton velocities and shapes do not change). When $q_{1,2} q_{2,2} < 0$, there is partial recombination. One soliton with charge $q = (q_{2,1}, q_{1,2} + q_{2,2})$ is created from the original pair. The collision is accompanied by intensive phonon emission. Formation of a breather-like localized torsional vibration in the chain is possible. Partial recombination also occurs for the collision of solitons with charges $q_1 = (\pm 1, 0)$ and $q_2 = (\mp 0.5, \pm 0.5)$ (see Fig. 6.21 right).

When they collide, solitons with charges $q_1 = (0, \pm 1)$ and $q_2 = (\pm 1, 0)$ form two solitons with charges $q = (\pm 0.5, \pm 0.5)$. The collision is accompanied by phonon emission and leads to a change of soliton type.

Let us test the stability of topological solitons with respect to thermal fluctuations in the chain. The dynamics of a thermalized chain consisting of N sites is described by the Langevin equations

$$
\left.
\begin{aligned}
M\ddot{u}_n &= -\frac{\partial H}{\partial u_n} + \xi_n - \Gamma M\dot{u}_n \ , \\[4pt]
M\ddot{v}_n &= -\frac{\partial H}{\partial v_n} + \eta_n - \Gamma M\dot{v}_n \ , \\[4pt]
M\ddot{w}_n &= -\frac{\partial H}{\partial w_n} + \zeta_n - \Gamma M\dot{w}_n \ ,
\end{aligned}
\right\} \quad n = 1,2,\dots,N \ , \qquad (6.44)
$$

where the Hamiltonian for the system H is given by (6.34), ξ_n, η_n, and ζ_n are normally distributed random forces describing the interaction of the nth molecule with a thermal bath, $\Gamma = 1/t_r$ is the friction coefficient, and t_r is the relaxation time of the molecular velocity. The random forces ξ_n, η_n, and ζ_n have correlation functions

$$
\langle \xi_n(t_1)\xi_m(t_2)\rangle = \langle \eta_n(t_1)\eta_m(t_2)\rangle = \langle \zeta_n(t_1)\zeta_m(t_2)\rangle = 2M\Gamma k_B T \delta_{nm}\delta(t_1 - t_2) \ ,
$$

$$
\langle \xi_n(t_1)\eta_m(t_2)\rangle = \langle \xi_n(t_1)\zeta_m(t_2)\rangle = \langle \eta_n(t_1)\zeta_m(t_2)\rangle = 0 \ , \quad n,m = 1,2,\dots,N,
$$

where k_B is Boltzmann's constant and T is the temperature of the thermal bath.

We integrate (6.44) numerically using the standard fourth-order Runge–Kutta method with a constant integration step Δt [37] and the lagged Fibonacci random number generator [43]. The delta-function is represented numerically as $\delta(t) = 0$ for $|t| > \Delta t/2$ and $\delta(t) = 1/\Delta t$ for $|t| < \Delta t/2$, i.e., the numerical integration step corresponds to the correlation time of the random force. In order to use the Langevin equation, one must have $\Delta t \ll t_r$. Therefore, we chose $\Delta t = 0.002\,\text{ps}$ and the relaxation time $t_r = 10 t_a = 0.1633\,\text{ps}$.

We now consider the soliton dynamics in the chain consisting of $N = 10^3$ sites with fixed ends. The initial conditions are taken so as to correspond to a stationary topological soliton. It is obvious that the solitons, which form due to the chain shift (the topological charge $q_1 \neq 0$), are stable with respect to thermal fluctuations. Their stability has a topological nature: to destroy them, one would have to move half of the chain sites by one ($q_1 = \pm 0.5$) or two ($q_1 = \pm 1$) periods. Numerical integration of the equations of motion (6.44) has shown that, at temperature $T = 300\,\text{K}$, thermal fluctuations can lead only to Brownian motion of the solitons along the chain.

Defects relating only to the chain torsion (the topological charges are $q_1 = 0$ and $q_2 = \pm 1$) can be unstable with respect to the thermal fluctuations. Indeed, the torsion of one site of the chain by $360°$ transfers it to the initial state, so to destroy this defect, it suffices to twist only a finite number of sites. Such defects in a thermalized chain have a finite life-time which decreases exponentially with increasing temperature of the thermal bath. Numerical modeling has shown that, at temperature $T = 100\,\text{K}$, the defect with topological charge $\mathbf{q} = (0, \pm 1)$ remains stable throughout the $t = 10^3\,\text{ps}$ of the numerical experiment, but at $T = 300\,\text{K}$ the lifetime of this defect is only $3\,\text{ps}$. The defect destruction process is shown

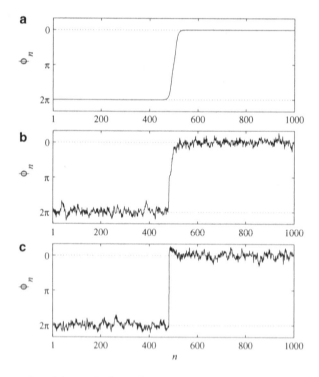

Fig. 6.22 Destruction of the topological soliton with charge $\mathbf{q} = (0, 1)$ in the thermalized chain (temperature of the thermal bath $T = 300$ K). Profiles of the soliton component ϕ_n at the initial time $t = 0$ (**a**), time $t = 2$ ps (**b**), and $t = 3$ ps (**c**) are shown

in Fig. 6.22. At the initial time $t = 0$, the defect has a smooth soliton-like form corresponding to the smooth torsion of the *trans* zigzag chain (see Fig. 6.22a). The thermal fluctuations lead to a gradual decrease in the soliton width. At $t = 2$ ps, the soliton width decreases by approximately a factor of 2 (see Fig. 6.22b), while at $t = 3$ ps, the width becomes equal to 1 (Fig. 6.22c). This means that the defect disappears (all the chain sites are in equivalent states).

To model topological soliton formation in an infinite thermalized chain, let us consider the cyclic chain of $N = 5,000$ sites. The ground state $u_n = 0$, $v_n = 0$, and $w_n = 0$, $n = 1, 2, \ldots, N$, is adopted as initial condition. The equations of motion (6.44) were integrated numerically during the time $t = 10^3$ ps.

It is convenient to describe the degree of chain thermalization by the instantaneous dimensionless temperature

$$c_T(t) = \sum_{n=1}^{N} M \left(\dot{u}_n^2 + \dot{v}_n^2 + \dot{w}_n^2 \right) / 3 N k_B T \, ,$$

Table 6.7 Dependencies of the average value of the dimensionless heat capacity C_E and dimensionless temperature C_T of the chain on the thermal bath temperature T

T (K)	1	100	200	300	400	425	450	475	500
C_E	0.994	0.995	0.999	1.013	1.032	1.033	1.045	1.093	1.135
C_T	0.994	0.994	0.994	1.002	1.003	0.994	0.994	0.994	1.002

the instantaneous dimensionless heat capacity $c_E(t) = H(t)/3Nk_BT$, and their averaged values

$$C_T = \lim_{t \to \infty} \frac{1}{t} \int_0^t c_T(\tau) d\tau , \qquad C_E = \lim_{t \to \infty} \frac{1}{t} \int_0^t c_E(\tau) d\tau .$$

The dependencies of C_E and C_T on the temperature of the thermal bath are presented in Table 6.7. The dimensionless temperature of the chain C_T practically always remains near 1, and this confirms the validity of the thermalization procedure. The dimensionless heat capacity increases monotonically with an increase in temperature. At small temperatures, the heat capacity is close to 1, so this emphasizes the 'harmonicity' of thermal vibrations in the chain. With an increase in temperature, the vibration amplitude also increases and the effect of nonlinearity ('anharmonicity') of the vibrations thus becomes stronger. A growth in C_E reflects the increasing contribution of the nonlinearity to the heat capacity of the system.

The chain vibration becomes essentially nonlinear at $T = 400$ K, and local topological defects of stretching and torsion (topological solitons with the charges $q = (\pm 0.5, \pm 0.5)$) appear at $T = 425$ K. A further increase in the temperature leads to an increase in the density of this type of defect (Fig. 6.23). Defects of other types do not arise. The absence of pure torsion defects in the chain can be explained by their short lifetime at high temperatures. The absence of defects of pure stretching (compression) in the chain is associated solely with the topology of the *trans* zigzag structure (note that the energy of defect pair formation for opposite-sign solitons exceeds that for same-sign solitons). We can conclude that the first type of soliton can only form through the interaction of the second type of soliton. Therefore, the density of the first type of defect has to be much less than that of the second.

The sharp exponential growth in the dimensionless heat capacity near $T = 425$ K is due to the accumulation of the second type of local topological defects in the chain. A phase transition of first order, i.e., a change in chain conformation, takes place. This temperature almost coincides with the melting point of the PE crystal, viz., $T_0 = 420$ K. Therefore, the accumulation of the second type of topological defects can be considered as the initial mechanism in the premelting of a PE crystal.

The results discussed above showed that three types of topological solitons can exist in the polyethylene crystal at realistic values of the interaction potential parameters:

- A two-dimensional soliton of stretching (compression),
- A soliton of stretching (compression) and torsion, and
- A torsional soliton.

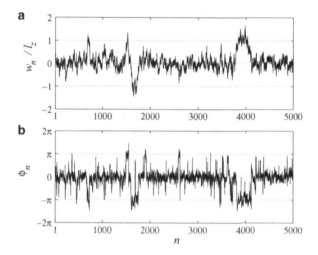

Fig. 6.23 Accumulation of topological solitons of the second type in the chain at the thermal bath temperature $T = 450$ K. The local displacement components of chain sites w_n (**a**) and ϕ_n (**b**) are shown at time $t = 10^3$ ps

All these solitons have a subsonic velocity spectrum. The most extended of them (40–50 chain segments) are the solitons of *trans* zigzag stretching. The solitons describing the local torsion of the *trans* zigzag backbone have half the length. The most massive quasi-particle (the rest mass is 1.4 proton mass) is the torsional soliton, while the other solitons have half the mass (0.72–0.8 proton mass). The torsional solitons also possess the highest energy of formation for an opposite-sign soliton pair, viz., ≥ 148 kJ mol^{-1}. The energy of formation of a stretching and compression soliton pair is ≥ 90 kJ mol^{-1}, and the energy of formation of a stretching and torsional soliton pair and a compression and torsional soliton pair is ≥ 81 kJ mol^{-1}. All the solitons have possible velocity bands in the subsonic range. They are dynamically stable and propagate with constant velocity while retaining their shapes.

Numerical modeling of the thermalized chain dynamics of crystalline PE has shown that thermal vibrations can lead to the formation of only one type of local mobile defect, namely, the topological solitons corresponding to stretching (compression) of the zigzag chain by a half of the chain period, followed by its torsion by 180°. Therefore, solitons with pure stretching (compression) of the *trans* zigzag backbone can be formed through interaction of same-sign solitons. Both these types of defect are topologically stable to thermal vibrations in the chain. In a thermalized chain, the defects are formed as pairs of opposite-sign topological solitons and they move as Brownian particles. The solitons corresponding to chain torsion by 360° are unstable to thermal vibrations. They have a finite lifetime which decreases exponentially as the temperature grows.

The concentration of topological defects grows sharply near the melting point. This allows us to consider their accumulation as the initial mechanism in the premelting of the PE crystal. A more detailed description of the results can be found in [7, 8, 44].

6.3.2 Topological Solitons in a Crystalline PTFE

The ground state of the PTFE molecule is the planar zigzag conformation of the chain (helix with symbol 1*2/1). The two-dimensional structure of this zigzag chain determines the significant features of the chain dynamics. However, most of the macromolecules belonging to this class have the three-dimensional helix conformation, rather than the two-dimensional zigzag conformation in the ground state. Therefore, it would be interesting to study topological solitons in a crystalline polymer consisting of macromolecules with the three-dimensional helix conformation in the ground state. Poly(tetrafluoroethylene) is the simplest example of such a polymer. The purpose of this section is to study the topological soliton dynamics in a helical polymer macromolecule through the example of PTFE. We will show that four types of topological defects can exist in the crystal formed by helical macromolecules, and that only one of them displays soliton dynamics.

At temperature $T < 19\,°C$, a molecule of PTFE in the crystalline state has the helical chain conformation of type 13/6, i.e., there are 13 CF_2 groups for every six turns of the molecule. The crystal lattice constants are $a = b = 5.59\,Å$ and $c = 16.88\,Å$ [26]. The helix has the helix angle $\Delta\phi = 12\pi/13 = 166.15°$ and the pitch $\Delta z = c/13 = 1.298\,Å$.

The helix radii defined by the carbon atoms ρ_C and fluorine ones ρ_F can be determined from the values of the C–C and C–F bond lengths and the FCF bond angle:

$$\rho_C = \left[(\rho_{CC}^2 - \Delta z^2)/2(1 - \cos\Delta\phi) \right]^{1/2} , \qquad (6.45)$$

$$\rho_F = \left[\rho_C^2 + 2\rho_C\rho_{CF}\cos(\alpha_{FCF}/2) + \rho_{CF}^2 \right]^{1/2} . \qquad (6.46)$$

According to [21], the bond lengths are $\rho_{CC} = 1.54\,Å$ and $\rho_{CF} = 1.34\,Å$, and the bond angle is $\alpha_{FCF} = 104.8°$. Thus, from (6.45) and (6.46), we obtain $\rho_C = 0.417\,Å$ and $\rho_F = 1.628\,Å$.

In the equilibrium state, the position vector of the nth carbon atom C_n is

$$\mathbf{R}_{C_n} = \left(\rho_C\cos(n\Delta\phi),\ \rho_C\sin(n\Delta\phi),\ n\Delta z \right) ,$$

and the position vectors of the first and second nth fluorine atoms $F_n^\pm$ are

$$\rho_{FCF}\mathbf{R}_{F_n^\pm} = \left(\rho_F\cos(n\Delta\phi \pm \beta),\ \rho_F\sin(n\Delta\phi \pm \beta),\ n\Delta z \right) ,$$

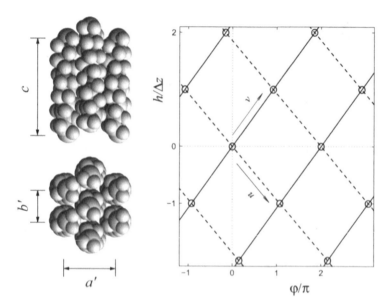

Fig. 6.24 *Left*: Arrangement of a helical molecular chain and its six neighbors in crystalline PTFE. *Right*: Construction of the dimensionless local coordinates u and v for the discrete helix. *Solid lines* correspond to the chain links and *circles* indicate sites of the left-handed helix

where

$$\beta = \arcsin\left(\frac{\rho_{CF}}{\rho_F}\sin\frac{\alpha_{FCF}}{2}\right) = 40.69° .$$

Thus, the PTFE molecule is described as three left-handed helices (one helix is formed by C_n atoms and the other two are formed by $F_n^{\pm}$ atoms).

In the crystal, each macromolecule is surrounded by six neighbor molecules (see Fig. 6.24 left). Let us consider the interaction between the corresponding chains. The four nearest molecular chains are right helices with position vectors

$$\mathbf{R}_{C_n} = \left(\pm\frac{a'}{2} + \rho_C\cos(\phi_0 - n\Delta\phi),\ \pm\frac{b'}{2} + \rho_C\sin(\phi_0 - n\Delta\phi),\ z_0 + n\Delta z\right),$$

$$\mathbf{R}_{F_n^{\pm}} = \left(\pm\frac{a'}{2} + \rho_F\cos(\phi_0 - n\Delta\phi \pm \beta),\ \pm\frac{b'}{2} + \rho_F\sin(\phi_0 - n\Delta\phi \pm \beta),\ z_0 + n\Delta z\right),$$

where ϕ_0 and z_0 are the relative angular and longitudinal shift of the helices, respectively. The two other nearest helices are obtained from the initial one by a shift by $\pm b'$ along the Y-axis.

Let us calculate the parameters of the crystal unit cell. For this purpose we minimize the energy of nonvalence interaction of the helices with respect to the four parameters a', b', ϕ_0, and z_0. The pair potentials of the interaction between atoms belonging to neighbor chains (CC, CF, and FF) were taken from [45]. We then obtained $a' = 9.41\,\text{Å}$, $b' = 5.72\,\text{Å}$, $\phi_0 = -110.77°$, and $z_0 = 1.95\,\text{Å}$. The relative arrangement of the helices is shown in Fig. 6.24 (left). Hereafter, in the dynamics simulation of a molecular chain, the six nearest neighbor chains will be assumed to be immobile. These chains comprise the immobile substrate for the molecular chain under consideration.

We now use the united atom approximation and consider the chain link CF_2 of the PTFE macromolecule as a single particle of mass $M = 50m_p$ (where m_p is the proton mass). In this case, the equilibrium position of the nth chain link is given by the position vector

$$\mathbf{R} = \left(R_0 \cos(n\Delta\phi),\ R_0 \sin(n\Delta\phi),\ n\Delta z\right),$$

where $R_0 = \rho_C$ is the helix radius. In the equilibrium state, the valence angle $\angle CCC$ is $\theta_0 = \arccos\left[-(\mathbf{e}_{n-1}, \mathbf{e}_n)/\rho_0^2\right]$, where the vector $\mathbf{e}_n = \mathbf{R}_{n+1} - \mathbf{R}_n$ determines the direction of the nth valence bond and $\rho_0 = \rho_{CC}$ is the equilibrium bond length. After elementary transformations, we obtain

$$\theta_0 = \pi - \arccos \frac{4R_0^2 \sin^2(\Delta\phi/2)\cos\Delta\phi + \Delta z^2}{\rho_0^2} = 116.30°.$$

The equilibrium value of the nth dihedral (torsion) angle is

$$\eta_n = \arccos \frac{(\mathbf{v}_{n-1}, \mathbf{v}_n)}{|\mathbf{v}_{n-1}||\mathbf{v}_n|}$$

$$= \eta_0 = \arccos \frac{h^2 \cos\Delta\phi + \sin^2\Delta\phi}{h^2 + \sin^2\Delta\phi},$$

where $\mathbf{v}_n = [\mathbf{e}_n, \mathbf{e}_{n+1}]$ is the vector product of $\mathbf{e}_n$ and $\mathbf{e}_{n+1}$, and $h = \Delta z/R_0$ is the dimensionless pitch of the helix. Hereafter, we shall use the rotation angle about the nth bond $\delta_n = \pi - \eta_n$, where η_n is the nth dihedral angle. The equilibrium rotation angle is $\delta_0 = \pi - \eta_0 = 16.32°$.

Let x_n, y_n, and z_n be the coordinates of the nth chain site. Transforming from Cartesian to cylindrical coordinates, we obtain

$$x_n = (R_0 + r_n)\cos(n\Delta\phi + \varphi_n),$$

$$y_n = (R_0 + r_n)\sin(n\Delta\phi + \varphi_n),$$

$$z_n = n\Delta z + h_n,$$

where r_n, φ_n, and h_n are the radial, angular, and longitudinal displacements of the nth chain site from its equilibrium position, respectively. The Hamiltonian of the chain is

$$H = \sum_n \left\{ \frac{1}{2} M \left[\dot{r}_n^2 + \dot{\varphi}_n^2 (R_0 + r_n)^2 + \dot{h}_n^2 \right] \right.$$

$$\left. + V(\rho_n) + U(\theta_n) + W(\delta_n) + S(r_n, u_n, v_n) \right\} . \tag{6.47}$$

Here the dot denotes the derivative with respect to time t, while ρ_n, θ_n, and δ_n are the length of the nth valence bond, the nth bond angle, and the nth rotation angle, respectively.

The potential of the nth valence bond is given by (6.2), where the length of the nth bond is $\rho_n = [a_{n,1} + b_n^2]^{1/2}$. Here,

$$a_{n,1} = d_n^2 + d_{n+1}^2 - 2d_n d_{n+1} c_{n,1} , \quad b_n = \Delta z + h_{n+1} - h_n , \quad d_n = R_0 + r_n ,$$

$$c_{n,1} = \cos(\Delta\phi + \varphi_{n+1} - \varphi_n) .$$

According to [21], the valence bond energy has the value $D_0 = 334.72 \, \text{kJ mol}^{-1}$ and $\alpha = 1.91 \, \text{Å}$. The deformation energy of the bond angle is given by the potential (6.3), where the nth bond angle is $\theta_n = \arccos[-(a_{n,2} + b_{n-1}b_n)/\rho_{n-1}\rho_n]$. Here,

$$a_{n,2} = d_{n-1}d_n c_{n-1,1} + d_n d_{n+1} c_{n,1} - d_n^2 - d_{n-1}d_{n+1}c_{n,2} ,$$

$$c_{n,2} = \cos(2\Delta\phi + \varphi_{n+1} - \varphi_{n-1}) ,$$

and the energy $K_\theta = 529 \, \text{kJ mol}^{-1}$ [22].

The internal rotation potential $W(\delta_n)$ describes the slowdown of chain links rotating about the nth valence bond. The nth rotational angle is

$$\delta_n = \arccos\left(-\frac{b_n b_{n+1} a_{n,2} + b_{n-1}b_n a_{n+1,2} - b_n^2 a_{n,4} - b_{n-1}b_{n+1}a_{n,1} + a_{n,3}a_{n+1,3}}{\sqrt{\beta_n \beta_{n+1}}} \right) .$$

where

$$a_{n,3} = d_{n-1}d_n s_{n-1,1} + d_n d_{n+1} s_{n,1} - d_{n-1}d_{n+1}s_{n,2} ,$$

$$a_{n,4} = d_n d_{n+2} c_{n+1,2} - d_n d_{n+1} c_{n,1} - d_{n-1}d_{n+2}c_{n,3} + d_{n-1}d_{n+1}c_{n,2} ,$$

$$s_{n,1} = \sin(\Delta\phi + \varphi_{n+1} - \varphi_n) , \quad s_{n,2} = \sin(2\Delta\phi + \varphi_{n+1} - \varphi_{n-1}) ,$$

$$c_{n,3} = \cos(3\Delta\phi + \varphi_{n+2} - \varphi_{n-1}) ,$$

and

$$\beta_n = a_{n-1,1}b_n^2 + a_{n,1}b_{n-1}^2 - 2a_{n,2}b_{n-1}b_n + a_{n,3}^2 .$$

The potential is given by (6.12) and its form is shown in Fig. 6.7. The potential has an absolute minimum at $\delta = \delta_0$, $2\pi - \delta_0$. A PTFE macromolecule has four rotation isomers per C–C bond. Two of them, *trans* (+) and *trans* (−), have equal minimum energies, viz., $\delta_1 = \delta_0$, $\delta_2 = 2\pi - \delta_0$, $W(\delta_1) = W(\delta_2) = 0$, and the other two, *gauche* (+) and *gauche* (−), have higher energies, viz., $\delta_3 \approx 2\pi/3$, $\delta_4 \approx 4\pi/3$, and $W(\delta_3) = W(\delta_4) > 0$.

The potential $S(r_n, u_n, v_n)$ describes the interaction of a molecular chain with the substrate formed by the six neighboring chains. This potential can be found numerically by calculating the energy of the chain interaction with the six immobile neighboring chains as a function of the chain displacement. For this purpose, it is necessary to calculate the sum of energies of nonvalence (van der Waals and Coulomb) interactions of all atoms, and to divide it by the number of chain sites.

Let us introduce the new local dimensionless coordinates

$$u_n = \frac{\varphi_n - h_n \Delta\phi/\Delta z}{2\pi}, \tag{6.48}$$

$$v_n = u_n + h_n/\Delta z, \tag{6.49}$$

The substrate potential is a two-dimensional periodic function of these variables: $S(r, u \pm 1, v \pm 1) \equiv S(r, u, v)$. The region $0 \le u \le 1$, $0 \le v \le 1$ is a unit cell of the discrete helix (Fig. 6.24 right). Indeed, each nth helix link occupies the position of the $(n-1)$th links at $u_n \equiv 1$ and the $(n+1)$th links at $v_n \equiv 1$. In both cases, the position of the infinite helix does not change. Therefore, it suffices to find numerically the substrate potential for the unit cell alone.

The substrate potential can be described analytically to high accuracy by the finite double Fourier series

$$S(r, u, v) = \frac{1}{2} K_r r^2 + b_{22} \cos\left[2\pi(u - v)\right]$$

$$+ \sum_{i=1}^{6}\sum_{j=1}^{6} a_{ij} \cos\left\{2\pi\left[(i-1)(u-1/2) + (j-1)(v-1/2)\right]\right\},$$

$$\tag{6.50}$$

where $K_r = 17.1\,\text{N/m}$ is the transverse rigidity, $b_{22} = -0.075878\,\text{kJ mol}^{-1}$, and the values of other Fourier coefficients are given in Table 6.8.

A two-dimensional plot of the substrate potential $E = S(0, u, v)$ is shown in Fig. 6.25. As can be seen, the substrate potential has a valley landscape. The potential minima $(u, v) = (0, 0)$, $(0, 1)$, $(1, 0)$, and $(1, 1)$ correspond to the ground state of the chain. The transition of the chain from the $(1, 0)$ ground state to the $(0, 1)$ state requires a minimum energy expenditure (this transition involves overcoming an energy barrier of height of $2.983\,\text{kJ mol}^{-1}$). In this case, the motion is along the bottom of the deepest valley. Hereafter, we will show that this transition corresponds to the topological defect of a chain, possessing soliton dynamics.

As shown in Sect. 2.2, there exist two sound velocities in an isolated PTFE chain: $v_1 = 6{,}978.6\,\text{m/s}$ is the velocity of the long-wave longitudinal phonons and $v_t =$

Table 6.8 Fourier coefficients $\{a_{ij}\}_{i=1, j=1}^{6,6}$ for the analytical representation of the substrate potential (6.50)

Value of a_{ij} (kJ mol^{-1}) at j					
i	1	2	3	4	5	6
1	2.642967	−0.405754	−0.000389	0.000112	−0.000004	−0.000006
2	0.840162	0.110066	−0.203920	0.004530	0.000084	0.000003
3	−0.001460	0.720333	−0.960938	−0.133891	0.000759	−0.000007
4	−0.000745	0.042676	0.828604	−0.184377	0.000844	0.000554
5	−0.000016	−0.001780	0.018256	−0.007373	−0.127483	0.011884
6	0.000001	−0.000012	−0.000185	0.028068	−0.149317	0.001723

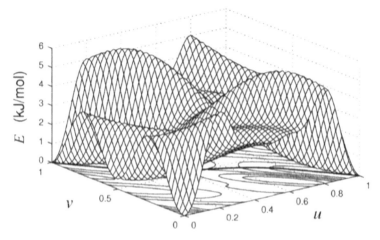

Fig. 6.25 Plot of the two-dimensional substrate potential $E = S(0, u, v)$. The *bold line* shows the most energetically preferred trajectory of particle motion in the unit cell of the helix

5,585.3 m/s is the velocity of the long-wave torsional phonons. The ratio of these velocities is $s_t = v_t/v_l = 0.80035$.

To find the stationary state of a topological defect (soliton) $\{r_n, \varphi_n, h_n\}_{n=1}^{N}$ in a helical PTFE macromolecule, one must solve numerically the constrained minimum problem

$$P = \sum_{n=1}^{N} \left[V(\rho_n) + U(\theta_n) + W(\delta_n) + S(r_n, u_n, v_n) \right] \longrightarrow \min_{r_1, \varphi_1, h_1, \dots, r_N, \varphi_N, h_N} \quad : \quad (6.51)$$

$$r_1 = 0, \quad \varphi_1 = \varphi_{-\infty}, \quad h_1 = h_{-\infty}. \quad (6.52)$$

$$r_N = 0, \quad \varphi_N = \varphi_{+\infty}, \quad h_N = h_{+\infty}. \quad (6.53)$$

The boundary conditions (6.52) and (6.53) for the problem (6.51) determine the type of topological defect. The number of chain units N must be sufficiently large for the defect shape to be independent of its value, e.g., $N = 500$.

Let us consider the possible local topological defects in a helical chain. For this purpose, we rewrite the boundary conditions (6.52) and (6.53) in the dimensionless helical coordinates (6.48) and (6.49):

$$r_1 = 0 , \quad u_1 = 0, 1 , \quad v_1 = 0, 1 , \tag{6.54}$$

$$r_N = 0 , \quad u_N = 0, 1 , \quad v_N = 0, 1 . \tag{6.55}$$

Each of these boundary conditions (6.54) and (6.55) corresponds to a certain ground state of the chain. Therefore, a solution of the constrained minimum problem given by (6.51)–(6.53) relates to the topological defect describing the transition of the helix from the ground state (6.54) to another (6.55). Let us define the topological charge of a defect as the two-dimensional vector $\mathbf{q} = (q_1, q_2)$, where $q_1 = u_N - u_1$ and $q_2 = v_N - v_1$. Thus, there are four possible types of elementary topological defects in the helical chain, with topological charges $\mathbf{q} = \pm(1, 0)$, $\pm(0, 1)$, and $\pm(1, 1)$, $\pm(-1, 1)$.

The distribution of chain deformations in the localization region of a topological defect with charge $\mathbf{q} = (1, 0)$ is shown in Fig. 6.26a. This defect is formed by rotating a chain segment through an angle of $2\pi - \Delta\phi$ and then shifting it along the z-axis by $-\Delta z$. As a result of this deformation, the nth link occupies the position of the $(n - 1)$th link. The chain deforms due to contraction of the bond angles and an increase in the torsion angles. As can be seen from Fig. 6.26a, the defect does not have a smooth profile along the radial r_n and angular φ_n components, so it cannot move as a smooth solitary wave.

The defect with topological charge $\mathbf{q} = (0, 1)$ (see Fig. 6.26b) is formed by rotating a chain segment through an angle $\Delta\phi$ and displacing it by Δz. As a result, the nth link takes the position of the $(n + 1)$th link. The chain deforms as a result of increasing the bond angles and diminishing the torsion angles. Once again, this defect does not have a smooth profile for the components r_n and φ_n.

The defect with charge $\mathbf{q} = (1, 1)$ (see Fig. 6.26c) is formed by rotating a helix segment through 360°. As a result, each helix unit returns to its initial position. This defect is characterized by the absence of a smooth profile for all three components r_n, φ_n, and h_n.

The defect with charge $\mathbf{q} = (-1, 1)$ has a smooth profile for all three components r_n, φ_n, and h_n (see Fig. 6.26d). This defect is formed by rotating a chain segment through an angle $2(\pi - \Delta\phi)$ and then displacing the chain along the z-axis by $2\Delta z$. As a result, the nth link occupies the position of the $(n + 2)$th link. The defect with the opposite charge $\mathbf{q} = (1, -1)$ also has a smooth profile for all three components. In this case, the chain deforms first of all due to contraction (extension) of the bond angles. These defects exhibit soliton dynamics, that is, they can move with a constant subsonic velocity $v < v_t$, retaining their shapes. In the unit cell $0 \leq u \leq 1$, $0 \leq v \leq 1$, the topological soliton is characterized by a smooth trajectory $\{(u(n), v(n))\}_{n=-\infty}^{+\infty}$, which connects the ground states $(1, 0)$ and $(0, 1)$. This trajectory, shown by the bold line in Fig. 6.25, corresponds to the deepest valley of the two-dimensional surface of the substrate potential $E = S(0, u, v)$.

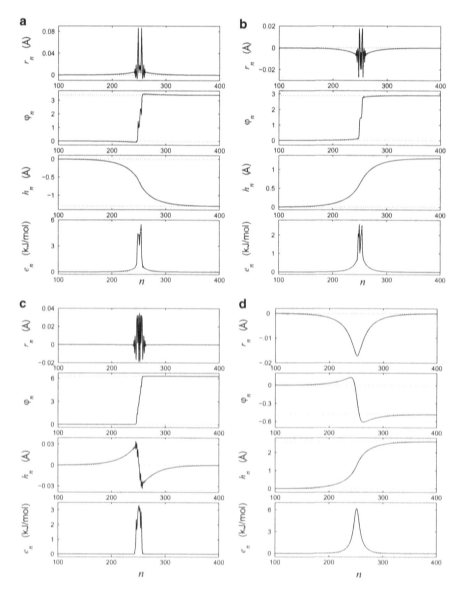

Fig. 6.26 Distribution of the radial r_n, angular φ_n, and longitudinal h_n displacements of the helix units, and the chain deformation energy e_n, where n is the number of the chain unit, in the localization region of the defect with topological charges $\mathbf{q} = (1,0)$ (**a**), $(0,1)$ (**b**), $(1,1)$ (**c**), and $(-1,1)$ (**d**). The energies of the defects are $E = 52.4$ (**a**), 33.6 (**b**), 28.9 (**c**), and $119.4\,\mathrm{kJ\,mol^{-1}}$ (**d**)

Numerical simulation of the dynamics of the topological solitons with charges $\mathbf{q} = \pm(-1,1)$ has shown that they are stable. Their motion is not accompanied by a change in shape. Let us consider the interaction between a soliton with charge $\mathbf{q} = (-1,1)$ and other stationary defects. For this purpose, we numerically integrate

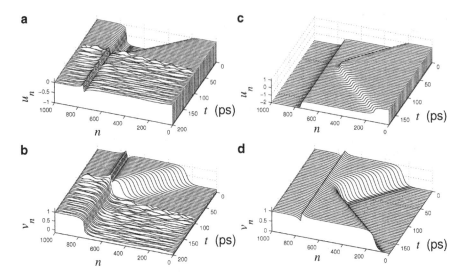

Fig. 6.27 Partial recombination of the topological soliton (charge $q_s = (-1, 1)$) with the topological defect (charge $q_d = (1, 0)$) as a result of their collision (**a**) and (**b**). Elastic reflection of the topological soliton (charge $q_s = (-1, 1)$) with the topological defect (charge $q_d = (-1, 0)$) (**c**) and (**d**)

the equations of motion

$$
\left.
\begin{aligned}
M\ddot{r}_n &= M(R_0 + r_n)\dot{\varphi}_n^2 - \frac{\partial}{\partial r_n} P \, , \\
M(R_0 + r_n)^2 \ddot{\varphi}_n &= -2M(R_0 + r_n)\dot{\varphi}_n \dot{r}_n - \frac{\partial}{\partial \varphi_n} P \, , \\
M\ddot{h}_n^2 &= -\frac{\partial}{\partial h_n} P \, ,
\end{aligned}
\right\} \quad n = 1, 2, \ldots, N \, ,
$$

$$(6.56)$$

with the initial condition corresponding to a soliton with charge $q_s = (-1, 1)$ and velocity $v = 0.1v_t$, as well as a stationary topological defect with another topological charge q_d.

When $q_d = (1, 0)$, the collision of the soliton and defect causes their partial recombination. This leads to the stationary defect with charge $q = (0, 1)$ remaining in the chain. The collision is accompanied by intense phonon emission (see Fig. 6.27a, b). When $q_d = (-1, 0)$, there is elastic reflection of the soliton from the defect. The defect itself remains immobile (see Fig. 6.27c, d). Elastic reflection also occurs when $q_d = (0, 1)$ and $(-1, -1)$. In these cases, the defect remains immobile. When $q_d = (0, -1)$, a stationary defect with $q = (-1, 0)$ remains in the chain as a result of partial relaxation. Partial recombination occurs also when $q_d = (1, 1)$. In this case, a stationary defect with the double topological charge $q = (0, 2)$ remains in the chain, i.e., stationary defects with multiple topological charges can exist in the chain.

The collision of two identical topological solitons always results in their elastic reflection, while the collision of two opposite-sign solitons leads to their complete recombination.

Numerical simulation of the topological defect dynamics in a thermalized chain shows that, for $T = 100$ K, all the defects are stable to thermal vibrations of the chain. For defects with the topological charge $\mathbf{q} \neq \pm(1, 1)$, this stability is topological in origin, i.e., the destruction of these defects involves moving half of the chain along the z-axis. The defect with charge $\mathbf{q} = \pm(1, 1)$ formed by local rotations of the helix units can be unstable to thermal vibrations. For $T = 100$ K, this defect remained stable over the whole integration time $t = 100$ ps. Under the action of thermal vibrations, the topological soliton, i.e., the defect with charge $\mathbf{q} = \pm(-1, 1)$, moves like a free Brownian particle. Other topological defects are pinned, so that their motion is reduced to rare thermally activated jumps to neighboring links.

Therefore, numerical investigation of the topological defects (solitons) in the PTFE crystal has shown that four types of defect can exist in a crystal composed of macromolecules in the conformation of a three-dimensional helix. All the defects are stable to thermal vibrations of the chain. Each defect can be characterized by a two-dimensional topological charge. One of the defects exhibits soliton dynamics, i.e., it can move along the chain as a solitary wave. The velocity spectrum of this soliton falls in the subsonic range. Other defects are pinned and their motion has the character of thermally activated random jumps.

A more detailed description of the results is given in [46, 47]. Numerical simulation carried out in [47] has also shown the existence of a phase transition in the PTFE crystal at temperature $T = 300$ K, characterized by the fast accumulation of topological defects and increasing heat capacity of the system.

6.4 Breathers in a PE Macromolecule

Localized excitations in nonlinear systems – such as solitons, polarons, and breathers – have become a subject of growing interest over the past few decades. Discrete breathers, localized nonlinear modes, are periodic, stable and localized vibrations in a system. An intensive study of them began with the pioneering work of Sievers and Takeno [48]. Furthermore, the existence of discrete breathers is guaranteed by a related theorem proven in [49, 50] and by numerous numerical investigations (see review [51]). Now the role of breathers in the mechanisms of energy transfer and relaxation in molecular systems has been well clarified [11–13].

In general, the investigation of breathers is carried out based on simple, one-dimensional models. However, these localized excitations can also exist in complex, discrete nonlinear systems of any dimension [52]. Necessary conditions for their existence include the discreteness of the system (boundedness of the phonon frequency spectrum) and the dependence of oscillation frequencies on the oscillation amplitudes (nonlinearity of interaction). Polymer molecules belong to such systems. The existence of high-frequency breathers in hydrocarbon macromolecules was shown in [52]. In these systems, there can exist localized high-frequency vibrations of C–H valence bonds due to anharmonic potential of these bonds. These vibrations are present in polyethylene (PE) molecule and have frequencies $\sim$3,100 cm^{-1}. In this section we shall show the existence of low-frequency localized vibrations of PE macromolecule zigzag backbone which result from coordinated changes in C–C valence bonds and angles.

The nonlinear dynamics of the *trans* zigzag chain of PE macromolecule was considered in Sects. 2.1 and 4.1. Here, we investigate the dynamics of a PE molecule using the united atom approximation, i.e. representing every CH$_2$ group as a single particle. In this approximation, the Hamiltonian of the chain has the form (6.36).

Small-amplitude vibrations can be divided into planar vibrations in the *trans* zigzag plane, and transverse vibrations in the plane transverse to the *trans* zigzag plane. Planar vibrations are further divided into low-frequency acoustic and high-frequency optical vibrations. Therefore, there are three dispersion curves which correspond to planar acoustic phonons $\omega = \omega_a(q)$, planar optical phonons $\omega = \omega_o(q)$, and torsional (transverse) phonons $\omega = \omega_t(q)$ in the chain. For the chain with substrate, these curves are shown in Fig. 6.28a (left).

For an isolated chain, i.e. when substrate potential $Z(u, v, w) \equiv 0$, the acoustic, torsional and optical phonons have the following frequency spectra:

$$0 \leq \omega_a \leq 228.75 \text{ cm}^{-1} , \quad 0 \leq \omega_t \leq 243.85 \text{ cm}^{-1} ,$$

$$838.47 \text{ cm}^{-1} \leq \omega_o \leq 1,168.11 \text{ cm}^{-1} ,$$

respectively. Consideration of the interaction with the chain substrate leads to a shift of the lower limit of the acoustic and transverse phonon spectra to high frequencies:

$$19.05 \text{ cm}^{-1} \leq \omega_a \leq 229.54 \text{ cm}^{-1} , \quad 52.43 \text{ cm}^{-1} \leq \omega_t \leq 249.43 \text{ cm}^{-1} ,$$

$$838.69 \text{ cm}^{-1} \leq \omega_o \leq 1,168.27 \text{ cm}^{-1} .$$

Consider a finite chain consisting of $N = 200$ links and introduce viscous friction at the chain ends which ensures adsorption of phonons. The equations of motion (6.41) are integrated numerically with $n = 1, 2, \ldots, N$ and a breather-like initial condition. If a discrete breather could exist in the chain, it should be obtainable. A superfluous 'non-breather' part of the initial condition will disappear through phonon emission.

Numerical modeling has shown that only one type of localized periodic vibrations exists in the zigzag chain, viz., the localized vibrations of valence C–C bonds. In the localization region of the excitation, there exist periodic compression and extension of valence bonds followed by a coordinated change in the valence angles (see Fig. 6.28b, c right). The vibrations occur in the plane of the *trans* zigzag chain,

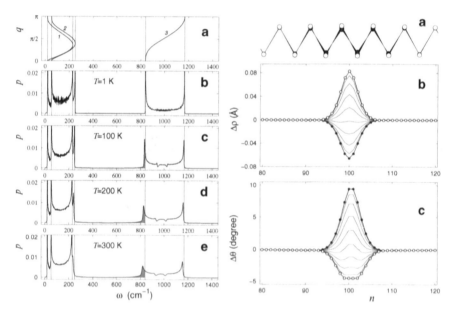

Fig. 6.28 *Left*: Dispersion curves $\omega = \omega_a(q)$, $\omega = \omega_t(q)$, and $\omega = \omega_o(q)$ (*lines 1–3*) for the *trans* zigzag chain interacting with the substrate (**a**). Dependence of the energy distribution density p on the frequencies of thermal vibrations at temperatures $T = 1\,\mathrm{K}$ (**b**), $T = 100\,\mathrm{K}$ (**c**), $T = 200\,\mathrm{K}$ (**d**), and $T = 300\,\mathrm{K}$ (**e**). The frequency region in which discrete breathers exist is *shaded*. *Right*: Localized planar periodic vibrations of the *trans* zigzag chain. Vibrations are shown schematically. *Line thickness* corresponds to the vibration amplitude (**a**). Dependence of valence bond length ρ_n (**b**) and angles θ_n (**c**) are given for ten different times. Breather frequency $\omega = 820.5\,\mathrm{cm}^{-1}$, energy $E = 26.4\,\mathrm{kJ/mol}$, and width $L = 4.28$

while zigzag segments oscillate perpendicularly to the main axis of the zigzag backbone (see Fig. 6.28a right).

These vibrations are stable excitations characterized by the frequency ω, energy E and dimensionless width $L = 2\big[\sum_{n=1}^{N}(n - n_c)^2 p_n\big]^{1/2}$. Here, the point $n_c = \sum_n n p_n$ determines the position of the vibration centre and the sequence $p_n = E_n/E$ describes the energy distribution density along the chain. The vibrations are nonlinear, i.e. its frequency decreases with increasing amplitude. Therefore, these vibrations are a truly discrete breather.

The dependence of the energy E and width L of the breather on its frequency ω is shown in Fig. 6.29 (left). The breather has the frequency spectrum near the lower limit of optical phonons. In essence, the breather is a modulated optical phonon with wave number $q = 0$, and so the chain substrate does not significantly affect it. The breather has frequency spectrum $\omega_b = 817\,\mathrm{cm}^{-1} \le \omega < \omega_o(0) = 838\,\mathrm{cm}^{-1}$ in an isolated chain and $\omega_b = 820\,\mathrm{cm}^{-1} \le \omega < \omega_o(0) = 839\,\mathrm{cm}^{-1}$ in a chain with substrate. When the frequency decreases, the breather energy increases steadily and its width decreases monotonically.

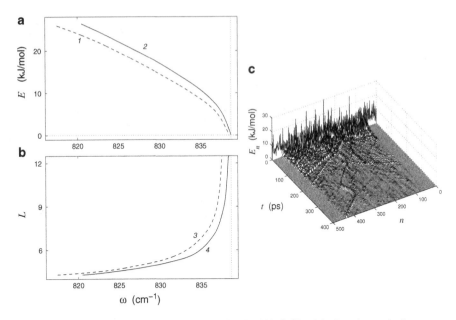

Fig. 6.29 *Left*: Dependence of the energy E (**a**) and width L (**b**) of the breather on its frequency ω in the isolated chain (*lines 1* and *3*) and in the chain with the substrate potential (*lines 2* and *4*). *Right*: Formation of discrete breathers from thermal vibrations of the zigzag chain ($N = 500$, $T = 200$ K). Absorbing ends are taken into account ($N_0 = 50$). The dependence of the energy distribution E_n in the chain on time t is shown

The breather frequency has to be separated from the frequencies of small-amplitude linear vibrations. One might also expect the existence of low-frequency breathers with frequencies near the upper limit of the torsion phonon spectrum, $\omega > \max \omega_t(q)$, and high-frequency breathers with frequencies near the upper limit of the optical phonon spectrum, $\omega > \omega_0(\pi)$. However, numerical analysis has shown the absence of these localized excitations.

Let us consider thermal vibrations of the *trans* zigzag chain. For this purpose, we analyse a finite chain, consisting of N links, of which N_0 links of both chain ends are attached to a thermal bath with temperature T. The dynamics of this system is described by the Langevin equations (6.44), where ξ_n, η_n, and $\zeta_n = 0$ are random forces and the friction coefficients $\Gamma_n = 0$ for $N_0 < n \leq N - N_0$. When $n \leq N_0$ and $N - N_0 < n \leq N$, $\Gamma_n = \Gamma$ and ξ_n, η_n, and ζ_n are normally distributed random forces which describe the interaction of the nth molecule with the thermal bath. The friction coefficient $\Gamma = 1/t_r$, where t_r is the relaxation time of molecular velocity.

Consider the frequency distribution of the kinetic energy of thermal vibrations in the molecule. To obtain these, Eq. (6.44) was numerically integrated for $N = 500$ and $N_0 = N/2$. The initial conditions were chosen to correspond to the ground state of the chain and the equations were integrated over time $t = 10t_r$ to bring the chain into thermal equilibrium with the thermal bath. The frequency density

$p(\omega)$ of the kinetic energy distribution in the molecule was then calculated. To improve accuracy, the distribution density was calculated using 1,000 independent realizations of the chain thermalization. The profile of the distribution density is shown in Fig. 6.28 (left) for different values of the temperature. The distribution density is normalized according to the condition $\int p(\omega)d\omega = 3$.

For temperature $T = 1$ K, the distribution density practically coincides with the corresponding distribution density of the linearized system. Here, the vibrations remain linear and only phonons are thermalized. With increasing temperature, the amplitude of the thermal vibrations also increases, and their anharmonicity begins to manifest itself. For $T = 100$ K, a shift in the density distribution beyond the lower limit of the optical phonon spectrum can be seen. This shift becomes more pronounced with further temperature increase. In this temperature region, high-frequency vibrations appear which are not optical phonons. The frequencies of these vibrations fall into the frequency interval of the discrete breather $(\omega_b, \omega_o(0))$, thus they can be identified as breathers. The part of the energy corresponding to the breathers can be determined as the integral

$$ p_b = \int_{\omega_b}^{\omega_o(0)} p(\omega)d\omega . $$

The contribution of the breathers to thermal energy increases with rising temperature ($p_b = 0.002$ and 0.106 for $T = 1$ and 100 K, respectively), reaching a maximum value $p_b = 0.115$ at $T = 200$ K, and then decreases (for $T = 300$ K, the breather contribution is $p = 0.083$).

To isolate breathers from thermal vibrations, consider a chain consisting of $N = 500$ links with end links ($N_0 = 50$) attached to a thermal bath with temperature T. After thermalization, we set the temperature of the thermal bath to $T = 0$ and analyse the process of heat energy sink from an internal region of the chain ($N_0 < n < N - N_0$). The relaxation process for $T = 200$ K is shown in Fig. 6.29 (right). The formation of several mobile localized excitations can be clearly visible in this figure. Detailed analysis has shown that these excitations are discrete breathers with frequencies $\omega \sim \omega_b$. Therefore, discrete breathers are present in thermal vibrations.

Consider the interaction of a discrete breather with thermal phonons. Take a finite chain ($N = 500$), with a stationary discrete breather with frequency $\omega = 820.5$ cm^{-1} placed in the centre of the chain with end links ($N_0 = 10$) attached to a thermal bath with temperature $T = 10$ K. Analysis of the breather dynamics has shown that the breather begins to collapse just as thermalization of the chain centre begins. The breather loses about half of its energy in 20 ps.

The probability of the thermally-activated formation of discrete breathers in the *trans* zigzag chain grows with increasing temperature. Therefore, concentration of breathers in the chain has to increase when the temperature increases. However, in a thermalized chain, breather has a finite lifetime which decreases with rising temperature. As a result, the dependence of the breather concentration p_b on the temperature T is non-monotonic. In fact, this concentration increases when $T \leq 200$ K and decreases when $T > 200$ K. Maximum value is reached for $T = 200$ K.

Numerical investigation shows that the breathers are best separated from thermal vibrations precisely at this temperature.

As can be seen in Fig. 6.28 (left), there are no other stable localized periodic vibrations in a thermalized chain except for the breathers considered above. As demonstrated previously, only vibrations with frequencies belonging to the linear vibration spectrum are thermalized. This confirms our conclusion regarding the existence of only one type of stable discrete breathers which correspond to the localized vibrations of valence C–C bonds. These excitations are present in a thermalized chain even at low temperatures.

The investigation carried out shows that, in linear polymer macromolecules, there can exist stable localized periodic vibrations, namely, discrete breathers. In a polyethylene macromolecule they represent planar vibrations of the *trans* zigzag chain with periodic compression and extension of valence C–C bonds in the localization region of the vibrations. Breathers are present in a thermalized chain and contribute to the heat capacity of the chains. A detailed description of the results obtained can be found in [53].

Note that the same breathers must also exist in PTFE molecules, where they will have the lower frequency spectrum. Recent studies have shown that breathers can exist in carbon nanotubes possessing the zigzag structure, where they correspond to localized torsion vibrations [54]. They can also exist at the edges of carbon nanotubes [55].

6.5   Conclusion

Numerical investigation of nonlinear dynamics of localized excitations showed that all types of excitations can exist in linear polymer molecules, viz., acoustic and topological solitons, as well as discrete breathers.

In a planar *trans* zigzag polyethylene macromolecule, there can exist dynamically stable, supersonic acoustic solitons corresponding to zigzag stretching. Furthermore, in a helix macromolecule of polytetrafluoroethylene, two types of supersonic acoustic solitons can exist: solitons of torsion and longitudinal compression of the helix. The solitons have finite supersonic velocity spectrum, they are stable for all permissible velocities, and display particle-like properties. On the other hand, in a carbon nanotube, acoustic solitons do not exist. There are only soliton-like excitations whose motion is always accompanied by longitudinal phonon emission. The amplitude of this emission decreases with increasing nanotube radius, but it is always nonzero. It can be said that an acoustic soliton in nanotube has a finite lifetime and therefore always possesses a finite free path.

In PE crystal, three types of topological solitons can exist:

• The two-dimensional soliton of stretching (compression) of the zigzag chain by one period,

- The soliton of stretching (compression) of the zigzag chain by a half of period, followed by torsion through 180°, and
- The soliton corresponding to torsion of the zigzag chain through 360°.

All these solitons have permissible velocity intervals in the subsonic range. They are dynamically stable, they move with constant velocities, and retain their shapes. In PTFE crystal, there are four types of localized topological defects of the helix macromolecule. All the defects are stable to thermal vibrations of the chain, but the defect of only one type is soliton, which can move as a solitary wave along the chain. The soliton velocity interval lies in the subsonic range. Other defects are pinned and their motion has the character of thermally-activated random hops.

In linear polymer molecules, there also exist breathers, i.e. stable localized periodic vibrations of macromolecule chains. In the PE macromolecule, these vibrations represent planar vibrations which correspond to periodic compression and extension of the valence bonds and angles in the *trans* zigzag chain.

References

1. Bishop, A.R.: Solitons and physical perturbations. In: Lanngren, K. and Scott, A. (eds.) Solutions in Action. Academic Press, New York (1978)
2. Collins, M.A.: A quasicontinuum approximation for solitons in an atomic chain. Chem. Phys. Lett. **77**, 342 (1981)
3. Collins, M.A.: Solitons and nonlinear phenomena. Adv. Chem. Phys. **53**, 225 (1983)
4. Kosevich, A.M.: Theory of Crystal Lattice. Vishcha Shkola, Kharkov (1988)
5. Gendelman, O.V., Manevich, L.I.: Exact soliton-like solutions in generalised dynamical models of quasi-one-dimensional crystal. Sov. Phys. JETP **102**, 511 (1992)
6. Manevitch, L.I., Ryvkina, N.G.: Mechanical Modelling of New Electromagnetic Materials. Elsevier, Amsterdam (1990)
7. Savin, A.V., Manevich, L.I.: Structural transformations in crystalline polyethylene: role of topological solitons in premelting. Vysokomol. Soedin. A **40**, 931 (1998)
8. Savin, A.V., Manevitch, L.I.: Solitons in crystalline polyethylene: a chain surrounded by immovable neighbors. Phys. Rev. B. **58**(17), 11386 (1998)
9. Reneker, D.H., Mazur, J.: Small defects in crystalline polyethylene. Polymer **29**, 3 (1988)
10. Ginzburg, V.V., Manevich, L., Ryvkina, N.G.: Dynamics of the polyethylene crystal. Mech. Compos. Mater. **27**, 167 (1991)
11. Kopidakis, G., Aubry, S.: Discrete breathers and delocalization in nonlinear disordered systems. Phys. Rev. Lett. **84**, 3236 (2000)
12. Kopidakis, G., Aubry, S., Tsironis, G.P.: Targeted energy transfer through discrete breathers in nonlinear systems. Phys. Rev. Lett. **87**, 165501 (2001)
13. Leitner, D.M.: Vibrational energy transfer in helices. Phys. Rev. Lett. **87**, 188102 (2001)
14. Christiansen, P.L., Zolotaryuk, A.V., Savin, A.V.: Solitons in an isolated helix chain. Phys. Rev. E **56**, 877 (1997)
15. Yomosa, S.: Solitary excitation in muscle proteins. Phys. Lett. A **32**, 1752 (1985)
16. Perez, P., Theodorakopoulos, N.: Solitary excitation in the α-helix: viscous and thermal effects. Phys. Lett. A **117**(8), 405 (1986)
17. Manevich, L.I., Savin, A.V.: Solitons of tension in polyethylene molecules. Polym. Sci. Ser. A **38**, 789 (1996)

18. Manevitch, L.I., Savin, A.V.: Solitons in crystalline polyethylene: isolated chains in the transconformation. Phys. Rev. E **55**, 4713 (1997)
19. Manevich, L.I., Ryapusov, S.V.: Nonlinear planar dynamics of the polyethylene molecule. Sov. Phys. Solid State **34**, 826 (1992)
20. Kirkwood, J.G.: The dielectric polarization of polar liquids. J. Chem. Phys. **7**, 506 (1939)
21. Sumpter, B.G., Noid, D.W., Liang, G.L., Wunderlich, B.: Atomistic dynamics of macromolecular crystals. Adv. Polym. Sci. **116**, 29 (1994)
22. Zhang, F.: Molecular-dynamics simulation of solitary waves in polyethylene. Phys. Rev. E **56**, 6077 (1997)
23. Christiansen, P.L., Savin, A.V., Zolotaryuk, A.V.: Soliton analysis in complex molecular systems: a zig-zag chain. J. Comput. Phys. **134**, 108 (1997)
24. Fletcher, R., Reeves, C.M.: Function minimization by conjugate gradients. Comput. J. **7**, 149 (1964)
25. Savin, A.V., Manevich, L.I., Hristiansen, P.L., Zolotaruk, A.V.: Nonlinear dynamics of zigzag molecular chains. Phys. Usp. **42**, 245 (1999)
26. Wunderlich, B.: Macromolecular Physics, vol. 1. Academic, New York (1973)
27. Savin, A.V., Manevitch, L.I.: Solitons in spiral polymeric macromolecules. Phys. Rev. E **61**(6), 7065 (2000)
28. Savin, A.V.: Nonlinear solitary waves in a polytetrafluoroethylene molecule. Vysokomol. Soedin. A **41**, 1416 (1999)
29. Radushkevich, L.V., Lukianovich, V.M.: About carbon structure, formed by thermal decomposition of carbon monoxide at iron contact. Zhurn. Fiz. Khim **26**(1), 88 (1952)
30. Iijima, S.: Helical microtubules of graphitic carbon. Nature **354**, 56 (1991)
31. Eletskii, A.V.: Carbon nanotubes and their emission properties. Phys. Usp. **45**, 369 (2002)
32. Astakhova, T.Yu., Gurin, O.D., Menon, M., Vinogradov, G.A.: Longitudinal solitons in carbon nanotubes. Phys. Rev. B. **64**, 035418 (2001)
33. Brenner, D.W.: Empirical potential for hydrocarbons for use in simulating the chemical vapor deposition of diamond films. Phys. Rev. B **42**(15), 9458 (1990)
34. Eilbeck, J.C., Flesh, R.: Calculation of families of solitary waves on discrete lattices. Phys. Lett. A **149**, 200 (1990)
35. Campbell, D.K., Peyrard, M., Sodano, P.: Kink–antikink interactions in the double sine–Gordon equation. Physica D **19**, 165 (1986)
36. Zolotaryuk, Y., Savin, A.V., Christiansen, P.L.: Solitary plane waves in an isotropic hexagonal lattice. Phys. Rev. B **57**(22), 14213 (1998)
37. Press, W.H., Teukolsky, S.A., Vetterling, W.T., Flannery, B.P.: Numerical Recipes in Fortran 77. The Art of Scientific Computing. Press Syndicate of the University of Cambridge, Cambridge/New York (1992)
38. Savin, A.V., Savina, O.I.: Nonlinear dynamics of carbon molecular lattices: soliton plane waves in graphite layers and supersonic acoustic solitons in nanotubes. Phys. Solid State **46**, 383 (2004)
39. Mazo, M.A., Balabaev, N.K., Oleinik, E.F.: Proceedings of International Symposium Molecular Mobility and Order in Polymer Systems, St. Petersburg, 98 (1994)
40. Noid, D.W., Sumpter, B.G., Wunderlich, B.: Molecular dynamics simulation of twist motion in polyethylene. Macromolecules **24**, 4148 (1991)
41. Christiansen, P.L., Savin, A.V., Zolotaryuk, A.V.: Zig-zag version of the Frenkel–Kontorova model. Phys. Rev. B **54**, 12892 (1996)
42. Zolotaryuk, A.V., Pnevmatikos, S.T., Savin, A.V.: Two-sublattice kink dynamics ionic defects in hydrogen-bonded chains. Physica D **51**, 407 (1991)
43. Petersen, W.P.: Lagged Fibonacci random number generators for the NEC SX-3. Int. J. High Speed Comput. **6**, 387 (1993)
44. Savin, A.V., Manevich, L.I.: Topological solitons in crystalline polyethylene. Polym. Sci. Ser. A **40**, 448 (1998)
45. Holt, D.B., Farmer, B.L.: Modeling of helix reversal defects in polytetrafluoroethylene. I. Force field development and molecular mechanics calculations. Polymer **40**, 4667 (1999)

46. Savin, A.V.: Topological solitons in crystalline poly(tetrafluoroethylene). Polym. Sci. Ser. A **43**, 860 (2001)
47. Savin, A.V., Manevitch, L.I.: Topological solitons in spiral polymeric macromolecules: a chain surrounded by immovable neighbors. Phys. Rev. B **63**, 224303 (2001)
48. Sievers, A.J., Takeno, S.: Intrinsic localized modes in anharmonic crystals. Phys. Rev. Lett. **61**, 970 (1988)
49. MacKay, R.S., Aubry, S.: Proof of existence of breathers for time-reversible or Hamiltonian networks of weakly coupled oscillators. Nonlinearity **7**, 1623 (1994)
50. Aubry, S.: Breathers in nonlinear lattices: existence, linear stability and quantization. Physica D **103**, 201 (1997)
51. Flach, S., Willis, C.R.: Discrete breathers. Phys. Rep. D **295**, 181 (1998)
52. Kopidakis, G., Aubry, S.: Discrete breathers in realistic models: hydrocarbon structures. Physica B **296**, 237 (2001)
53. Savin, A.V., Manevitch, L.I.: Discrete breathers in a polyethylene chain. Phys. Rev. B **67**, 144302 (2003)
54. Savin, A.V., Kivshar, Y.S.: Discrete breathers in carbon nanotubes. EPL **82**, 66002 (2008)
55. Savin, A.V., Kivshar, Y.S.: Surface solitons at the edges of graphene nanoribbons. EPL **89**, 46001 (2010)

Chapter 7
Autolocalization of Quantum Particles

In this chapter we discuss the autolocalized state (soliton) dynamics of a quantum particle (intermolecular excitation) in a molecular chain.

One of the central concepts in bioenergetics is the question of how the chemical energy released by the hydrolysis of adenosine triphosphate (ATP) transforms and transfers along the protein molecules, and over significant distances compared to the molecular scale. The difficulty in the physical explanation for this transfer is associated with the fact that the energy released by ATP hydrolysis, equal to 0.42 eV, is only 20 times higher than the average energy of thermal fluctuations under physiological conditions. This energy is insufficient to excite electronic states. Therefore, it is normally assumed that the energy of ATP hydrolysis is transferred along the protein molecules in the form of the vibrational energy of the C=O group belonging to the peptide groups (PGs) of all proteins [1].

Among the interpeptide vibrations, the longitudinal amide-I vibrations of the double bond C=O are the most intensive. They manifest themselves through infrared absorption near $1,650 \, cm^{-1}$ (the single-quantum excitation energy is 0.205 eV) in all proteins. These vibrations have a relatively large electric dipole transition moment of 0.35 D. This leads to the resonance interaction between neighboring PGs, causing the vibration to hop from one PG to the next. On the other hand, the dependence of the energy of this vibration on the distance between neighboring PGs results in the appearance of exciton–phonon interactions in the chain. These interactions are determined by the coupling between the amide-I vibration and the displacement of the PG.

Given the regularity of the helical protein structure, Davydov and Kislukha [2] suggested a nonlinear model for vibration localization based on the exciton concept from solid-state physics [3–5]. The model consists of a one-dimensional deformable chain of harmonically-coupled molecules (PGs), each having an internal vibrational degree of freedom described by the exciton Hamiltonian. Using this model, they established that, in α-helical protein molecules, nonlinear collective autolocalized excitations can move without loss of energy or change in shape.

© Springer International Publishing Switzerland 2015
L.N. Lupichev et al., *Synergetics of Molecular Systems*, Springer Series
in Synergetics, DOI 10.1007/978-3-319-08195-3_7

Later, these autolocalized excitations received the name of *Davydov solitons* [6–10]. The balance between the resonant and exciton–phonon interactions discussed above leads to stabilization of the soliton motion. Here, the resonant interaction between neighboring molecules plays the role of dispersing small-amplitude waves (free excitons), while the exciton–phonon interaction (EPI) represents the nonlinearity causing localization of the intermolecular excitation.

The Davydov soliton model is based on the polaron concept, first introduced by Landau [11] and developed further by Pekar [12, 13] to describe electron autolocalization in a polarized field induced by the electron itself. Today, the term *polaron* has lost its original meaning (electron interacting with long-wave polarized optical vibrations) and it is mainly used for a much wider class of autolocalized states when a quantum particle (or quasi-particle) interacts with a field of any nature [14]. Recently, in connection with the development of the soliton theory [15–25] and its wide application to various areas in physics [26–40], autolocalized states, capable of moving, were called solitons. In addition, the term *soliton* is used not only in the strict mathematical sense, i.e., in the case of completely integrable Hamiltonian systems [15, 17, 19–21, 24], but also to describe dynamically stable nonlinear collective structures [22, 23, 25, 28–40], and in particular, autolocalized states [26, 27, 38–40]. Therefore, in 'polaron language', the Davydov soliton can also be called a one-dimensional acoustic polaron.

The dynamical equations describing the motion of a one-dimensional acoustic polaron (autolocalized exciton or polaron) in the continuum approximation constitute a self-consistent system which includes the time-dependent Schrödinger equation with a deformation potential and an inhomogeneous linear wave equation for this potential. The right-hand side of the wave equation corresponds to the second spatial derivative of the wave function satisfying the Schrödinger equation. This system, called the *Zakharov system*, has an important significance in physics and generally describes the nonlinear interaction of two physical subsystems: fast and slow.

The Zakharov equations were first obtained in plasma physics in the description of the interaction of Langmuir (high-frequency) waves with ion sound waves [28]. For different purposes and applications in condensed matter physics, equations of this type, describing self-trapping (autolocalization) of a quantum particle by the acoustic mode of a one-dimensional crystal lattice, were established by Kosevich [41] in the investigation of one-dimensional dynamics problems in nonlinear crystal mechanics. Similar results were obtained by Petrina and Epolskiy [42] in the investigation of molecular chain dynamics based on the Bogolubov equations [42]. Davydov and Kislukha developed this approach to model the soliton energy transfer of the interpeptide amid-I excitation [2]. In contrast to [43], the work described in [2] considers the excitation motion in a deformable molecular chain without taking the kinetic energy of the molecules into account. In this approximation, the equations of motion reduce to a completely integrable nonlinear Schrödinger equation (NSE) with cubic nonlinearity [18].

The Zakharov equations have a well-known soliton solution in hyperbolic secant form, describing the envelope profile of the high-frequency vibrations of a fast subsystem which can propagate with any subsonic velocity. In the stationary case, a solution of this type was first obtained by Rashba [44, 45] when investigating the interactions of intermolecular excitations with acoustic phonons in molecular crystals. These and later works [2, 46–51] studied the conditions for formation of bound states of intramolecular excitation and deformation in a molecular crystal without considering the excitation motion. The autolocalized state turned out to have the lowest energy, but in the case of the short-range coupling of the excitation with phonons in a three-dimensional crystal, a metastable exciton state is also possible, separated from the autolocalized state by an energy barrier.

The Zakharov equations constitute a non-integrable system. This was discovered by Degtyarev, Makhankov, and Rudakov [52] when they numerically modeled the interaction of Langmuir solitons with each other and with ion sound. This interaction turned out to be inelastic, and this was proved in the strict mathematical sense by Schulman [53]. The interaction of these solitons with sound waves was also investigated by Malomed [54] using the inverse scattering method [20, 21]. It has been shown that a soliton situated in a sound field gradually collapses, emitting high-frequency waves.

The equations described above, both from the point of view of plasma physics and the theory of molecular systems, provide the simplest model. To describe realistic cases, this system of equations requires refinements and generalizations. First, in the model of a molecular chain with harmonic intermolecular potential, a soliton solution has physical meaning only for velocities significantly less than the longitudinal sound velocity. When the soliton velocity approaches the sound velocity in a chain, the amplitude of local compression tends to infinity. As a result, there appears an effect of neighboring molecules passing through each other, which has no physical meaning. To avoid this, Davydov and Zolotaryuk [55–57] introduced positive intermolecular anharmonicity, enhancing the repulsion of neighboring molecules in a chain if they approach one another. In [57], an algorithm has been developed to obtain successive approximations to a soliton solution. It turns out that including positive anharmonicity also extends both the limits of applicability of the continuum approximation and the range of permissible velocities of soliton solutions to include the sound velocity. In the case of cubic anharmonicity, Karbovskiy and Kislikha have found approximate soliton solutions using a variational approach [58]. Further study of the model with cubic anharmonicity was carried out in [59, 60]. It should be emphasized that the dispersion law for acoustic phonons was assumed to be linear in all these studies [55–60].

Second, any real molecular chain is a discrete system that leads to certain peculiarities in the localized state dynamics. The dynamical equations in the discrete case, as determined in [42, 43], will hereafter be called the discrete acoustic self-trapping equations. They do not admit exact soliton solutions, even if one neglects the inertia of molecular motion, when these equations reduce to the corresponding discrete NSE. Note that exact soliton solutions exist only in the case of another type of NSE discretization scheme, as discussed by Ablowitz and Ladik [61, 62], who demonstrated its complete integrability.

The manifestation of discreteness in the case of stationary and moving autolocalized states in a harmonic chain was investigated by Kuprievich using variational [63] and numerical [64] approaches, by Kislukha in a study of the limit case of strong autolocalization [65, 66], Vakhnenko and Gaididei [67] who used the exact results and perturbation theory [61, 62], and also by other authors [68–74]. In [75, 76], the dynamic stability of Davydov solitons was established by numerical modeling. In [76], the velocity spectrum of stable solitons was found. The question of the existence and dynamics of autolocalization states in an α-helix protein molecule, consisting of three hydrogen bonded chains, was studied in [7, 8, 77, 78]. In this chapter we consider the features of the autolocalized state dynamics in inhomogeneous molecular chains and two-dimensional structures including β-layers of protein molecules.

Acoustic solitons in the anharmonic chain are stabilized by the balance between the effects of nonlinearity due to the positive intermolecular anharmonicity and the dispersion of acoustic waves caused by the chain discreteness [79]. In the continuum approximation, this dispersion can be taken into account if the expansion of difference derivatives with respect to molecule displacements is limited to terms with derivatives up to fourth order. As a result, for cubic anharmonicity, the Boussinesq equation is derived, and this admits only supersonic soliton solutions. These solutions (acoustic solitons) describe local compression regions moving along the chain. On the other hand, the Davydov soliton in a harmonic chain can propagate only with subsonic velocities. Therefore, the propagation of supersonic two-component solitons is expected to be possible in a molecular chain with intermolecular excitation if both intermolecular anharmonicity and acoustic wave dispersion are taken into account.

A soliton solution of the Schrödinger equation with the self-consistent potential satisfied by the Boussinesq equation was investigated in the series of studies [80–87], which are devoted to various applications. For example, in [80], devoted to the dynamics of near-sound Langmuir solitons, a supersonic solution is found, while it is asserted in [81] that supersonic soliton solutions do not exist at all. In [82, 83], this system of equations was solved using a non-self-consistent method, and as a result, solutions were found which describe, not autolocalization, but the capture of a quantum quasi-particle (external excess electron) by the supersonic acoustic soliton. The complete integrability of this system was investigated in [86].

The most comprehensive study of soliton solutions of the coupled Schrödinger and Boussinesq equations is given in [85, 87], where it has been suggested that two soliton modes can exist: the Davydov soliton (autolocalized state of a quantum quasi-particle) and a supersonic acoustic mode (coupled state of a quantum quasi-particle with an acoustic soliton). The dynamical stability of these solutions was investigated in [76]. The discrete equations of acoustic self-trapping in a context with intermolecular anharmonicity were studied in [88, 89], but the two-component soliton solutions with supersonic velocities were not examined. In this chapter we will consider in detail the coupled states of a quantum quasi-particle with an acoustic soliton.

An essential complement to the research discussed above was a study of the soliton-induced transfer of intermolecular vibrational energy in a situation where each chain molecule is in an external single-well potential which describes a chain substrate [90]. From the study of this model, it turned out that two stable states exist for a definite range of the coupling constant between chain molecules and substrate: exciton and soliton states separated by an energy barrier. The exciton state is stable for large values of the coupling constant, while the soliton state is stable for small values. This model can also be used to describe the dynamics of autolocalized states in a three-dimensional crystal, where the existence of a similar barrier was established by other methods [50, 51]. In this chapter we will obtain numerically a parameter range for which the stable soliton state of a quasi-particle exists.

The most convincing experimental evidence for the existence of solitons in proteins was obtained by looking at the absorption spectrum of crystalline acetanilide (ACN) [91]. The chains of hydrogen bonds in the ACN crystal have a surprising structural similarity to the chains of hydrogen bonds in the peptide groups of α-helix proteins [92]. This similarity was noticed by Careri and prompted him to experimental studies in the hope of finding new physical properties of ACN which might be of a certain biological interest. Experimentally, the amide-I excitation in crystalline ACN is observed as the maximum of infrared absorption near $1,667\,\mathrm{cm}^{-1}$. In addition, another anomalous band has been found $17\,\mathrm{cm}^{-1}$ below the maximum of the amide-I vibration. Analysing the experimental data, Careri, Scott, and others [91, 92] have concluded that this anomalous line is due to a new type of soliton which emerges as a result of the interaction between the amide-I vibration and the displacement of a hydrogen-bonded proton, but not the displacement of a PG as a whole. With increasing temperature, the magnitude of the second (soliton) maximum in the absorption spectrum decreases. This is accounted for by the destabilizing effect of thermal vibrations of peptide groups in the crystal.

Therefore, in order to describe the soliton transport of the amide-I excitation along the chain of hydrogen-bonded PGs, one must consider both the interaction of the intramolecular excitation with deformation of hydrogen bonds (acoustic vibrations) and the nonlinear interaction of the excitation with intrapeptide proton displacement (optical vibration of chain molecules).

A mathematical treatment of the modified soliton theory for the ACN crystal is similar to the original Davydov model which describes acoustic self-trapping of the intramolecular excitation. The corresponding equations, which can be called the discrete self-trapping equations, describe the interaction of the excitation with optical vibrations of a one-dimensional lattice [93]. In fact, they coincide with the equations obtained by Holstein in the study of one-dimensional polaron dynamics [94, 95]. Later, this model of the dynamics of quantum quasi-particles, nonlinearly interacting with optical lattice vibrations, was investigated in many other studies, e.g., see the reviews [39, 96] and the references therein. Davydov and Enolski have established the existence of stationary soliton excitations in the presence of the dispersion of optical vibrations [97]. It has been shown by numerical modeling that these solitons exist and that they are dynamically stable formations at any (even zero) magnitude of the dispersion of optical vibrations [98, 99].

A classical vibrational model of the interpeptide excitation dynamics, alternative to the quantum Davydov model, was suggested by Taneko [100–102]. This model can describe the transport of vibrational energy in the form of solitary waves (vibron solitons). The main condition for emergence of these waves, which are in essence the envelope solitons, is the interaction between high-frequency vibrations of the oscillators, modeling the intrapeptide C=O bonds, and one of the modes (acoustic [101] or optical [102]) of another one-dimensional lattice, viz., a PG chain. Autolocalization, induced by the intrinsic anharmonicity of the high-frequency oscillators, was studied by Kosevich and Kovalev [103], and this was subsequently used by Oraevsky and Sudakov to model the transport of vibrational energy in proteins [104].

The successful application of the concept of the Davydov soliton in biology depends on its stability at physiological temperatures $T \approx 300$ K. The influence of thermal motion of the chain molecules on the soliton properties was studied in a series of works [105–112, 116–118]. Davydov investigated the role of thermal effects analytically, using the theory he developed himself [105–107], which takes into account the impact of quantum fluctuations in the equilibrium positions of molecules and their thermal vibrations relative to the new equilibrium positions. It follows from his analysis that the soliton exists at physiological temperatures, but its degree of localization decreases with increasing temperature. Numerical simulation of the soliton dynamics carried out by the authors with quantum consideration of thermal vibrations has shown that thermal vibrations contribute to the localization of the intramolecular excitation, leading to an effective decrease in the resonant interaction of neighboring molecules [108–115]. A similar result was obtained by Cruseiro et al. [116].

On the other hand, Lomdahl and Kerr [117, 118] (see also the work of Laurence et al. [119]) have shown that classical consideration of thermal vibrations based on the Langevin equation leads to the instability of Davydov solitons at physiological temperatures. It follows from these results that the amplitude of thermal vibrations in the molecule is more than one order of magnitude greater than the amplitude of the chain's local deformation caused by intramolecular excitation during its autolocalization. Thermal vibrations induced in the molecule by random external forces (white noise) lead to a breakdown of the coupling between the excitation and the local deformation of the chain, thereby destroying the soliton.

Today, the question of the stability of the Davydov soliton to thermal fluctuations in protein macromolecules is still open. It is obvious that the interaction of the amide-I excitation with other, lower frequency intermolecular vibrations must increase the soliton stability. The frequencies of these vibrations significantly exceed acoustic phonon frequencies, so they are not completely thermalized. In this case, application of the Langevin equation requires the use of colored noise rather than white, which only leads to complete thermalization of low-frequency vibrations. The periods of optical vibrations are much shorter than the time required by the excitation to pass over one chain link, so the excitation is sensitive to an averaged magnitude of the vibrations, responding only to a change in the equilibrium positions of the vibrations. In this chapter, these issues will be discussed in detail.

7.1 Davydov Soliton

Let us consider autolocalized states of a quantum quasi-particle (the intramolecular amide-I excitation of an external electron) in molecular systems of hydrogen-bonded chains. Two of the most significant examples of such systems are the α-helix and β-sheet protein macromolecules. To describe their dynamics, we will use the semi-classical approximation, in which a quasi-particle is treated quantum-mechanically and the displacements of the system monomers and the intramonomer displacements of atoms (i.e., acoustic and optical vibrations of the molecular system) are considered within the framework of classical mechanics [26].

7.1.1 General Dynamical Equations: Semi-classical Approximation

One of the secondary structures of a protein is the α-helix. The interactions between amino acid residues are regular and determine the periodicity of the secondary structure. The stability of the α-helix is ensured by the hydrogen bonds between (NH)- and (CO)-groups of the main chain. The atom groups NHCO (the peptide group) form three regular chains of bonds. The α-helix has the form of a rod, and its inner part consists of a tightly twisted backbone with the radicals directed outwards. The charge distribution over the atoms involved in the peptide groups is such that each peptide group has dipole moment equal to approximately 3.5 D, directed along the hydrogen bond [26].

Let us consider a chain of hydrogen-bonded peptide groups HNCO of the α-helix protein molecule:

$$\cdots \text{H–N–C=O} \cdots \text{H–N–C=O} \cdots \text{H–N–C=O} \cdots .$$

To describe the dynamics of a quantum of the intramolecular amide-I excitation (vibration of the double valence bond C=O), the exciton Hamiltonian [3, 26] can be used:

$$H = \sum_n \left[\varepsilon_n B_n^+ B_n - J_n (B_n^+ B_{n-1} + B_{n-1}^+ B_n) \right] , \tag{7.1}$$

where ε_n is the energy of the amide-I excitation at the nth molecule (PG), depending on the relative displacements of the nth molecule, its neighbors, and intramolecular atom displacements, J_n is the energy of the resonant dipole–dipole interaction between the excitations at the nth and $(n + 1)$th molecules, depending generally on the relative distances between these molecules, and B_n^+ and B_n are the creation and annihilation operators of the excitation at the nth molecule, respectively, which satisfy the usual commutation relations

$$[B_n, B_m^+] = \delta_{nm} , \quad [B_n, B_m] = 0 , \quad [B_n^+, B_m^+] = 0 .$$

Hereafter, we will replace the magnitudes ε_n and J_n by the linear terms in their power series expansions with respect to small displacements:

$$\varepsilon_n = \varepsilon_0 + \chi_L(y_n - y_{n-1}) + \chi_R(y_{n+1} - y_n) + \chi_O z_n + \cdots , \tag{7.2}$$

$$J_n = J_n^0 - \chi_{D,n}(y_n - y_{n-1}) + \cdots , \tag{7.3}$$

where ε_0 is the energy of the amide-I excitation at a single molecule, J_n^0 is the energy of the resonant interaction between excitations at the $(n-1)$th and nth molecules in a non-deformed chain, $\chi_L \geq 0$ ($\chi_R \geq 0$) is the parameter of interaction between the excitation and deformation of the left (right) hydrogen bond of the molecule, y_n is the displacement of the nth molecule from its equilibrium position, χ_O is the parameter of interaction between the excitation and the intramolecular atom displacement z_n, and $\chi_{D,n}$ is the parameter (whose sign coincides with the sign of J_n^0) describing a decrease in the resonant interaction with increasing distance between the neighboring $(n-1)$th and nth molecules. The intrapeptide displacement of the hydrogen atom is usually considered as the intramolecular displacement of the PG. Then $z_n = y_n - x_n$, where x_n is the displacement of the hydrogen atom from its equilibrium position in the nth PG.

In the lattice representation, the wave function of a quantum of the intramolecular excitation can be written as

$$|\Psi\rangle = \sum_n \phi_n(t) B_n^+ |0\rangle , \tag{7.4}$$

where $|0\rangle$ is the ground state (the exciton vacuum) and $\phi_n(t)$ is the complex-valued probability amplitude for finding a quantum of excitation at the nth molecule. The normalization condition $\langle \Psi(t)|\Psi(t)\rangle = 1$ implies the normalization relationship for the state vector (7.4):

$$\sum_n |\phi_n(t)|^2 = 1 . \tag{7.5}$$

The time dependence of the wave function (7.4) is given by the Schrödinger equation

$$i\hbar \frac{\partial}{\partial t}|\Psi\rangle = H|\Psi\rangle ,$$

with the Hamiltonian (7.1). The displacements $y_n(t)$ and $z_n(t)$ in (7.2) and (7.3) are the classical variables. Thus, the following Lagrange function can be introduced:

$$\mathscr{L} = \left\langle \Psi \left| i\hbar \frac{\partial}{\partial t} - H \right| \Psi \right\rangle + \mathscr{L}_{\text{lat}} , \tag{7.6}$$

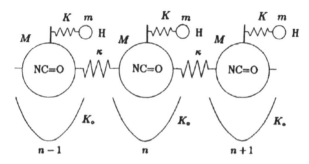

Fig. 7.1 Schematic model of the hydrogen bond chain HNC=O⋯HNC=O⋯HNC=O, where n is the number of the PG

where

$$\mathcal{L}_{\text{lat}} = \sum_n \left[\frac{1}{2} m \dot{x}_n^2 + \frac{1}{2} M_n \dot{y}_n^2 - V(y_{n+1} - y_n) - \frac{1}{2} K(y_n - x_n)^2 - \frac{1}{2} K_o y_n^2 \right].$$

Here, the dot denotes differentiation with respect to time t, while m and M_n are the reduced masses of hydrogen and the nth PG, respectively, and K is the elastic constant of the intrapeptide displacement of hydrogen atom. The Morse potential

$$V(\rho) = \epsilon_o \left[\exp(-b\rho) - 1 \right]^2$$

describes the hydrogen bond between neighboring PGs and ϵ_o is the energy of the hydrogen bond. For small deformation of the bond, we have $V(\rho) \approx \kappa \rho^2 / 2$, where $\kappa = V''(0)$ is the stiffness constant of the bond. The phenomenological parameter of the Morse potential is $b = \sqrt{\kappa / 2\epsilon_o}$. The interaction of the nth PG with the chain substrate is described by the potential $V_o(y_n) = K_o y_n^2 / 2$, where K_o is the stiffness coefficient of the interaction. The structure of the chain under consideration is shown schematically in Fig. 7.1.

Let us calculate the expectation value appearing in the Lagrangian function (7.6):

$$\left\langle \Psi \left| i\hbar \frac{\partial}{\partial t} - H \right| \Psi \right\rangle = \sum_n \phi_n^* (i\hbar \dot{\phi}_n - \varepsilon_n \phi_n + J_n \phi_{n-1} + J_{n+1} \phi_{n+1}).$$

Then, from the least action principle

$$\delta \int_{t_1}^{t_2} \mathcal{L}(\dots, \dot{\phi}_n, \phi_n, \dot{y}_n, y_n, \dot{x}_n, x_n, \dots) dt = 0,$$

we find three differential difference equations

$$i\hbar \dot{\phi}_n = \varepsilon_n \phi_n - (J_n \phi_{n-1} + J_{n+1} \phi_{n+1}), \tag{7.7}$$

$$M_n \ddot{y}_n = V'(y_{n+1} - y_n) - V'(y_n - y_{n-1}) - K(y_n - x_n) - K_o y_n$$

$$+ \chi_L(|\phi_{n+1}|^2 - |\phi_n|^2) + \chi_R(|\phi_n|^2 - |\phi_{n-1}|^2)$$
$$+ 2\chi_{D,n+1}\text{Re}(\phi^*_{n+1}\phi_n) - 2\chi_{D,n}\text{Re}(\phi^*_n\phi_{n-1}) - \chi_0|\phi_n|^2 , \qquad (7.8)$$

$$m\ddot{x}_n = K(y_n - x_n) + \chi_0|\phi_n|^2 , \qquad (7.9)$$

where the coefficients ε_n and J_n are given by (7.2) and (7.3).

The following momenta correspond to the coordinates ϕ_n, y_n, and x_n:

$$P^{\text{ex}}_n = \frac{\partial \mathscr{L}}{\partial \dot{\phi}_n} = i\hbar\phi^*_n , \quad P^{\text{ac}}_n = \frac{\partial \mathscr{L}}{\partial \dot{y}_n} = M_n\dot{y}_n , \quad P^{\text{op}}_n = \frac{\partial \mathscr{L}}{\partial \dot{x}_n} = m\dot{x}_n .$$

The equations of motion (7.7)–(7.9) can be obtained from the Hamiltonian function in the usual way:

$$\mathscr{H} = \sum_n (i\hbar\phi^*_n\dot{\phi}_n + M_n\dot{y}^2_n + m\dot{x}^2_n) - \mathscr{L}_{\text{lat}}$$

$$= \sum_n \phi^*_n(\varepsilon_n\phi_n - J_n\phi_{n-1} - J_{n+1}\phi_{n+1}) + \mathscr{H}_{\text{lat}} , \qquad (7.10)$$

where

$$\mathscr{H}_{\text{lat}} = \sum_n \left[\frac{1}{2}m\dot{x}^2_n + \frac{1}{2}M_n\dot{y}^2_n + V(y_{n+1} - y_n) + \frac{1}{2}K(y_n - x_n)^2 + \frac{1}{2}K_\circ y^2_n \right] .$$
$$(7.11)$$

It is convenient to write the Hamiltonian function (7.10) in the symmetric form

$$\mathscr{H} = \sum_n \left[\varepsilon_n\phi_n\phi^*_n - J_n(\phi^*_n\phi_{n-1} + \phi_n\phi^*_{n-1}) \right] + \mathscr{H}_{\text{lat}} . \qquad (7.12)$$

7.1.2 Soliton Dynamics in an Inhomogeneous Chain

Let us consider the dynamics of the intramolecular excitation in a single free molecular chain ($K_\circ = 0$) with an inhomogeneous distribution of masses and energies of the resonant interaction of the chain PGs. Excluding the interaction of the excitation with the intrapepride displacement and considering (7.2) and (7.3), the chain Hamiltonian function (7.12) can be written in the form

$$\mathscr{H} = \sum_n \left\{ |\phi_n|^2 \left[\varepsilon_0 + \chi_L(y_n - y_{n-1}) + \chi_R(y_{n+1} - y_n) \right] \right.$$
$$- (\phi_n\phi^*_{n-1} + \phi^*_n\phi_{n-1}) \left[J_n - \chi_{D,n}(y_n - y_{n-1}) \right]$$
$$\left. + \frac{1}{2}M_n\dot{y}^2_n + \frac{1}{2}\kappa(y_n - y_{n-1})^2 \right\} . \qquad (7.13)$$

Here, J_n and $\chi_{D,n}$ are now the energy of the resonant dipole–dipole interaction and the parameter of the resonant exciton–phonon interaction (EPI) of the nth and $(n-1)$th molecules, respectively. When measuring the energy from the bottom of the exciton band, the term in ε_0 can be omitted in (7.13).

The Hamiltonian function (7.13) gives the equations of motion

$$i\hbar\dot{\phi}_n = -J_{n+1}\phi_{n+1} - J_n\phi_{n-1} + \phi_n\big[\chi_R(y_{n+1} - y_n) + \chi_L(y_n - y_{n-1})\big]$$
$$+ \chi_{D,n}(y_n - y_{n-1})\phi_{n-1} + \chi_{D,n+1}(y_{n+1} - y_n)\phi_{n+1} \, . \tag{7.14}$$

$$M_n\ddot{y}_n = \kappa(y_{n-1} - 2y_n + y_{n+1}) + \chi_L(|\phi_{n+1}|^2 - |\phi_n|^2) + \chi_R(|\phi_n|^2 - |\phi_{n-1}|^2)$$
$$+ 2\chi_{D,n+1}\mathrm{Re}(\phi_n\phi_{n+1}^*) - 2\chi_{D,n}\mathrm{Re}(\phi_{n-1}\phi_n^*) \, . \tag{7.15}$$

The values of the PG masses M_n depend significantly on the distribution of amino acid residues in the α-helix molecule. The resonant interaction energy J_n depends weakly on the distribution of amino acid residues, but depends heavily on the relative orientation of the transition dipole moments of the PGs:

$$J_n = -\frac{d^2}{a^3}\Big[(\mathbf{e}_n, \mathbf{e}_{n-1}) - 3(\mathbf{l}, \mathbf{e}_n)(\mathbf{l}, \mathbf{e}_{n-1})\Big] \, ,$$

where $d = 0.35\,\mathrm{D}$ is the transition dipole moment of the amide-I excitation, $\mathbf{l}$ is the unit vector giving the direction along the chain, $a = 4.5 \times 10^{-10}$ m is the chain spacing, and $\mathbf{e}_n$ and $\mathbf{e}_{n-1}$ are the unit vectors giving the directions of the dipole moments of the nth and $(n-1)$th PGs, respectively. If all the dipole moments are directed along the chain, then $J_n = J = 2d^2/a^3 = 1.55 \times 10^{-22}$ J for all n [120, 121]. In general, $J_n = J(2\cos\alpha_{n-1}\cos\alpha_n - \sin\alpha_{n-1}\sin\alpha_n)/2$, where α_n is the angle between the vectors $\mathbf{e}_n$ and $\mathbf{l}$.

The excitation dynamics in an infinite chain is conveniently modeled in a finite cyclic chain consisting of N PGs. For this purpose, in the Hamiltonian (7.13) and the equations of motion (7.14) and (7.15), one must take $n = 1, 2, \ldots, N$ and put $n + 1 = 1$ for $n = N$ and $n - 1 = N$ for $n = 1$. The Davydov soliton dynamics were investigated numerically in a cyclic chain consisting of 100 and 200 PGs and the following values were taken for the parameters: $\kappa = 13$ N/m, average mass of the PGs $M = 190.7 \times 10^{-27}$ kg [7], and average tilt angle of the PG dipole moment $\alpha = 0$. The EPI parameters $\chi_L = 0$, $\chi_R = 5 \times 10^{-11}$ N, and $\chi_{D,n} = 3J_n/a$ were used. These were obtained from quantum mechanical calculations of the electronic structure of the formamide dimer (the EPI asymmetry is explained by the asymmetry in the distribution of the amide-I excitation inside the PG) [122].

Equations (7.14) and (7.15) were integrated numerically using the standard fourth-order Runge–Kutta method with constant grid spacing $\Delta t = \sqrt{M/k}/12 = 0.1$ ps [123]. The accuracy of the integration was estimated through the conservation of the integrals of motion for the system. For a given value of Δt, the total probability integral of (7.5) and the total energy integral (7.13) are conserved with an accuracy to seven and six significant figures, respectively. The initial condition was chosen to correspond to the exact solution of the nonlinear Schrödinger equation, which is the continuum approximation to the discrete equations of motion. The following initial conditions correspond to the Davydov soliton [2, 26, 43], located at the n_0th PG ($n_0 = 10$) and moving with velocity $V = a\sqrt{\kappa/M}/4 = 0.928 \times 10^3$ m/s:

$$\phi_n(0) = D_n^{-1/2} \frac{\exp[ik(n - n_0)]}{\cosh \xi_n} \, ,$$

$$y_n(0) - y_{n-1}(0) = \chi \frac{Q}{2\kappa(1 - s^2)\cosh^2 \xi_n} \, ,$$

$$\dot{y}_n(0) - \dot{y}_{n-1}(0) = \chi \frac{Q^2 s \sinh \xi_n}{\kappa(1 - s^2)\cosh^3 \xi_n} \, ,$$

where

$$\xi_n = Q(n - n_0) \, , \quad Q = -\frac{\chi^2}{4\kappa J(1 - s^2)} \, , \quad \chi = \chi_L + \chi_R + 6J/a \, ,$$

and $k = \arcsin(\hbar V/2aJ)$ is the wave number, $s = V/V_{\text{ph}} = 1/4$ is the dimensionless velocity, $V_{\text{ph}} = a\sqrt{\kappa/M}$ is the sound velocity, and $D_N = \sum_{n=1}^{N} 1/\cosh^2 \xi_n$ is the normalization coefficient. For the chosen parameters of the chain, the soliton width is approximately equal to ten chain spacings, thereby justifying the continuum approximation.

Numerical integration has shown that, in a homogeneous chain ($M_n = M, \alpha_n = 0, n = 1, 2, \ldots, N, N = 100$), the soliton moves with constant velocity $s = 0.21$, keeping its profile (it passes 422 PGs of the chain in $t = 240 \times 10^{-12}$ s). The soliton also moves along a chain with a random inhomogeneous mass distribution, where M_n is chosen from the interval $[M/2, 3M/2]$ (see Fig. 7.2a), as well as along a chain with a random inhomogeneous distribution of dipole orientation, where $M_n = M$ and the angles α_n are taken randomly from the interval $[-\beta, \beta]$, with $\beta = 5°$ and $\beta = 10°$. The soliton is stable to mass distribution inhomogeneity and does not collapse when moving into a chain region with different density. In a chain with $M_1 = M_2 = \ldots = M_{50} = M, M_{51} = M_{52} = \ldots = M_N = M'$ for $M' = 4M$ and $M' = M/4$, the passage of the soliton between two regions of different densities causes only insignificant periodic changes in its profile (soliton 'breathing'). The soliton is stable with respect to certain fixed PGs of a chain. It is important in this case that the distance between fixed PGs is not less than twice the soliton width (the soliton width is approximately equal to ten chain links, so it does not collapse when moving along a chain where every 20th PG is fixed, but it collapses when every 10th PG is fixed). The soliton does not collapse when passing two and even three fixed neighboring GPs. The stability of the Davydov soliton to inhomogeneity in the mass distribution is accounted for by its subsonic velocity, and the fact that the chain deformation energy of the soliton is a few hundredths of a percent of the

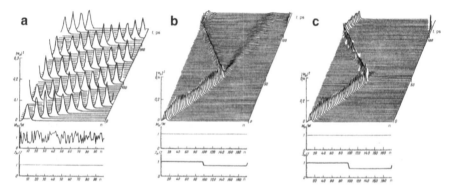

Fig. 7.2 Dynamics of the Davydov soliton in a chain with a random mass distribution (**a**). Decay of the soliton when encountering a boundary of chain regions with different dipole orientations (**b**), $\beta = 25°$. Soliton reflection from the boundary (**c**), $\beta = 30°$

intramolecular amide-I excitation energy, which means that its energy loss due to its energy loss due to inhomogeneities in the phonon system is insignificant.

The soliton is more sensitive to inhomogeneities in the distribution of the PG transition dipole moment orientation, which lead to an inhomogeneous distribution of the resonant interaction energy. In a chain with a random inhomogeneous distribution of the angles α_n, where $M_n \equiv M$ and $\{\alpha_n\}_{n=1}^{N=200}$ are chosen randomly from the interval $[-\beta, \beta]$, the soliton is already destroyed when $\beta \geq 15°$. When the soliton encounters a boundary of chain regions with different dipole orientations ($M_n = M$, $n = 1, 2, \ldots, N$, $N = 200$, $\alpha_1 = \alpha_2 = \ldots = \alpha_{100} = 0$, $\alpha_{101} = \alpha_{102} = \ldots = \alpha_{200} = \beta$), three cases are observed:

- The soliton does not collapse and keeps the same direction of motion (its velocity only decreases) for $\beta \leq 20°$.
- The soliton decays into two oscillating wave packets moving in opposite directions for $\beta = 25°$.
- The soliton reflects from the boundary of two regions with $\beta \geq 30°$ (see Fig. 7.2a, b).

When the soliton encounters a differently oriented PG (all $M_n = M$ and $\alpha_n = 0$ except $\alpha_{100} = \beta$, $n = 1, 2, \ldots, N$, $N = 200$), the same three cases are observed:

- The soliton passes through the inhomogeneity, not collapsing and losing only a small portion of its energy, when $\beta \leq 35°$.
- The soliton collapses when $\beta = 40°$.
- The soliton reflects from the inhomogeneity when $\beta \geq 45°$.

The stability of the Davydov soliton to inhomogeneities in the PG mass distribution in a chain makes it an effective carrier of energy along the α-helix segments of protein molecules. Note that supersonic acoustic solitons are unstable to these inhomogeneities [124–126]. They emit phonons passing through an inhomogeneity and can decay into several solitons moving with different velocities. Therefore, acoustic solitons, suggested as an alternative to the Davydov soliton [127, 128], cannot serve as an energy carrier over large distances along the α-helix segments of protein molecules. The sensitivity of the Davydov soliton to inhomogeneities in the distribution of the resonant interaction energy allows one to affect the soliton transport of energy by changing the local orientation of chain PGs. For example, the binding of barbiturate to an α-helical protein molecule should lead, not only to weakening of the hydrogen bond and an increase in the distance between two nearest neighbor PGs [129, 130], but also to a perturbation of their mutual orientation, which causes an additional decrease in the resonant interaction energy in this region of the α-helix molecule.

This dynamical modeling shows that the Davydov soliton is stable to inhomogeneities in the α-helix protein molecules and so can serve as an effective energy carrier. The results of this section were first obtained in [131, 132].

7.2 Autolocalization of the Excitations in a β-Sheet

Here we consider the transport of the intrapeptide excitation in a β-sheet structure of a protein molecule. In contrast to the α-helical structure, which is quasi-one-dimensional, the β-sheet structure is two-dimensional. In the continuum approximation, the excitation dynamics of the β-sheet structure are described by the two-dimensional Schrödinger equation. It is still not known [25] whether this equation has two-dimensional soliton solutions.

7.2.1 Discrete Model of Protein β-Sheet

The structure of a β-sheet protein molecule is shown in Fig. 7.3. The protein β-sheet is a polypeptide chain, folded in a zigzag plane structure that is stabilized by the hydrogen bonds between peptide groups. The PGs themselves form a two-dimensional lattice with two periods a and b. We will treat the β-sheet as a plane on which the chain of hydrogen-bonded PGs are laid parallel with period b (a is the period of the hydrogen bond chain, m is the chain number, and n is the number of the PG in the chain). Hereafter we will take into account the resonant interaction of the intrapeptide amide-I excitation only with the ten nearest PGs. The coefficients $C_1, C_2, \ldots, C_8$ in Fig. 7.3 correspond to the energies of the resonant interaction between the PG with numbers (m, n) and neighboring PGs.

There exist two β-sheet packings of a protein molecule: parallel and antiparallel. In the model under consideration, they differ only in the values of parameters b, C_1, $C_2, \ldots, C_8$. Parameter values for the parallel β-sheet **p** and antiparallel β-sheet **ap** are given in Table 7.1 [121, 133, 134].

Fig. 7.3 Structure of a β-sheet protein molecule. The polypeptide chain and hydrogen bonds between the peptide groups are shown by *solid* and *dashed lines*, respectively. *Ovals* and *arrows* indicate the directions of the PG dipole moments

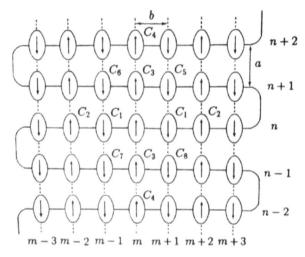

Table 7.1 Values of the parameters a and b (nm), and interaction coefficients $C_1, \ldots, C_8$ (cm^{-1}) for parallel **p** and antiparallel **ap** β-sheets

	a	b	C_1	C_2	C_3	C_4	C_5	C_6	C_7	C_8
p	0.473	0.334	1.0	2.3	−13.6	−1.7	−2.1	−2.1	2.1	2.1
ap	0.473	0.345	−0.1	1.8	−13.0	−1.6	15.3	−6.7	2.5	1.4

The total energy of a quantum of the amide-I excitation in the β-sheet can be written in the form

$$\mathcal{H} = \mathcal{H}_{ex} + \mathcal{H}_{int} + \mathcal{H}_{ph} . \tag{7.16}$$

Here, the energy of the resonant interaction of the excitation is

$$\mathcal{H}_{ex} = \varepsilon_0 + \sum_{m,n} 2\mathrm{Re}\left\{ \phi_{m,n}^* \left[C_1 \phi_{m+1,n} + C_2 \phi_{m+2,n} + C_3 \phi_{m,n+1} + C_4 \phi_{m,n+2} + \frac{1}{2} R_{m,n} \right] \right\}. \tag{7.17}$$

where m is the number of the hydrogen bond chain ($m = 0, \pm 1, \pm 2, \ldots$), n is the number of the PG in a chain ($n = 0, \pm 1, \pm 2, \ldots$),

$$R_{m,n} = C_5 \phi_{m+1,n+1} + C_6 \phi_{m-1,n+1} + C_7 \phi_{m-1,n-1} + C_8 \phi_{m+1,n-1} .$$

if m is an even number and

$$R_{m,n} = C_7 \phi_{m+1,n+1} + C_8 \phi_{m-1,n+1} + C_5 \phi_{m-1,n-1} + C_6 \phi_{m+1,n-1} .$$

if m is odd number, and $\varepsilon_0 = 0.205\,\mathrm{eV}$ is the energy of one quantum excitation in an isolated PG. The complex functions $\phi_{m,n}$ in the lattice representation specify the wave function of the excitation in the β-sheet. They are normalized by the condition

$$\sum_{m,n} |\phi_{m,n}|^2 = 1 , \tag{7.18}$$

where the magnitude $|\phi_{m,n}|^2$ gives the probability for finding a quantum of excitation in the PG with the numbers (m, n).

The energy of the EPI excitation has the form

$$\mathscr{H}_{\text{int}} = \sum_{m,n} |\phi_{m,n}|^2 \left[\chi_{m,1}(y_{m,n+1} - y_{m,n}) + \chi_{m,2}(y_{m,n} - y_{m,n-1}) \right] , \tag{7.19}$$

where $y_{m,n}$ is the displacement of hydrogen bonds along the chain from the equilibrium position of the PG with the numbers (m, n). If m is an even number, $\chi_{m,1} = \chi$ and $\chi_{m,2} = 0$, while if m is an odd number, $\chi_{m,1} = 0$ and $\chi_{m,2} = \chi$, where χ is the EPI parameter (hereafter the value of χ will be varied).

The deformation energy of the β-sheet has the form

$$\mathscr{H}_{\text{ph}} = \sum_{m,n} \left[\frac{M}{2} \dot{y}_{m,n}^2 + \frac{\kappa}{2}(y_{m,n+1} - y_{m,n})^2 \right] , \tag{7.20}$$

where M is the reduced mass of the PG and κ is the stiffness of the hydrogen bond ($M = 114.2 m_\text{p}$ [7], where m_p is the proton mass and $\kappa = 13\,\text{N/m}$).

The stationary state of this system can be obtained from the minimum total energy condition (7.16). By virtue of (7.19) and (7.20), necessary conditions for stationarity are

$$\dot{y}_{m,n} = 0 , \quad \frac{\partial \mathscr{H}}{\partial y_{m,n}} = \kappa(y_{m,n} - y_{m,n-1}) - \kappa(y_{m,n+1} - y_{m,n}) + f_{m,n} = 0 ,$$

where $f_{m,n} = \chi(|\phi_{m,n-1}|^2 - |\phi_{m,n}|^2)$ for even m and $f_{m,n} = \chi(|\phi_{m,n}|^2 - |\phi_{m,n+1}|^2)$ for odd m. It follows that $y_{m,n} - y_{m,n-1}$ is equal to $\chi|\phi_{m,n-1}|^2/\kappa$ for even m and $-\chi|\phi_{m,n}|^2/\kappa$ for odd m. Therefore, from (7.16), (7.19), and (7.20), we obtain

$$\mathscr{H} = \tilde{\mathscr{H}} = \mathscr{H}_{\text{ex}} - \frac{\chi^2}{2\kappa} \sum_{m,n} |\phi_{m,n}|^4 . \tag{7.21}$$

To find a stationary state, it is sufficient to solve the minimum problem for the reduced Hamiltonian $\tilde{\mathscr{H}}$ with the constraint (7.18)

$$\tilde{\mathscr{H}} \to \min : \sum_{m,n} |\phi_{m,n}|^2 = 1 . \tag{7.22}$$

If we consider the interaction of the excitation with other low-frequency intrapeptide vibrations [92, 93] (e.g., with the hydrogen atom) rather than with the longitudinal deformation of hydrogen bonds, the problem of finding the stationary state will have the same form (7.22). Indeed, in this case

$$\mathscr{H}_{\text{int}} = \sum_{m,n} \chi |\phi_{m,n}|^2 x_{m,n} , \qquad \mathscr{H}_{\text{ph}} = \sum_{m,n} \frac{1}{2} m_p \dot{x}_{m,n}^2 + \frac{1}{2} K y_{m,n}^2 ,$$

where $x_{m,n}$ is the normal coordinate of the vibration and K is the stiffness coefficient of the system. Necessary conditions for stationarity are

$$\dot{x}_{m,n} = 0 , \qquad \frac{\partial \mathscr{H}}{\partial x_{m,n}} = \chi |\phi_{m,n}|^2 + K x_{m,n} = 0 ,$$

which lead to the relations $x_{m,n} = -\chi |\phi_{m,n}|^2 / K$ and $\tilde{\mathscr{H}} = \mathscr{H}$.

Up to now we have been considering the infinite β-sheet. Below we analyse a finite β-sheet consisting of M chains, each of which contains N PGs ($m = 1, 2, \ldots, M$ and $n = 1, 2, \ldots, N$). We shall consider a β-sheet with free edges (in which case all terms in (7.17) with indices $m \pm k$ greater than M or less than 1, together with all terms with indices $n \pm k$ greater than N or less than 1 should be omitted) and a β-sheet with periodic boundary conditions (in which case the sums over $m \pm k$ and $n \pm k$ in (7.17) should be carried out modulo M and N, respectively).

7.2.2 Stationary States of Intrapeptide Excitation in β-Sheet Structures

Let us find the stationary excitation states in a β-sheet by the numerical method of steepest descent [123] for the minimum problem (7.22). We will consider a β-sheet with periodic boundary conditions and a β-sheet with free edges ($M = 20$ and $N = 20$). It is convenient to characterize the degree of excitation localization in the stationary state by the magnitude $D = \sum_{m=1}^{M} \sum_{n=1}^{N} |\phi_{m,n}|^4$, so that $D = 1$ if the excitation is localized at a single PG and $D = 1/MN$ if the excitation is uniformly distributed throughout the entire sheet. Here, we choose the zero level of the state energy as the energy ε_0 of one quantum excitation in an isolated PG and give it in units of 10^{-3} eV.

Numerical solution of the problem (7.22) has shown that, in the β-sheet with periodic boundary conditions, two types of stationary states exist, depending on the value of the EPI interaction: the exciton state (the excitation is uniformly distributed over the sheet) and the strongly localized soliton state (the excitation is virtually localized on one PG).

There are two exciton stationary states in the parallel β-sheet with periodic boundary conditions. For the first state,

$$\phi_{m,n} = (-1)^{\lfloor m/2 \rfloor} \frac{1 + (-1)^m}{\sqrt{2MN}} , \tag{7.23}$$

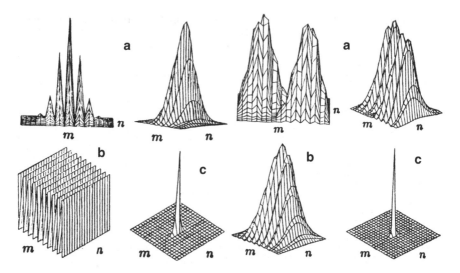

Fig. 7.4 *Left*: Stationary profiles of the excitation distribution in the parallel β-sheet ($M = N = 20$) with free edges for $\chi = 1.1 \times 10^{-10}$ N (**a**) and with periodic boundary conditions for $\chi = 1.1 \times 10^{-10}$ N (**b**) and $\chi = 1.15 \times 10^{-10}$ N (**c**). *Right*: Stationary profiles of the excitation distribution in the antiparallel β-sheet ($M = N = 20$) with free edges: the two-humped stationary state for $\chi = 1.45 \times 10^{-10}$ N (**a**), the ground stationary state for $\chi = 1.4 \times 10^{-10}$ N (**b**), and the ground stationary state for $\chi = 1.45 \times 10^{-10}$ N (**c**)

where $m = 1, 2, \ldots, M$, $n = 1, 2, \ldots, N$, $[p]$ is the integer part of p, $D = 2/MN$, and $E_1 = 2(-C_2 + C_3 + C_4) - \chi^2/\kappa MN$ is the state energy. In this case, the excitation is localized only on chains with even number m (see Fig. 7.4b left). For the second state,

$$\phi_{m,n} = 1/\sqrt{MN} \,, \tag{7.24}$$

where $D = 1/MN$ and the energy $E_2 = 2(C_1 + C_2 + C_3 + C_4) - \chi^2/2\kappa MN > E_1$. When $\chi \geq \chi_p = 1.15 \times 10^{-10}$ N, in the parallel β-sheet, along with the stationary exciton states discussed above, there exists a strongly localized stationary soliton state if the excitation is mainly localized on a single PG (Fig. 7.4c left). When $\chi < \chi_p$, this state was not found. The dependence of the stationary state energy on the parameter χ for the parallel β-sheet with periodic boundary conditions is given in Table 7.2.

In the antiparallel β-sheet, two stationary exciton states also exist. For the first,

$$\phi_{m,n} = (-1)^{[m/2]}/\sqrt{MN} \,, \tag{7.25}$$

and for the second,

$$\phi_{m,n} = (-1)^m/\sqrt{MN} \,, \tag{7.26}$$

Table 7.2 Dependence of the stationary state energy (given in 10^{-3} eV) on the value of χ (given in 10^{-10} N) for the parallel β-sheet with periodic boundary conditions for (a) the main exciton state (7.23), (b) the exciton state (7.24), and (c) the localized state

χ	0.0	1.0	1.1	1.15	1.2	1.3	1.4
a	−4.364	−4.376	−4.379	−4.380	−4.381	−4.384	−4.389
b	−2.976	−2.982	−2.983	−2.984	−2.984	−2.986	−2.987
c	−	−	−	−4.341	−4.496	−4.912	−5.426

Table 7.3 Dependence of the stationary state energy (given in 10^{-3} eV) on the EPI parameter χ (given in 10^{-10} N) for the antiparallel β-sheet with periodic boundary conditions for (a) the main exciton state (7.25), (b) the exciton state (7.26), and (c) the localized state

χ	0.0	1.0	1.4	1.45	1.5	1.6	1.7
a	−6.666	−6.672	−6.678	−6.678	−6.679	−6.681	−6.683
b	−4.699	−4.705	−4.711	−4.712	−4.713	−4.714	−4.716
c	−	−	−	−6.195	−6.445	−7.032	−7.706

where $m = 1, 2, \ldots, M$ and $n = 1, 2, \ldots, N$. The energy of the first state, viz.,

$$E_1 = 2(-C_2 + C_3 + C_4) - C_5 + C_6 + C_7 - C_8 - \chi^2/2\kappa MN ,$$

is less than the energy of the second, viz.,

$$E_2 = 2(-C_1 + C_2 + C_3 + C_4) - C_5 - C_6 - C_7 - C_8 - \chi^2/2\kappa MN .$$

When $\chi \geq \chi_p = 1.45 \times 10^{-10}$ N, there is also a soliton state (Fig. 7.4c right). The dependence of the stationary state energy on the EPI parameter χ is given in Table 7.3.

The soliton state is not always energetically more favorable than the exciton state. When $\chi = \chi_p$, the first exciton state turns out to be more energetically favorable, but with increasing χ, the soliton state becomes more favorable.

In a β-sheet with free edges, the stationary state has the exciton form for $\chi < \chi_p$ (see Fig. 7.4a (left) and a, b (right)) and the strongly localized form for $\chi > \chi_p$ (Fig. 7.4c (left) and c (right)).

When $\chi < \chi_p$, the stationary distribution of the excitation for the parallel β-sheet with free edges has a bell-shaped profile, touching the sheet edges, and the excitation is mainly localized at the chains with even number m (Fig. 7.4a left). The dependencies of the energy E and the localization degree D of the stationary excitation state on the EPI parameter χ for the parallel β-sheet with free edges are given in Table 7.4.

When $\chi < \chi_p$, the stationary distribution of the excitation for the antiparallel β-sheet with free edges also has a bell-shaped profile, touching the sheet edges, and the excitation is mainly localized at the chains with even number m (Fig. 7.4b left).

Table 7.4 Dependencies of the energy E (given in 10^{-3} eV) and the localization degree D of the stationary excitation state on the EPI parameter χ (given in 10^{-10} N) for the parallel β-sheet with free edges

χ	0.0	1.0	1.1	1.15	1.2	1.3	1.4
E	−4.301	−4.325	−4.332	−4.341	−4.496	−4.912	−5.426
D	0.008	0.013	0.015	0.463	0.613	0.753	0.828

Table 7.5 Dependencies of the energy E (given in 10^{-3} eV) and the localization degree D of the stationary excitation state on the EPI parameter χ (given in 10^{-10} N) for the antiparallel β-sheet with free edges

χ	0.0	1.0	1.4	1.45	1.5	1.6	1.7
E	−6.565	−6.579	−6.593	−6.195	−6.445	−7.032	−7.706
D	0.005	0.006	0.007	0.667	0.742	0.825	0.872

Table 7.6 Dependencies of the energy E (given in 10^{-3} eV) and the localization degree D of the ground stationary excitation state on the sheet width M for the β-sheet with free edges (χ values are given in 10^{-10} N)

	Parallel β-sheet				Antiparallel β-sheet			
	$\chi = 1.1$		$\chi = 1.15$		$\chi = 1.4$		$\chi = 1.45$	
M	E	D	E	D	E	D	E	D
1	−4.116	0.368	−4.245	0.558	−5.314	0.857	−5.611	0.878
5	−4.269	0.131	−4.339	0.473	−5.973	0.531	−6.185	0.676
9	−4.312	0.022	−4.341	0.468	−6.439	0.011	−6.190	0.667
13	−4.335	0.014	−4.341	0.467	−6.499	0.007	−6.190	0.667
21	−4.350	0.009	−4.341	0.467	−6.535	0.004	−6.190	0.667
29	−4.355	0.007	−4.341	0.467	−6.544	0.003	−6.190	0.667

Note that, in the antiparallel β-sheet with free edges, in addition to the one-humped stationary states, there are two-humped stationary states which are less energetically favorable (Fig. 7.4a right). The dependencies of the energy E and the localization degree D of the stationary excitation state on the EPI parameter χ for the antiparallel β-sheet with free edges are given in Table 7.5.

To study the dependence of the stationary state profile on the sheet width for the β-sheet with free edges, we considered the stationary states for the sheet with $N = 30$ and $M = 1, 5, 9, 13, 17, 21, 25$, and 29 when $\chi < \chi_p$ ($\chi = 1.1 \times 10^{-10}$ N for the parallel β-sheet and $\chi = 1.4 \times 10^{-10}$ N for the antiparallel one) and when $\chi = \chi_p$. The dependencies of the energy E and the localization degree D of the ground stationary state on its width M for the β-sheet with free edges are given in Table 7.6.

The investigation carried out in this section allows us to conclude that, with increasing sheet size, the stationary state profile becomes broader when $\chi < \chi_p$, but it does not come off the sheet edges. When $\chi \geq \chi_p$, the stationary state still remains highly localized and does not depend on the sheet width. Therefore, in the β-sheet with $\chi < \chi_p$, only delocalized exciton-like stationary states are possible, but when $\chi \geq \chi_p$, there is a highly localized stationary state. This property is a manifestation of the two-dimensionality of the system (in an isolated chain, stationary soliton states exist for any value of the parameter χ, and only the soliton width depends on χ).

7.2.3   Intrapeptide Excitation Dynamics in a β-Sheet

The dynamics of one quantum amide-I excitation in a β-sheet is described by the Hamiltonian equations

$$i\hbar\dot{\phi}_{m,n} = \frac{\partial \mathcal{H}}{\partial \phi_{m,n}} , \qquad M \ddot{y}_{m,n} = -\frac{\partial \mathcal{H}}{\partial y_{m,n}} , \qquad (7.27)$$

with the Hamiltonian (7.16), where $m = 1, 2, \ldots, M$ and $n = 1, 2, \ldots, N$.

Let us consider the excitation dynamics in a β-sheet with free edges ($M = 20$ and $N = 30$) at the initial induction of the excitation at one PG with numbers (m_0, n_0). For this purpose, one must solve the equations of motion (7.27) with the initial condition $\phi_{m_0,n_0}(0) = 1$, $\phi_{m,n}(0) = 0$ for $m \neq m_0$ or $n \neq n_0$, and $y_{m,n}(0) = 0$, $\dot{y}_{m,n}(0) = 0$ for all m and n.

Equations (7.27) were solved numerically by the standard fourth-order Runge–Kutta method [123]. Numerical integration has shown that, for values of the EPI parameter χ below a certain threshold value χ_p, the excitation propagates through the sheet as an oscillating, spreading two-dimensional wave packet. For $\chi \geq \chi_p$, the excitation hangs up at one PG and an immobile autolocalized stationary state is formed (see Fig. 7.5 left). The dependence of the threshold value χ_p on the position of the excited PG at the initial time is given in Table 7.7.

This behavior of the excitation is in good agreement with the results of a study of the stationary states. One can say that the soliton transport of the excitation can only occur in narrow β-sheets consisting of several chains of hydrogen bonds. In broad sheets, transport of the excitation can occur either by two-dimensional excitons or by spreading of two-dimensional wave packets. Therefore, quasi-one-dimensionality of molecular structure is a necessary condition for soliton transport of the intramolecular excitation.

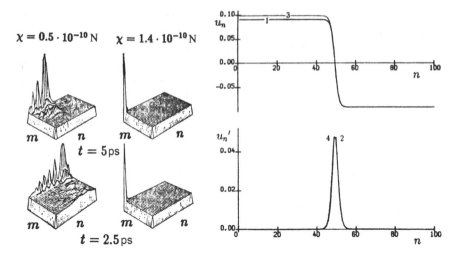

Fig. 7.5 *Left*: Excitation dynamics in a parallel β-sheet ($M = 20, N = 30$) with free edges ($m_0 = 20, n_0 = 1$). *Right*: Profiles of an acoustic soliton in a chain with cubic anharmonicity ($\gamma = 3$) at the initial time $\tau = 0$, obtained as the solution to the problem (7.41) with initial velocity $s = 1.05$ (*lines 1* and *2*) and at time $\tau = 1,911.9$ after the passage of 2,000 chain links (*lines 3* and *4*). The soliton velocity at the final time is $s = 1.0461$

Table 7.7 Dependence of the threshold value of the EPI parameter χ_p on the position of the initially excited PG (m_0, n_0)

	Parallel β-sheet				Antiparallel β-sheet			
m_0	10	10	1	20	10	10	1	20
n_0	1	15	15	1	1	15	15	1
$\chi_p(10^{-10})$ N	1.3	1.35	1.3	1.25	1.55	1.6	1.45	1.35

7.3 Soliton Dynamics of a Quasi-particle in an Anharmonic Chain

In this section we restrict ourselves to the dynamics of a quantum quasi-particle (a single quantum of intramolecular excitation of an extra electron) in a free anharmonic chain. We only take into account the interaction of the quasi-particle with acoustic vibrations of the molecular chain.

7.3.1 Acoustic Solitons

Let us consider a free anharmonic molecular chain [135–137]. Without taking into account the intramolecular displacements, the Hamiltonian function (7.11) has the form

$$\mathcal{H}_{\text{lat}} = \sum_n \left[\frac{1}{2} M' \dot{y}_n^2 + V(y_{n+1} - y_n) \right], \tag{7.28}$$

where $M' = M + m$ is the total molecular mass, y_n is the displacement of the nth molecule from its equilibrium position, and $V(r) = \kappa r^2/2 - \beta_1 r^3/3 + \beta_2 r^4/4 + \cdots$ is the anharmonic on-site interaction potential.

For convenience, we introduce the following dimensionless variables: displacement $u_n = y_n/a$ (a is the chain spacing), time $\tau = v_0 t/a$ ($v_0 = a\sqrt{\kappa/M'}$ is the sound velocity in the chain), and energy $E = \mathcal{H}/\kappa a^2$. Then the dimensionless Hamiltonian function of the chain (7.28) takes the form

$$E = \sum_n \left[\frac{1}{2} u'_n + U(u_{n+1} - u_n) \right], \tag{7.29}$$

where the prime denotes differentiation with respect to the dimensionless time τ and the dimensionless anharmonic potential is

$$U(u) = \frac{1}{\kappa a^2} V(au) = \frac{1}{2} u^2 - \frac{\gamma_1}{3} u^3 + \frac{\gamma_2}{4} u^4 - \cdots .$$

Here $\gamma_1 = \beta_1 a/\kappa$, $\gamma_2 = \beta_2 a^2/\kappa, \ldots$ are the dimensionless anharmonicity parameters.

The Hamiltonian (7.29) implies the discrete equations of motion

$$u''_n = F(u_{n+1} - u_n) - F(u_n - u_{n-1}), \quad n = 0, \pm 1, \pm 2, \ldots, \tag{7.30}$$

where $F(u) = dU/du = u - \gamma_1 u^2 + \gamma_2 u^3 - \cdots$. Let us search for a solution of (7.30) in the form of a traveling smooth solitary wave with constant profile (acoustic soliton). For this purpose, we set $u_n(\tau) = u(n - s\tau)$, where s is the dimensionless wave velocity and the function u depends smoothly on the discrete variable n. To use this continuum approximation, the parameter $\mu = \max_n |du(n)/dn|$ which describes the reciprocal soliton width, must be small. It is obvious that all derivatives are equal to zero: $d^m u/dn^m = O(\mu^m)$. Therefore, in the continuum approximation the following partial differential equation corresponds to the discrete equations (7.30):

$$0 = (1 - s^2)u_{xx} + \frac{1}{12} u_{xxxx} + \frac{1}{360} u_{xxxxxx}$$

$$- \gamma_1 \left(2u_x u_{xx} + \frac{1}{3} u_{xx} u_{xxx} + \frac{1}{6} u_x u_{xxxx} \right)$$

$$+ \gamma_2 \left(3u_x^2 u_{xx} + u_x u_{xxx} + \frac{1}{4} u_{xx}^2 + \frac{1}{4} u_x^2 u_{xxxx} \right) - 4\gamma_3 u_{xx} u_x^3 + O(\mu^7).$$

where x is the continuous variable approximating the discrete one n. Taking into account only terms up to the order of μ^5, this equation takes the form

$$(1 - s^2)u_{xx} + \frac{1}{12}u_{xxxx} - 2\gamma_1 u_x u_{xx} + 3\gamma_2 u_x^2 u_{xx} = 0 . \tag{7.31}$$

We now make the replacement $\rho = u_x$ in (7.31) and integrate once. Then, taking into account the boundary conditions $\rho, \rho_x, \rho_{xx} \to 0$ as $x \to \pm\infty$, we obtain the well known Boussinesq equation

$$(1 - s^2)\rho + \frac{1}{12}\rho_{xx} - \gamma_1\rho^2 + \gamma_2\rho^3 = 0 , \tag{7.32}$$

which has an exact solution in only two situations. Firstly, when $\gamma_2 = 0$ (cubic anharmonicity) and

$$\rho(x) = \frac{A}{\cosh^2(\mu x)} , \qquad A = -\frac{3(s^2 - 1)}{2\gamma_1} , \qquad \mu = \sqrt{3(s^2 - 1)} , \tag{7.33}$$

and secondly, when $\gamma_1 = 0$ (quartet anharmonicity) and

$$\rho(x) = \frac{A}{\cosh(\mu x)} , \qquad A = \pm\sqrt{\frac{2(s^2 - 1)}{\gamma_2}} , \qquad \mu = \sqrt{12(s^2 - 1)} . \tag{7.34}$$

The solution (7.33) describes the motion of the compression region ($A < 0$) of a constant profile with supersonic velocity ($s > 1$) along the chain. The solution (7.34) corresponds to the motion of the compression region ($A < 0$) and expansion region ($A < 0$) with supersonic velocity ($s > 1$).

For definiteness, we restrict ourselves below to the case of cubic anharmonicity ($\gamma_1 = \gamma > 0$, $\gamma_2 = 0$, $\gamma_3 = 0, \ldots$). In this case, the acoustic soliton is characterized by the amplitude A, the energy

$$E = \int_{-\infty}^{+\infty} \left[\frac{1}{2}(1 + s^2)\rho^2 - \frac{1}{3}\gamma\rho^3 \right] dx = \frac{1}{\mu}\left[\frac{2}{3}(1 + s^2)A^2 + \frac{16}{45}\gamma A^3 \right] , \tag{7.35}$$

the total compression of the chain

$$R = \int_{-\infty}^{+\infty} \rho\, dx = 2A/\mu , \tag{7.36}$$

and the root-mean-square diameter

$$L = 2\left(\frac{1}{R}\int_{-\infty}^{+\infty} x^2 \rho\, dx \right)^{1/2} = \pi/\sqrt{3}\mu , \tag{7.37}$$

given in units of the chain period.

7.3.2 Numerical Method for Finding Acoustic Soliton Profile

A soliton solution of the discrete equations of motion (7.30) can be found as the extremum of the Lagrangian

$$\mathcal{L} = \sum_n \left\{ -\frac{s^2}{24}[16(u_{n+1} - u_n)^2 - (u_{n+2} - u_n)^2] + U(u_{n+1} - u_n) \right\} , \qquad (7.38)$$

which is determined by the discrete equations

$$\mathcal{L}_{u_n} = \frac{s^2}{12}[16(u_{n+1} - 2u_n + u_{n-1}) - (u_{n+2} - 2u_n + u_{n-2})]$$

$$- F(u_{n+1} - u_n) + F(u_n - u_{n-1}) = 0 , \qquad n = 0, \pm 1, \pm 2, \dots . \quad (7.39)$$

In the continuum approximation, the discrete equations (7.39) coincide with (7.31) up to terms of the order of μ^5. Introducing the notation $u_{n+1} - u_n = \rho_n$, (7.39) takes the form

$$\mathcal{L}_{u_n} = \frac{s^2}{12}[16(\rho_n - \rho_{n-1}) - (\rho_{n+1} + \rho_n - \rho_{n-1} - \rho_{n-2})]$$

$$- F(\rho_n) + F(\rho_{n-1}) = 0 , \qquad n = 0, \pm 1, \pm 2, \dots . \quad (7.40)$$

An acoustic soliton corresponds to a saddle point of the Lagrangian, so a soliton solution of the discrete equations (7.40) should be sought numerically by minimizing the functional $\mathcal{F}_1$:

$$\mathcal{F}_1 = \frac{1}{2} \sum_{n=3}^{N-1} \mathcal{L}_{u_n}^2 \longrightarrow \min_{\rho_2,\dots,\rho_{N-1}} \quad : \rho_1 = 0 , \quad \rho_N = 0 , \qquad (7.41)$$

where N is the number of chain sites.

If one makes the replacement $u_{n+1} - u_n = \rho_n$ in the Lagrangian (7.38), it takes the form

$$\mathcal{L} = \sum_n \left\{ -\frac{s^2}{24}[16\rho_n^2 - (\rho_{n+1} + \rho_n)^2] + U(\rho_n) \right\} .$$

Then an extremum of the Lagrangian is defined by the discrete equations

$$\mathcal{L}_{\rho_n} = \frac{s^2}{12}(\rho_{n+1} - 14\rho_n + \rho_{n-1}) + F(\rho_n) = 0 , \qquad n = 0, \pm 1, \pm 2, \dots , \quad (7.42)$$

for which, in the continuum approximation, we can write the modified Boussinesq equation

$$(1 - s^2)\rho + \frac{s^2}{12}\rho_{xx} - \gamma\rho^2 = 0 .$$ (7.43)

This has the soliton solution

$$\rho(x) = \frac{A}{\cosh^2(\mu x)} , \quad A = -\frac{3(s^2 - 1)}{2\gamma} , \quad \mu = \sqrt{\frac{3(s^2 - 1)}{s^2}} .$$ (7.44)

It is convenient to solve the discrete equations (7.42) numerically by minimizing the functional $\mathscr{F}_2$:

$$\mathscr{F}_2 = \frac{1}{2}\sum_{n=2}^{N-1}\mathscr{L}_{\rho_n}^2 \longrightarrow \min_{\rho_2,\dots,\rho_{N-1}} : \rho_1 = 0 , \quad \rho_N = 0 .$$ (7.45)

The constrained minimum problems (7.41) and (7.45) were solved numerically by the conjugate gradient method [138]. For the minimization procedure, $N = 100$ chain links were taken. In the conjugate gradient method, initial points $\{\rho_n^0\}_{n=1}^N$ with $\rho_n^0 = 0$ for $n \neq N/2$ and $\rho_{N/2}^0 = -1/\gamma$ were chosen. The localized solution $\{\rho_n\}_{n=1}^N$ with the bell-shaped form corresponds to an acoustic soliton. The soliton is characterized by the energy

$$E = \sum_{n=1}^N\left[\frac{1}{2}s^2\rho_n^2 + U(\rho_n)\right] ,$$

the amplitude $A = \min_n \rho_n$, the total compression $R = \sum_{n=1}^N \rho_n$, the position of its centre $m = \sum_{n=1}^N n\rho_n/R$, and the root-mean-square diameter

$$L = 2\sqrt{\sum_{n=1}^N (m - n)^2\rho_n/R} ,$$

given in units of the chain period.

7.3.3 Modeling the Dynamics of the Acoustic Soliton

The acoustic soliton dynamics in an infinite chain are conveniently modeled numerically in a finite chain with free edges. This dynamics is described by the equations of motion

$$u''_1 = F(u_2 - u_1) ,$$

$$\vdots$$

$$u''_n = F(u_{n+1} - u_n) - F(u_n - u_{n-1}) , \tag{7.46}$$

$$\vdots$$

$$u''_M = -F(u_M - u_{M-1}) ,$$

with the energy integral

$$E = \sum_{n=1}^{M} \frac{1}{2} u'^2_n + \sum_{n=1}^{M-1} U(u_{n+1} - u_n) . \tag{7.47}$$

Equations (7.46) were solved numerically by the standard fourth-order Runge–Kutta method with the constant grid spacing $\Delta\tau$ [123]. The accuracy of the integration can be estimated through the conservation of the energy integral (7.47). For the value of $\Delta\tau = 0.1$ used in the calculation, the integral (7.47) is conserved to an accuracy of five significant figures throughout the time of numerical integration.

The molecule number M in the chain was taken equal to $N + 100$, where $N = 100$ is the number of sites used in the solution of the minimum problems (7.41) and (7.45). The following initial conditions correspond to the acoustic soliton:

$$u_n(0) = -\frac{R}{2} + \sum_{k=1}^{n} \rho_n , \quad u'_n(0) = -s\rho_n , \quad \text{for} \quad n = 1, 2, \dots, N ,$$

$$u_n(0) = \frac{R}{2} , \quad u'_n(0) = 0 , \quad \text{for} \quad n = N + 1, \dots, M . \tag{7.48}$$

With this choice of initial conditions, the soliton position is characterized by its centre, which is conveniently defined as the intersection point of the broken line sequentially connecting the points (n, u_n) $(n = 1, 2, \dots, M)$ with the n-axis.

To model the soliton dynamics in an infinite chain, we shift the soliton to the left through 100 chain sites as soon as it passes through 100 chain sites, i.e., we make the replacements $u_n(\tau) = u_{n+100}(\tau)$, $u'_n(\tau) = u'_{n+100}(\tau)$ for $n = 1, 2, \dots, N$ and $u_n(\tau) = u_N(0)$, $u'_n(\tau) = 0$ for $n = N + 1, \dots, M$. This method of numerical simulation of the dynamics allows us to integrate the equations of motion for the relatively small dimension $M = 200$, with a considerable saving in computer time. The method is especially efficient for analysis of the supersonic soliton dynamics. As a result of each shift, a subsonic phonon tail emitted by the soliton is cut off. Note that, when using a cyclic chain, the soliton moves in a background of phonons emitted by itself.

Table 7.8 Dependence of the soliton velocities s_1 and s_2, obtained in numerical simulation of the soliton dynamics, on the velocity value s_i used in the solution of the minimum problems (7.41) and (7.45)

s_i	1.01	1.02	1.03	1.04	1.05	1.06	1.07	1.08	1.09	1.10
s_1	1.0098	1.0194	1.0287	1.0376	1.0461	1.0540	1.0612	1.0679	1.0743	1.0806
s_2	1.0098	1.0193	1.0286	1.0375	1.0459	–	–	–	–	–

The results of the numerical minimization show that the problem (7.41) has a localized soliton solution only for velocities $1 < s \leq 1.05$ (if $s \geq 1.06$, the problem has no solution when a minimum value of the functional $\mathscr{F}_1 \approx 0$ is reached). The problem (7.45) has a localized solution with $\mathscr{F}_2 \approx 0$ for all $s > 1$. With increasing s, the accuracy of the soliton solution, obtained by minimizing the functionals $\mathscr{F}_1$ and $\mathscr{F}_2$, decreases. Table 7.8 presents the velocity values s_i used to solve the minimum problems (7.41) and (7.45), together with the velocity values s_1 and s_2 obtained by numerical simulation of the soliton dynamics with the initial conditions (7.48), which correspond to the solutions for these problems. The anharmonicity parameter $\gamma = 3$ is used in the simulation.

The decrease in accuracy of the obtained soliton solutions with increasing velocity s is accounted for by the increasing influence of chain discreteness on the dynamics. With increasing s, the soliton diameter $L \approx \pi/\sqrt{3}\mu$, where $\mu = \sqrt{3(s^2 - 1)}$, decreases proportionally to s^{-1}. For $s > 1.05$, the diameter already becomes $L < 3.27$, precluding the use of the continuum approximation. The method using the functional $\mathscr{F}_2$ is more stable to chain discreteness in comparison with the method using the functional $\mathscr{F}_1$. The dynamics of the acoustic soliton found by solving the problem (7.45) for $s = 1.05$ is shown in Fig. 7.5 (right). As can be seen from this figure, the numerical method for finding a soliton profile using the functional $\mathscr{F}_2$ slightly underestimates the soliton amplitude. Figure 7.6 (left) shows the dependencies of the energy E, diameter L, total chain compression R, and amplitude A of the acoustic soliton on s, as obtained by different methods: numerical simulation of the dynamics, solution of the modified Boussinesq equation (7.43), solution of the Boussinesq equation (7.32) ($\gamma_2 = 0$), and numerical minimization of the functional $\mathscr{F}_2$. As can be seen from this figure, the three latter methods, based on the use of the continuum approximation, give appropriate results only for $1 < s \leq 1.05$, when the soliton has sufficient width to justify use of the continuum approximation.

It follows from the foregoing numerical analysis that a much more efficient method for finding a soliton profile is the numerical method based on minimization of the functional $\mathscr{F}_2$, which allows a good approximation of the soliton profile when $1 < s \leq 1.05$. We now apply this method to find a bound state of the acoustic soliton with a quantum quasi-particle.

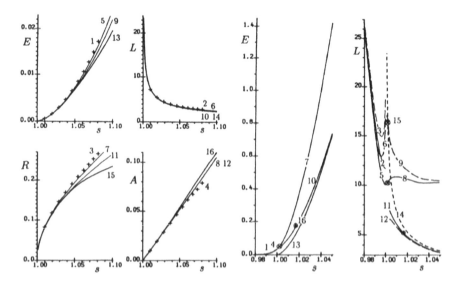

Fig. 7.6 *Left*: Dependencies of the energy E, diameter L, total compression of the chain R, and amplitude A on the velocity s in the chain with cubic anharmonicity ($\gamma_1 = 3$), as obtained by the following methods: numerical simulation of the dynamics (*lines 1–4*), solution of the modified Boussinesq equation (7.43) (*lines 5–8*), solution of the Boussinesq equation (7.32) (*lines 9–12*), and numerical minimization of the functional $\mathscr{F}_2$ (*lines 13–16*). *Right*: Dependencies of the energy E and diameters L_1 and L_2 on the velocity s for the following soliton types: (1) the soliton **d**, found using the Lagrangians $\mathscr{L}_1$ (*lines 1–3*) and $\mathscr{L}_e$ (*lines 4–6*), (2) the soliton a_1 (*lines 7–9*), (3) the soliton a_2 (*lines 10–12*), and (4) the purely acoustic soliton (*lines 13 and 14*). Points with markers (*15*) and (*16*) correspond to the solutions (7.64) and (7.67), respectively

7.3.4 Davydov Soliton and Bound States of a Quantum Quasi-particle with an Acoustic Soliton

Here we consider the dynamics of a quantum quasi-particle in a free anharmonic chain. Taking into account only the interaction of a quasi-particle with acoustic phonons, the Hamiltonian function of the chain (7.12) can be written in the form

$$\mathscr{H} = \sum_n \left\{ [\varepsilon_0 + \chi(y_{n+1} - y_{n-1})]\phi_n\phi_n^* - J(\phi_n^*\phi_{n+1} + \phi_n\phi_{n+1}^*) \right.$$

$$\left. + \frac{1}{2}M'\dot{y}_n^2 + V(y_{n+1} - y_n) \right\} . \tag{7.49}$$

Here, as before, $\{\phi_n\}_{n=-\infty}^{n=+\infty}$ is the complex-valued wave function of a quasi-particle satisfying the normalization condition (7.5), ε_0 is the energy of a quasi-particle located at one molecule of an undeformed chain, J is the energy of a quasi-particle hopping on a neighboring molecule (the same for all molecules), and χ is the EPI parameter (for definiteness, the symmetrical EPI is considered, so $\chi_L = \chi_R = \chi$).

For convenience, we introduce the dimensionless displacement $u_n = y_n/a$, time $\tau = v_0 t/a$, energy $E = \mathcal{H}/\kappa a^2$, and renormalized wave function $\psi_n = \sqrt{\alpha}\phi_n$, with $\alpha = \hbar/aM'v_0$ and $\hbar$ the Planck constant (see also the derivation of the dimensionless Hamiltonian (7.28)). Then the dimensionless Hamiltonian function (7.49) takes the form

$$
E = \sum_n \left\{ \left[\lambda_0 + \frac{1}{2}g(u_{n+1} - u_{n-1}) \right] \psi_n \psi_n^* - D(\psi_n^* \psi_{n+1} + \psi_n \psi_{n+1}^*) \right.
$$
$$
\left. + \frac{1}{2}u_n'^2 + U(u_{n+1} - u_n) \right\} ,   \tag{7.50}
$$

where $\lambda_0 = a\varepsilon_0/\hbar v_0$, $g = 2\chi a^2/\hbar v_0$, and $D = aJ/\hbar v_0$ are the dimensionless parameters, and $\{\psi_n\}_{n=-\infty}^{n=+\infty}$ is the wave function satisfying the normalization condition

$$
\sum_n |\psi_n|^2 = \alpha .   \tag{7.51}
$$

$U(u) = u^2/2 - \gamma u^3/3$ is the dimensionless potential of the intermolecular interaction with cubic anharmonicity and the prime denotes differentiation with respect to the dimensionless time τ.

The dimensionless Hamiltonian (7.50) yields the equations of motion

$$
i\psi_n' = \lambda_0 \psi_n - D(\psi_{n-1} + \psi_{n+1}) + \frac{1}{2}g(u_{n+1} - u_{n-1})\psi_n ,
$$
$$
u_n'' = F(u_{n+1} - u_n) - F(u_n - u_{n-1}) + \frac{1}{2}g(|\psi_{n+1}|^2 - |\psi_{n-1}|^2) ,   \tag{7.52}
$$

where $n = 0, \pm 1, \pm 2, \ldots$, and $F(u) = dU/du = u - \gamma u^2$. We now change from the absolute displacements u_n to the relative ones $\rho_n = u_{n+1} - u_n$, whence (7.52) takes the form

$$
i\psi_n' = \lambda_0 \psi_n - D(\psi_{n-1} + \psi_{n+1}) + \frac{1}{2}g(\rho_n + \rho_{n-1})\psi_n ,
$$
$$
\rho_n'' = F(\rho_{n+1}) - 2F(\rho_n) + F(\rho_{n-1})
$$
$$
+ \frac{1}{2}g(|\psi_{n+2}|^2 - |\psi_{n+1}|^2 - |\psi_n|^2 + |\psi_{n-1}|^2) ,   \tag{7.53}
$$

where $n = 0, \pm 1, \pm 2, \ldots$.

We represent the wave function of a quasi-particles in the form of the ansatz

$$
\psi_n(\tau) = \varphi_n(\tau) \exp \left\{ i[nk - (\lambda + \lambda_0 - 2D\cos k)\tau] \right\} ,   \tag{7.54}
$$

where $\varphi_n(\tau)$ is the real-valued wave function which, by virtue of (7.51), satisfies the normalization condition

$$\sum_n \varphi_n^2(\tau) = \alpha . \tag{7.55}$$

Then (7.53) takes the form

$$\varphi_n' = D(\varphi_{n-1} - \varphi_{n+1}) \sin k ,$$

$$\lambda \varphi_n = -D \cos k (\varphi_{n+1} - 2\varphi_n + \varphi_{n-1}) + \frac{1}{2} g(\rho_{n-1} + \rho_n)\varphi_n , \tag{7.56}$$

$$\rho_n'' = F(\rho_{n+1}) - 2F(\rho_n) + F(\rho_{n-1}) + \frac{1}{2} g(\varphi_{n+2}^2 - \varphi_{n+1}^2 - \varphi_n^2 + \varphi_{n-1}^2) ,$$

where $n = 0, \pm 1, \pm 2, \ldots$.
  In the continuum approximation,

$$\varphi_n(\tau) = \varphi(n - s\tau) = \varphi(\xi) , \qquad \rho_n(\tau) = \rho(n - s\tau) = \rho(\xi) ,$$

where s is the velocity of a wave with constant profile and ξ is the wave variable. Then the discrete equations (7.56) take the form of ordinary differential equations with respect to the wave variable

$$0 = (-s + 2D \sin k)\varphi_\xi , \tag{7.57}$$

$$\lambda \varphi = -D \cos k \, \varphi_{\xi\xi} + g\rho\varphi , \tag{7.58}$$

$$s^2 \rho_{\xi\xi} = \left(1 - \frac{1}{12} \frac{d^2}{d\xi^2}\right)^{-1} F(\rho)_{\xi\xi} + g(\varphi^2)_{\xi\xi} , \tag{7.59}$$

and the normalization condition (7.55) has the form

$$\int_{-\infty}^{+\infty} \varphi^2 d\xi = \alpha . \tag{7.60}$$

Equation (7.57) determines the wave number $k = \arcsin(s/s_e)$, where $s_e = 2D$ is the dimensionless maximum velocity of a quasi-particle. We seek a solution of (7.58) and (7.59) in the form of a solitary wave (soliton), i.e., we assume that the functions φ, ρ, and their derivatives satisfy the asymptotic behavior

$$\varphi , \; \varphi_\xi , \; \varphi_{\xi\xi} , \; \rho , \; \rho_\xi , \; \rho_{\xi\xi} \longrightarrow 0 , \quad |\xi| \to \infty . \tag{7.61}$$

We integrate (7.59) twice. Then, taking into account the boundary conditions (7.61), we get the equation

$$(1 - s^2)\rho + \frac{1}{12} s^2 \rho_{\xi\xi} - \gamma\rho^2 + g\varphi^2 = 0 . \tag{7.62}$$

Thus the functions ρ and φ are the solutions of the system comprising the stationary Schrödinger equation (7.58) and the modified Boussinesq equation (7.62), with the normalization condition (7.60) for the wave function φ. This system of equations is the Lagrange–Euler system for the constrained extremum problem:

$$\mathscr{L} = \int_{-\infty}^{+\infty} \left[-\sqrt{s_{\mathrm{e}}^2 - s^2}\varphi\varphi_\xi + \frac{1}{2}s^2\rho^2 + U(\rho) + g\rho\varphi^2 \right]\mathrm{d}\xi$$

$$\longrightarrow \text{ extremum}: \int_{-\infty}^{+\infty} \varphi^2\mathrm{d}\xi = \alpha . \tag{7.63}$$

Equations (7.58), (7.60), and (7.62) have two explicit solutions. The first is

$$\varphi(\xi) = \frac{\sqrt{3\mu_1\alpha/4}}{\cosh^2(\mu_1\xi)} , \qquad \rho(\xi) = \frac{A_1}{\cosh^2(\mu_1\xi)} , \tag{7.64}$$

with

$$A_1 = -3\mu_1^2\sqrt{s_{\mathrm{e}}^2 - s^2}/g , \qquad \mu_1 = \sqrt{3(s^2 - 1)/s^2} , \tag{7.65}$$

for the anharmonicity parameter

$$\gamma = \gamma_1^0 = \frac{g}{6\sqrt{s_{\mathrm{e}}^2 - s^2}} + \frac{\alpha g^3}{12(s_{\mathrm{e}}^2 - s^2)}\left[\frac{s^2}{3(s^2 - 1)} \right]^{3/2} . \tag{7.66}$$

The second solution is

$$\varphi(\xi) = \frac{\sqrt{\mu_2\alpha/2}}{\cosh(\mu_2\xi)} , \qquad \rho(\xi) = \frac{A_2}{\cosh^2(\mu_2\xi)} , \tag{7.67}$$

for another value of the anharmonicity parameter, viz.,

$$\gamma = \gamma_2^0 = \frac{gs^2}{2\sqrt{s_{\mathrm{e}}^2 - s^2}} . \tag{7.68}$$

The amplitude of this solution is

$$A_2 = -\mu_2^2\sqrt{s_{\mathrm{e}}^2 - s^2}g , \tag{7.69}$$

where the reciprocal width of the soliton μ_2 is a positive root of the cubic equation

$$\frac{s^2\mu^3}{3} + (1 - s^2)\mu - \frac{g^2\alpha}{2\sqrt{s_{\mathrm{e}}^2 - s^2}} = 0 , \tag{7.70}$$

which always has exactly one positive root for $s < s_{\mathrm{e}}$.

By virtue of the conditions (7.65) and (7.66), the first solution (7.64) exists only
for the single value of the anharmonicity parameter $\gamma = \gamma_1^0(s)$ $(1 < s < s_e)$. Due to
the conditions (7.68)–(7.70), the second solution (7.67) already exists already at the
value $\gamma = \gamma_2^0(s)$ of the anharmonicity parameter $(0 \leq s < s_e)$. Thus, both soliton
solutions (7.64) and (7.67) can describe supersonic soliton states of a quasi-particle.
These solutions were first found by Davydov and Zolotaryuk in [85,87], where they
put forward the hypothesis that there were two soliton modes of a quasi-particle in
an anharmonic chain: the Davydov mode with velocity spectrum $0 \leq s < s_e$ and the
supersonic acoustic mode (a bound state of a quasi-particle and an acoustic soliton)
with velocity spectrum $1 < s < s_e$.

It is convenient to characterize a two-component soliton by the energy

$$
E = \int_{-\infty}^{+\infty} \left[-\sqrt{s_e^2 - s^2}\, \varphi \varphi_\xi + \frac{1}{2}s^2 \rho^2 + U(\rho) + g\rho\varphi^2 \right] d\xi ,
$$

the root-mean-square diameter with respect to the first component, viz.,

$$
L_1 = 2 \left[\alpha^{-1} \int_{-\infty}^{+\infty} \xi^2 \varphi^2(\xi) d\xi \right]^{1/2} ,
$$

and the root-mean-square diameter with respect to the second component, viz.,

$$
L_2 = 2 \left[R^{-1} \int_{-\infty}^{+\infty} \xi^2 \rho(\xi) d\xi \right]^{1/2} ,
$$

where $R = \int_{-\infty}^{+\infty} r(\xi) d\xi$ is the total compression of the chain induced by the quasi-
particle soliton state. For the first soliton (7.64) (we denote the characteristics of this
soliton by the additional index $i = 1$), the energy is

$$
E_1 = -\frac{3}{4}\alpha\mu_1 + \frac{4}{5}g\alpha A_1 + \frac{A_1^2}{\mu_1}\left[\frac{2}{3}(1 + s^2) - \frac{16}{45}\gamma A_1 \right] , \tag{7.71}
$$

the first diameter is

$$
L_{1,1} = \sqrt{2(\pi^2/6 - 1)/\mu_1} , \tag{7.72}
$$

and the second diameter is

$$
L_{2,1} = \pi/\sqrt{3}\mu_1 . \tag{7.73}
$$

For the second soliton (7.67) (we also denote the characteristics of this soliton by
the additional index $i = 2$), the energy is

$$
E_2 = -\frac{1}{2}\alpha\mu_2 + \frac{2}{3}g\alpha A_2 + \frac{A_2^2}{\mu_2}\left[\frac{2}{3}(1 + s^2) - \frac{16}{45}\gamma A_2 \right] , \tag{7.74}
$$

and the diameters are

$$L_{1,2} = L_{2,2} = \pi/\sqrt{3}\mu_2 . \tag{7.75}$$

For the pure acoustic soliton (7.44), the energy E_0 and diameter L_0 are determined by (7.35) and (7.37), respectively.

For definiteness, we set $\alpha = 0.01$, $g = 1.0$, and $s_e = 2.0$. For these values, the system of two equations $\gamma = \gamma_1^0(s)$ and $\gamma = \gamma_2^0(s)$ has one solution $s = 1.002129$ and $\gamma = 0.29$. Thus, for $\gamma = 0.29$, (7.58), (7.60), and (7.62) have two supersonic two-component soliton solutions ($s = 1.002129$). For the first soliton, we found $E_1 = 0.052963$, $L_{1,1} = 10.065$, $L_{2,1} = 16.074$, and $A_1 = -0.0661171$, while for the second, we obtained $E_2 = 0.047728$, $L_{1,2} = L_{2,2} = 8.035$, and $A_2 = -0.0882036$, and for the pure acoustic soliton, we obtained $E_0 = 0.0057657$, $L_0 = 16.074$, and $A_0 = -0.0220475$.

As can be seen from these data, the amplitudes A_1 and A_2 of the two-component solitons significantly exceed, in absolute magnitude, the amplitude of the acoustic soliton A_0 and the diameters with respect to the second component $L_{2,1} = L_0$ and $L_{2,2} < L_0$. The energy difference $E_1 - E_2 = 0.0052353$ is approximately equal to the energy of one acoustic soliton. All this suggests that, for $s > 1$, the first soliton solution (7.64) describes the bound state of a quasi-particle with two acoustic solitons, and the second solution (7.67) describes the bound state of a quasi-particle with one acoustic soliton.

The quasi-particle interaction with one acoustic soliton causes an increase in the soliton amplitude, $|A_2| > |A_0|$, and a decrease in its diameter, $L_{2,2} < L_0$. The quasi-particle interaction with two acoustic solitons leads to soliton pairing, while in the absence of a quasi-particle, these solitons interact with each other as elastic particles. Together they form a broader ($L_{2,1} > L_{2,2}$), but less deep ($|A_1| < |A_2|$), potential well for a quasi-particle than is formed by one soliton. Numerical investigations confirmed this finding.

7.3.5 Numerical Methods for Finding Quasi-particle Autolocalized States in an Anharmonic Chain

The following Lagrangian corresponds to the dimensionless Hamilton function (7.50):

$$\mathscr{L} = \sum_n \left\{ i\psi_n^* \psi'_n - \left[\lambda_0 + \frac{1}{2} g(\rho_n + \rho_{n-1}) \right] \psi_n \psi_n^* \right.$$

$$\left. + D(\psi_n^* \psi_{n+1} + \psi_n \psi_{n+1}^*) + \frac{1}{2} u'^2_n - U(\rho_n) \right\} . \tag{7.76}$$

Substituting the ansatz (7.54) into the Lagrangian (7.76) and taking into account (7.55) and (7.56), we rewrite the Lagrangian in the form

$$
\mathscr{L} = -\sum_n \left[\frac{1}{2}\sqrt{s_e^2 - s^2}(\varphi_{n+1} - \varphi_n)^2 + \frac{1}{2}g(\rho_n + \rho_{n-1})\varphi_n^2 - \frac{1}{2}u_n'^2 + U(\rho_n) \right].
$$

(7.77)

Let us further assume that the discrete equations of motion (7.52) have smooth soliton solutions $\varphi_n(\tau) = \varphi(n - s\tau)$ and $u_n(\tau) = u(n - s\tau)$, where $s = 2D \sin k = s_e \sin k$. Then finally, making the replacement $u_n' = -s(u_{n+1} - u_n) = -s\rho_n$, the Lagrangian (7.77) can be represented in the form

$$
\mathscr{L}_1 = -\sum_n \left[\frac{1}{2}\sqrt{s_e^2 - s^2}(\varphi_{n+1} - \varphi_n)^2 + \frac{1}{2}g(\rho_n + \rho_{n-1})\varphi_n^2 - \frac{1}{2}s^2\rho_n^2 + U(\rho_n) \right].
$$

(7.78)

The quasi-particle localized state corresponds to the extremum of the Lagrangian $\mathscr{L}_1$, viz.,

$$
\mathscr{L}_1 \longrightarrow \text{extremum}: \sum_n \varphi_n^2 = \alpha ,
$$

(7.79)

which is defined by the Lagrange–Euler equations

$$
-\mathscr{L}_{1\,\varphi_n} = -\sqrt{s_e^2 - s^2}(\varphi_{n+1} - 2\varphi_n + \varphi_{n-1}) + g(\rho_{n-1} + \rho_n)\varphi_n - \lambda\varphi_n = 0,
$$

(7.80)

$$
-\mathscr{L}_{1\,\rho_n} = (1 - s^2)\rho_n - \gamma\rho_n^2 + \frac{1}{2}g(\varphi_n^2 + \varphi_{n+1}^2) = 0,
$$

(7.81)

where the Lagrange multiplier λ is found from the condition $\sum_n \mathscr{L}_{1\,\varphi_n}\varphi_n = 0$:

$$
\lambda = \alpha^{-1}\sum_n \left[\sqrt{s_e^2 - s^2}(\varphi_{n+1} - \varphi_n)^2 + g(\rho_{n-1} + \rho_n)\varphi_n^2 \right].
$$

From (7.81), we have

$$
\rho_n = \frac{1}{2\gamma}\left[1 - s^2 - \sqrt{(1 - s^2)^2 + 2g\gamma(\varphi_n^2 + \varphi_{n+1}^2)} \right].
$$

(7.82)

As $\varphi_n \to 0$, the chain deformation $\rho_n \to 0$ only for $s \leq 1$. Therefore, using the Lagrangian $\mathscr{L}_1$ allows us to obtain only subsonic solitons.

For a harmonic chain ($\gamma = 0$), it follows from (7.81) that

$$
\rho_n = -g\frac{\varphi_n^2 + \varphi_{n+1}^2}{2(1 - s^2)} ,
$$

and with this, (7.80) thus reduces to the discrete Schrödinger equation

$$\sqrt{s_e^2 - s^2}(\varphi_{n+1} - 2\varphi_n + \varphi_{n-1}) + \frac{g^2}{2(1-s^2)}(\varphi_{n-1}^2 + 2\varphi_n^2 + \varphi_{n+1}^2)\varphi_n + \lambda\varphi_n = 0 \, .$$

In the continuum approximation, this equation goes over into the well known nonlinear Schrödinger equation (NSE)

$$\sqrt{s_e^2 - s^2}\varphi_{\xi\xi} + \frac{2g^2}{(1-s^2)}\varphi^3 + \lambda\varphi = 0$$

with the wave function φ satisfying the normalization condition (7.60). The width of the soliton solution of the NSE (Davydov soliton)

$$\varphi(\xi) = \frac{\sqrt{\mu\alpha/2}}{\cosh(\mu\xi)} \, , \quad \mu = \frac{\alpha g^2}{2(1-s^2)\sqrt{s_e^2 - s^2}} \, ,$$

tends to zero as $s \to 1$ (the soliton exists only for $s < 1$). Considering anharmonicity ($\gamma > 0$) leads to an increase in the soliton width, and it will then also exist for $s \leq 1$.

A soliton solution of the discrete equations (7.80) and (7.81) corresponds to the minimum of the function $-\mathscr{L}_1$, so this solution can be sought numerically by solving the constrained minimum problem using the descent method:

$$-\mathscr{L}_1 = \sum_n \left[\frac{1}{2}\sqrt{s_e^2 - s^2}(\varphi_{n+1} - \varphi_n)^2 + \frac{1}{2}g(\rho_n + \rho_{n-1})\varphi_n^2 - \frac{1}{2}s^2\rho_n^2 + U(\rho_n)\right]$$

$$\longrightarrow \min : \sum_{n=1}^N \varphi_n^2 = \alpha \, , \tag{7.83}$$

where $n + 1 = 1$ for $n = N$ and $n - 1 = N$ for $n = 1$. The number of sites N used in the calculation is chosen experimentally. N should be chosen in such a way that the periodic boundary conditions do not affect the soliton shape. To meet this requirement, it suffices to take N ten times greater than the soliton diameter.

The minimum of the function $-\mathscr{L}_1$ can be sought by the more time consuming method of sequential self-consistency for (7.80) and (7.81) ($n = 1, 2, \ldots, N$) [64, 72]. Putting $\varphi_n = 0$ for $n \neq N/2$ and $\varphi_{N/2} = \sqrt{\alpha}$ at the initial stage of the self-consistency procedure, we find the vector $\{\rho_n\}_{n=1}^N$ and then the vector $\{\varphi_n\}_{n=1}^N$ as eigenvectors of the tridiagonal symmetric $N \times N$ matrix A with diagonal elements

$$A_{n,n} = 2\sqrt{s_e^2 - s^2} + g(\rho_{n-1} + \rho_n)$$

and subdiagonal elements

$$A_{n,n+1} = A_{n+1,n} = -\sqrt{s_e^2 - s^2} \, ,$$

which corresponds to the smallest eigenvalue. Then once again, using $\{\varphi_n\}_{n=1}^N$, we find $\{\rho_n\}_{n=1}^N$ from (7.82) and the new value of $\{\varphi_n\}_{n=1}^N$, and so on, until complete self-consistency for (7.80) and (7.81) is achieved, i.e., when the sum $\sum_{n=1}^N (\mathscr{L}_{1\varphi_n}^2 + \mathscr{L}_{1\rho_n}^2)$ reaches its minimum value. Complete self-consistency for the equations usually requires 30–40 iterations.

The Lagrangian $\mathscr{L}_1$ does not take into account the acoustic phonon dispersion, so it does not allow one to obtain supersonic solitons by the minimization procedure. For this purpose, as shown in Sect. 7.3.3, the following Lagrangian should be used:

$$\mathscr{L}_2 = -\sum_n \left[\frac{1}{2}\sqrt{s_e^2 - s^2}(\varphi_{n+1} - \varphi_n)^2 + \frac{1}{2}g(\rho_n + \rho_{n-1})\varphi_n^2 \right.$$
$$\left. - \frac{1}{2}s^2\rho_n^2 - \frac{1}{24}s^2(\rho_{n+1} - \rho_n)^2 + U(\rho_n) \right], \qquad (7.84)$$

which has extrema that satisfy the Lagrange–Euler equations

$$-\mathscr{L}_{2\varphi_n} = -\sqrt{s_e^2 - s^2}(\varphi_{n+1} - 2\varphi_n + \varphi_{n-1}) + g(\rho_{n-1} + \rho_n)\varphi_n - \lambda\varphi_n = 0,$$
$$\qquad (7.85)$$

$$-\mathscr{L}_{2\rho_n} = -s^2\rho_n + \frac{1}{12}s^2(\rho_{n+1} - 2\rho_n + \rho_{n-1}) + \frac{1}{2}g(\varphi_n^2 + \varphi_{n+1}^2) + U(\rho_n) = 0.$$
$$\qquad (7.86)$$

In the continuum approximation, these discrete equations coincide with the coupled Schrödinger and Boussinesq equations (7.58) and (7.62), with the normalization condition (7.60) for the wave function.

When $s > 1$, the soliton solution of the discrete equations (7.85) and (7.86) corresponds to a saddle point of the Lagrangian (7.84). Hence, it is impossible to find a soliton solution by minimizing the Lagrangian $-\mathscr{L}_2$. Here, the self-consistency method can be used for (7.85) and (7.86). The vector $\{\varphi_n\}_{n=1}^N$ is most advantageously sought by solving the constrained minimum problem numerically using the simple method of gradient descent:

$$\sum_{n=1}^N \left[\frac{1}{2}\sqrt{s_e^2 - s^2}(\varphi_{n+1} - \varphi_n)^2 + \frac{1}{2}g(\rho_n + \rho_{n-1})\varphi_n^2 \right] \longrightarrow \min_{\varphi_1,\dots,\varphi_N} : \sum_{n=1}^N \varphi_n^2 = \alpha.$$

In order to find the displacements $\{\rho_n\}_{n=1}^N$, we can use the conjugate gradient method [138] to solve the minimum problem

$$\frac{1}{2}\sum_{n=1}^N \mathscr{L}_{2\rho_n} \longrightarrow \min_{\rho_1,\dots,\rho_N}.$$

To find a bound state of a quasi-particle with two acoustic solitons, it is convenient to use the solution $\{\varphi_n, \rho_n\}_{n=1}^N$ of the problem (7.83) with $s = 1$ as starting point for

the self-consistency procedure (it can be found by minimizing the functional $\mathscr{L}_1$).
To find a bound state with one acoustic soliton, the solutions

$$\left\{\varphi_n = c/\cosh\left[\mu_0(n - N/2)\right], \; \rho_n = A_0/\cosh^2\left[\mu_0(n - N/2)\right]\right\}_{n=1}^{N}$$

can be used, where $A_0 = 3(1 - s^2)/2\gamma$, $\mu_0 = \sqrt{3(s^2 - 1)/s^2}$, and the constant c is
determined from the normalization condition (7.55). The self-consistency procedure
is considered complete when the functional

$$\mathscr{F} = \frac{1}{2}\sum_{n=1}^{N}(\mathscr{L}_{2\,\varphi_n}^2 + \mathscr{L}_{2\,\rho_n}^2)$$

reaches its minimum.

7.3.6 Two-Component Soliton Dynamics

The accuracy of the two-component solitons can be evaluated by simulating their
dynamics numerically. Let us consider the soliton motion in a finite chain with free
ends, where the soliton dynamics is determined by the equations of motion

$$i\psi_1' = -\frac{1}{2}s_e\psi_2 + \frac{1}{2}g(u_2 - u_1)\psi_1 ,$$

$$u_1'' = F(u_2 - u_1) + \frac{1}{2}g(|\psi_2|^2 + |\psi_1|^2) ,$$

$$\vdots$$

$$i\psi_n' = -\frac{1}{2}s_e(\psi_{n+1} + \psi_{n-1}) + \frac{1}{2}g(u_{n+1} - u_{n-1})\psi_n , \qquad (7.87)$$

$$u_n'' = F(u_{n+1} - u_n) - F(u_n - u_{n-1}) + \frac{1}{2}g(|\psi_{n+1}|^2 - |\psi_{n-1}|^2) ,$$

$$\vdots$$

$$i\psi_M' = -\frac{1}{2}s_e\psi_{M-1} + \frac{1}{2}g(u_M - u_{M-1})\psi_M ,$$

$$u_M'' = -F(u_M - u_{M-1}) - \frac{1}{2}g(|\psi_{M-1}|^2 + |\psi_M|^2) ,$$

with the integral of total probability

$$P = \alpha^{-1}\sum_{n=1}^{M}|\psi_n|^2 \qquad (7.88)$$

and the integral of the total energy

$$E = -\frac{1}{2}s_e \sum_{n=1}^{M-1}(\psi_n\psi_{n+1}^* + \psi_n^*\psi_{n+1})$$

$$+\frac{1}{2}g\left[\sum_{n=1}^{M-1}(u_{n+1}-u_n)|\psi_n|^2 + \sum_{n=2}^{M}(u_n - u_{n-1})|\psi_n|^2\right]$$

$$+\sum_{n=1}^{M}\frac{1}{2}u_n' + \sum_{n=1}^{M-1}U(u_{n+1}-u_n) .$$ (7.89)

Equations (7.87) were solved numerically using the standard fourth-order Runge–Kutta method with constant grid spacing $\Delta\tau$ [123]. The accuracy of the numerical integration was estimated through the conservation of the integrals of motion (7.88) and (7.89). For the values $\alpha = 0.01$, $g = 1.0$, $\gamma = 0.3$, and $s_e = 2.0$, the grid spacing $\Delta\tau = 0.005$ was used. This ensures conservation of the integrals of motion to an accuracy of seven significant figures during the total time of the numerical simulations.

The quasi-particle dynamics was considered in a chain consisting of $M = N + 50$ molecules, where N (the site number used in the self-consistency procedure for (7.85) and (7.86)) was taken approximately ten times greater than the soliton diameter. The following initial conditions correspond to the two-component soliton

$$\psi_n(0) = \varphi_n^0 \exp(ink) , \quad u_n(0) = \frac{1}{2}R + \sum_{k=1}^{n}\rho_k^0 , \quad u_n'(0) = -s\rho_n^0 ,$$

for $n = 1, 2, \ldots, N$, and

$$\psi_n(0) = 0 , \quad u_n(0) = -\frac{R}{2} , \quad u_n'(0) = 0 ,$$

for $n = N + 1, \ldots, M$, where $R = -\sum_{n=1}^{N}\rho_n^0$ and $\{\varphi_n^0, \rho_n^0\}_{n=1}^{N}$ is the extremum point of the Lagrangian (7.84) which corresponds to the soliton. With this choice of the initial conditions, the soliton position in the chain is characterized by its centre, which is conveniently defined as an intersection point of the broken line, sequentially connecting the points (n, u_n), $n = 1, 2, \ldots, M$, with the n-axis. At the initial time, the soliton is centred at $m = N/2$. To model the soliton dynamics in an infinite chain, we shift the soliton to the left through 50 sites as soon as it passes through 50 chain sites to the right, i.e., we make the replacements $\psi_n(\tau) = \psi_{n+50}(\tau)$, $u_n(\tau) = u_{n+50}(\tau)$, and $u_n'(\tau) = u'_{n+50}(\tau)$ for $n = 1, 2, \ldots, N$, and $\psi_n(\tau) = 0$, $u_n(\tau) = u_N(0)$, and $u_n'(\tau) = 0$ for $n = N + 1, \ldots, M$.

7.3.7 Results of Numerical Investigation

The two-component soliton $\{\varphi_n^0, \rho_n^0\}_{n=1}^N$, found by the method of self-consistency for (7.85), is characterized by the energy

$$
E = \sum_{n=1}^N \left[\frac{1}{2}\sqrt{s_e^2 - s^2}(\varphi_{n+1}^0 - \varphi_n^0)^2 + \frac{1}{2}g(\rho_n^0 + \rho_{n-1}^0)\varphi_n^{0^2} + \frac{1}{2}s^2\rho_n^{0^2} + U(\rho_n^0) \right],
$$

the centre of the distribution of the quasi-particle wave function

$$
m_1 = \sum_{n=1}^N n\varphi_n^{0^2}/\alpha ,
$$

the centre of the chain deformation distribution

$$
m_2 = \sum_{n=1}^N n\rho_n^0/R , \qquad R = \sum_{n=1}^N \rho_n^0 ,
$$

the root-mean-square diameter with respect to the first component

$$
L_1 = 2\left[\sum_{n=1}^N (m_1 - n)^2 \varphi_n^{0^2}/\alpha \right]^{1/2} ,
$$

and the diameter with respect to the second component

$$
L_2 = 2\left[\sum_{n=1}^N (m_2 - n)^2 \rho_n^0/R \right]^{1/2} .
$$

Numerical solution of the discrete equations (7.85) and (7.86) ($N = 200$, $g = 1.0$, $\alpha = 0.01$, $\gamma = 0.3$, and $s_e = 2.0$) has shown the existence of three soliton modes: the Davydov mode $\mathbf{d}$ (subsonic self-trapping mode of a quasi-particle), the first supersonic acoustic mode a_1 (the bound state of a quasi-particle with two acoustic solitons), and the second supersonic acoustic mode a_2 (the bound state of a quasi-particle with one acoustic soliton). The dependencies of E, L_1, and L_2 on the soliton velocity s for each mode are shown in Fig. 7.6 (right). Numerical modeling of the dynamics has shown that all three soliton modes are dynamically stable.

The Davydov mode has velocity spectrum $0 \leq s \leq 1$. The Davydov soliton $\mathbf{d}$ moves along the chain with constant velocity and unchanged shape for $s = 1$. The profile of the soliton after passing over 1,000 links coincides completely with the initial profile (see Fig. 7.7 left).

The soliton of the first acoustic mode has a two-humped profile with respect to each component. This two-humped profile is barely noticeable for s close to 1

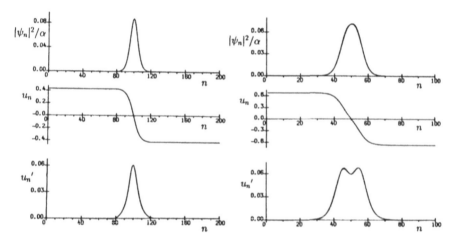

Fig. 7.7 *Left*: Profile of the soliton **d** for $s = 1.0$ at the initial time $\tau = 0$ and time $\tau = 1,000.05$, after passing over 1,000 chain links. *Right*: Profile of the soliton a_1 for $s = 1.004$ at the initial time $\tau = 0$ and time $\tau = 995.825$, after passing 1,000 chain links

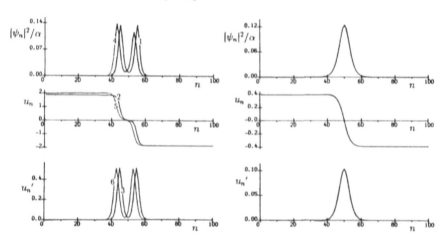

Fig. 7.8 *Left*: Profile of the soliton a_1 at the initial time $\tau = 0$ (*lines 1–3*) and at $\tau = 954.05$ (*lines 4–6*), after passing over 1,000 chain links at $s = 1.05$. At the final time the soliton has velocity $s = 1.046$. *Right*: Profile of the soliton a_2 at the initial time $\tau = 0$ and at $\tau = 996.05$ after passing over 1,000 chain links at $s = 1.004$

(Fig. 7.7 right), but it becomes more pronounced with increasing velocity (Fig. 7.8 left). This can be accounted for only by the pairing of two acoustic solitons due to their interaction with a quasi-particle. The solitons of this mode are dynamically stable. At the initial velocity $s = 1.004$, the soliton a_1 moves with constant velocity and keeps its shape. Its profile at the initial time $\tau = 0$ coincides completely with its profile at time $\tau = 995.825$, after passing over 1,000 chain links (Fig. 7.7 right). With increasing velocity s, the acoustic soliton diameter decreases, and

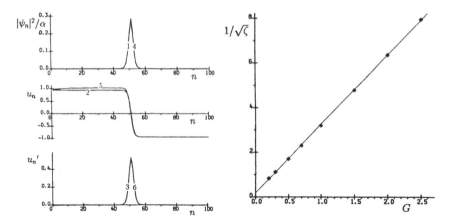

Fig. 7.9 *Left*: Profile of the soliton a_2 at the initial time $\tau = 0$ (*lines 1–3*) and at time $\tau = 956.05$ (*lines 4–6*), after passing over 1,000 chain links at $s = 1.05$. At the final time the soliton has the velocity $s = 1.046$. *Right*: Dependence of the threshold $\zeta_p^{-1/2}$ on G found numerically (*points*) and its linear approximation $\zeta_p^{-1/2} = a_1 + a_2 G$ (*line*) found by the method of least squares ($a_1 = 0.16606$ and $a_2 = 3.08087$)

this leads to decreasing accuracy of the soliton solutions obtained by the self-consistency procedure for the discrete equations (7.85) and (7.86). At the initial velocity $s = 1.05$, the soliton a_1 moves along the chain keeping its shape, but its velocity becomes $s = 1.046$ (Fig. 7.8 left).

The soliton of the mode a_2 has a one-humped profile with respect to both components (Figs. 7.8 right and 7.9 left). The solitons are dynamically stable. For $s = 1.004$, the soliton a_2 propagates along the chain keeping its shape. Its profile at the initial time $\tau = 0$ coincides completely with its profile at time $\tau = 996.05$, after passing over 1,000 chain links (Fig. 7.8 right). For $s = 1.05$, the soliton moves with velocity $s = 1.046$, keeping its shape (Fig. 7.9 left).

The difference between the energies of the solitons a_1 and a_2 at all velocities is approximately equal to the energy of one pure acoustic soliton (see Fig. 7.6 right). This supports the idea that the solitons a_1 and a_2 are the bound states of a quasi-particle with two and one acoustic solitons, respectively.

For $\gamma = 0.3$, $\alpha = 0.01$, $g = 1.0$, and $s_e = 2$, (7.66) has solution $s = 1.002057$. At this velocity, by virtue of (7.65) and (7.71)–(7.73), the two-component soliton corresponding to the solution (7.64) has energy $E = 0.0503459$ and diameters $L_1 = 10.238809$ and $L_2 = 16.351806$. Equation (7.68) has the solution $s = 1.01658$. At this velocity, by virtue of (7.69), (7.70), (7.74), and (7.75), the two-component soliton corresponding to the solution (7.67) has energy $E = 0.1742091$ and diameters $L_1 = L_2 = 5.2087173$. The values of E, L_1, and L_2 obtained for the solution (7.64) lie exactly on the curves $E(s)$, $L_1(s)$, and $L_2(s)$ corresponding to the soliton a_1, while those obtained for the solution (7.67) lie on the curves $E(s)$, $L_1(s)$, and $L_2(s)$ corresponding to the soliton a_2 (see Fig. 7.6 right). Therefore, the solution (7.64) corresponds to the soliton mode a_1, the solution (7.67) for $s > 1$, i.e., for $\gamma > g/2\sqrt{s_e^2 - 1}$, corresponds to the soliton mode a_2, while for $s \leq 1$, i.e., for $\gamma \leq g/2\sqrt{s_e^2 - 1}$, it refers to the soliton mode $\mathbf{d}$. Note that the mode $\mathbf{d}$ at $s = 1$ changes directly to the mode a_1, and there is a gap in the velocity spectrum between the modes $\mathbf{d}$ and a_2 (see Fig. 7.6 right).

The Lagrangian $\mathcal{L}_1$ for $s \leq 0.98$ gives almost the same profile for the soliton $\mathbf{d}$ as the Lagrangian $\mathcal{L}_2$. The difference appears only in the range of $s = 1$, when consideration of acoustic phonon dispersion is already required (Fig. 7.6 right). Therefore, use of the Lagrangian $\mathcal{L}_2$ is justified only for transonic and supersonic velocities. Thus, for $s_e < 1$, when only solitons $\mathbf{d}$ with velocity spectrum $0 \leq s < s_e$ exist, it makes sense to use only the Lagrangian $\mathcal{L}_1$ and search for the soliton profile numerically by solving the constrained minimum problem (7.83). We will apply this method to further investigate Davydov solitons in a chain with substrate.

The results of this section were published in [139, 140].

7.4   Quasi-particle Autolocalization in a Chain with Substrate

Let us consider the quantum quasi-particle dynamics in a harmonic chain with an external substrate. When we take into account the quasi-particle interaction with acoustic phonons alone, the Hamiltonian function of the chain given by (7.11) and (7.12) has the form

$$\mathcal{H} = \sum_n \left\{ [\varepsilon_0 + \chi(y_{n+1} - y_{n-1})]\phi_n \phi_n^* - J(\phi_n^* \phi_{n+1} + \phi_n \phi_{n+1}^*) \right.$$
$$\left. + \frac{1}{2}M'\dot{y}_n^2 + \frac{\kappa}{2}(y_{n+1} - y_n)^2 + \frac{K_o}{2}y_n^2 \right\} . \tag{7.90}$$

Here, as before, $\{\phi_n\}_{n=-\infty}^{n=+\infty}$ is the complex-valued quasi-particle wave function which satisfies the normalization condition (7.5), ε_0 is the energy of a quasi-particle, located at one molecule of the non-deformed chain, J is the energy of the quasi-particle hopping onto a neighboring molecule (the same for all molecules), χ is the EPI parameter (for definiteness, the symmetrical EPI is considered), M' is the total mass, κ is the stiffness coefficient of the intermolecular interaction, and K_o is the stiffness coefficient of the interaction between the chain molecules and the substrate.

The Hamiltonian function (7.90) yields the equations of motion

$$\left. \begin{array}{l} i\hbar\dot{\phi}_n = -J(\phi_{n-1} + \phi_{n+1}) + [\varepsilon_0 + \chi(y_{n+1} - y_{n-1})]\phi_n, \\[2mm] M'\ddot{y}_n = \kappa(y_{n+1} - 2y_n + y_{n-1}) - K_o y_n + \chi(|\phi_{n+1}|^2 - |\phi_{n-1}|^2), \end{array} \right\} \quad n = 0, \pm 1, \pm 2, \ldots .$$
$$\tag{7.91}$$

For convenience, we introduce the dimensionless displacement $u_n = \kappa y_n / 2\chi$ and time $\tau = v_0 t / a = t\sqrt{\kappa/M'}$. Then (7.91) takes the form

$$
\left.
\begin{aligned}
i\phi_n' &= -D(\phi_{n-1} + \phi_{n+1}) + \left[\lambda_0 + \frac{1}{2}g(u_{n+1} - u_{n-1})\right]\phi_n, \\
u_n'' &= u_{n+1} - 2u_n + u_{n-1} - \eta u_n + \frac{1}{2}(|\phi_{n+1}|^2 - |\phi_{n-1}|^2), \\
&\quad u_n + \frac{1}{2}(|\phi_{n+1}|^2 - |\phi_{n-1}|^2),
\end{aligned}
\right\} \quad n = 0, \pm 1, \pm 2, \ldots ,
$$

$$(7.92)$$

where

$$
D = \frac{J}{\hbar\sqrt{\kappa/M'}} = \frac{aJ}{\hbar v_0}, \quad
\lambda_0 = \frac{\varepsilon_0}{\hbar\sqrt{\kappa/M'}} = \frac{a\varepsilon_0}{\hbar v_0}, \quad
g = \frac{4\chi^2}{\kappa\hbar\sqrt{\kappa/M'}} = \frac{4\chi^2 a}{\kappa\hbar v_0},
$$

and $\eta = K_0/\kappa$ are the dimensionless coefficients, and the prime denotes differentiation with respect to the dimensionless time τ.

We represent the quasi-particle wave function in the form of the ansatz

$$
\phi_n(\tau) = \varphi_n(\tau)\exp\left\{i[nk - (\lambda + \lambda_0 - 2D\cos k)\tau]\right\} , \tag{7.93}
$$

where $\{\varphi_n(\tau)\}_{n=-\infty}^{+\infty}$ is the real-valued wave function which, by virtue of (7.5), satisfies the normalization condition

$$
\sum_n \varphi_n^2 = 1 . \tag{7.94}
$$

Then (7.92) takes the form

$$
\varphi_n' = D(\varphi_{n-1} - \varphi_{n+1})\sin k ,
$$

$$
\lambda\varphi_n = -D(\varphi_{n+1} - 2\varphi_n + \varphi_{n-1})\cos k + \frac{1}{2}g(u_{n+1} - u_{n-1})\varphi_n , \tag{7.95}
$$

$$
u_n'' = u_{n+1} - 2u_n + u_{n-1} - \eta u_n + \frac{1}{2}(\varphi_{n+1}^2 - \varphi_{n-1}^2) ,
$$

for $n = 0, \pm 1, \pm 2, \ldots..$ In the continuum approximation, we have

$$
\varphi_n(\tau) = \varphi(n - s\tau) = \varphi(\xi) , \qquad u_n(\tau) = u(n - s\tau) = u(\xi) ,
$$

where s is the dimensionless velocity of the wave with constant profile, ξ is the wave variable, and the derivatives are

$$
u_n'' = s^2\frac{d^2u}{dn^2} = s^2(u_{n+1} - 2u_n + u_{n-1}) ,
$$

$$\varphi_n{}' = -s\frac{d\varphi}{dn} = -\frac{s}{2}(\varphi_{n+1} - \varphi_{n-1}) .$$

Therefore the discrete equations (7.95) can be rewritten in the form

$$s = 2D \sin k ,$$

$$\lambda\varphi_n = -D(\varphi_{n+1} - 2\varphi_n + \varphi_{n-1})\cos k + \frac{1}{2}g(u_{n+1} - u_{n-1})\varphi_n , \qquad (7.96)$$

$$0 = (1 - s^2)(u_{n+1} - 2u_n + u_{n-1}) - \eta u_n + \frac{1}{2}(\varphi_{n+1}^2 - \varphi_{n-1}^2) ,$$

for $n = 0, \pm 1, \pm 2, \ldots$.

Let us introduce new dimensionless displacements $v_n = (1 - s^2)u_n$. Then the discrete equations (7.96) can be rewritten as

$$\left.\begin{aligned}
G(\varphi_{n+1} - 2\varphi_n + \varphi_{n-1}) + (v_{n+1} - v_{n-1})\varphi_n + \tilde{\lambda}\varphi_n = 0 , \\
-v_{n+1} + 2v_n - v_{n-1} + \zeta v_n + \frac{1}{2}(\varphi_{n-1}^2 - \varphi_{n+1}^2) = 0 ,
\end{aligned}\right\} \quad n = 0, \pm 1, \pm 2, \ldots ,$$

$$(7.97)$$

where the new dimensionless coefficients are

$$G = \frac{2D(1 - s^2)\cos k}{g} = \frac{\kappa\hbar(v_0^2 - v^2)\sqrt{v_{\mathrm{ex}}^2 - v^2}}{4a\chi^2 v_0^2} ,$$

$$\zeta = \frac{K_\circ}{\kappa(1 - s^2)} , \qquad \tilde{\lambda} = -\frac{2\lambda(1 - s^2)}{g} . \qquad (7.98)$$

Here, $v_0 = a\sqrt{\kappa/M'}$ is the sound velocity in the chain, $v_{\mathrm{ex}} = 2aJ/\hbar$ is the maximum quasi-particle velocity, and $v = sv_0$ is the soliton velocity.

The system of discrete equations (7.97) is the Lagrange–Euler system for the constrained minimum problem

$$\mathscr{F} = \frac{1}{2}\sum_n \left[G(\varphi_{n+1} - \varphi_n)^2 + (v_{n+1} - v_{n-1})\varphi_n^2 + (v_{n+1} - v_n)^2 + \zeta v_n^2\right]$$

$$\longrightarrow \min : \quad \sum_n \varphi_n^2 = 1 . \qquad (7.99)$$

The Lagrange multiplier $\tilde{\lambda}$ is found from the normalization condition (7.94), whence $\tilde{\lambda} = -\sum_n \left[G(\varphi_{n+1} - \varphi_n)^2 + (v_{n+1} - v_{n-1})\varphi_n^2\right]$, since $\sum_n \mathscr{F}_{\varphi_n}\varphi_n = 0$.

Therefore, a soliton solution to the equations of motion (7.92) can be found numerically by solving the constrained minimum problem for a finite cyclic chain consisting of N molecules (7.99) using a descent method. We define the soliton

Table 7.9 Dependencies of ζ_p and $L_p(\zeta_p)$ on the parameter G

G	0.2	0.3	0.5	0.7	1.0	1.5	2.0	2.5
ζ_p	1.460	0.802	0.346	0.190	0.0979	0.044	0.025	0.016
$L(\zeta_p)$	4.57	5.70	6.76	8.63	12.3	14.8	19.1	23.1

diameter $\{\varphi_n^0, v_n^0\}_{n=1}^N$ by the equation $L = 2\left[\sum_n (m-n)^2 \varphi_n^{0\,2}\right]^{1/2}$, where $m = \sum_n n\varphi_n^{0\,2}$ is the soliton centre. To obtain the soliton solution, one must take the number N of molecules in the chain to be ten times greater than the soliton diameter L (in this case, the periodic boundary conditions do not influence the soliton shape).

Numerical solution of the problem (7.98) has shown that soliton solutions exist only for parameter values ζ less than the threshold value $\zeta_p(G)$, which depends on the parameter G. For $\zeta \leq \zeta_p(G)$, a stable localized soliton solution exists, while for $\zeta > \zeta_p(G)$, only the stable exciton solution $\varphi_n^0 \equiv 1/\sqrt{N}$, $v_n^0 \equiv 0$ exists. Note that, in a chain without substrate ($\zeta = 0$), the soliton solution exists for all values of the parameter G. Introducing the substrate ($\zeta > 0$) increases the soliton diameter $L(\zeta)$ and decreases the energy of the quasi-particle coupling with chain deformation. With increasing ζ, $L(\zeta)$ increases steadily as well. The values of ζ_p and $L_p(\zeta_p)$ for specific values of the parameter G are given in Table 7.9.

Consider now the position of the points $\left(G, 1/\sqrt{\zeta_p(G)}\right)$ found on the plane $(G, 1/\sqrt{\zeta})$. Using the method of least squares, we obtain the line which gives the best fit for the function $1/\sqrt{\zeta_p(G)}$. As can be seen from Fig. 7.9 (right), this line $1/\sqrt{\zeta_p(G)} = a_1 + a_2 G$, with $a_1 = 0.16606$ and $a_2 = 3.080865$, passes through almost all these points. Therefore, the Davydov soliton in the chain with the substrate, i.e., the soliton solution of (7.92), exists only if

$$(3.080865G + 0.16606)^2 \zeta \leq 1, \tag{7.100}$$

where the dimensionless parameters G and ζ are defined according to (7.98). The dependence of the root-mean-square diameter of the soliton on the parameters G and ζ in the soliton existence range is shown in Fig. 7.10 (left). Note that all the results obtained are also valid for a chain with non-symmetric EPI, and in this case one must put $\chi = (\chi_L + \chi_R)/2$ in (7.98).

Consider now the dynamics of one quantum of the intramolecular amide-I excitation in a cyclic chain with a substrate where the chain dynamics is described by the equations of motion (7.91) with $n = 1, 2, \ldots, N$ ($n + 1 = 1$ for $n = N$ and $n - 1 = N$ for $n = 1$). We choose the parameter values corresponding to a chain of hydrogen-bonded peptide groups in the α-helix protein molecule: $\varepsilon_0 = 0.22\,\text{eV}$, $a = 4.5 \times 10^{-10}\,\text{m}$, $J = 1.55 \times 10^{-22}\,\text{J}$, $M' = 114.2m_p$, $\kappa = 13\,\text{N/m}$, $\chi = 0.4 \times 10^{-10}\,\text{N}$, where m_p is the proton mass. At these values, the sound velocity $v_0 = 3{,}712.5\,\text{m/s}$ is greater than the maximum velocity of excitation motion $v_{ex} = 1{,}322.9\,\text{m/s}$ ($s_{ex} = v_{ex}/v_0 = 0.356338$). A soliton solution will be found numerically, solving the constrained minimum problem (7.99) by a simple descent method.

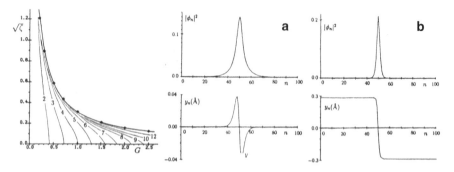

Fig. 7.10 *Left*: Dependence of the root-mean-square diameter of the soliton L on the dimensionless parameters $\sqrt{\zeta}$ and G in the range of soliton existence, which lies under the hyperbola $\sqrt{\zeta} = 1/(a_1 + a_2 G)$ (*thick line*). *Points* show the values found numerically. *Thin lines* show the level lines of the function $L(G, \sqrt{\zeta})$, and numbers on the lines give the function values. *Right*: Profile of the soliton in the chain with the substrate ($K_o = 5$ N/m) (**a**) and without ($K_o = 0$) (**b**)

The condition (7.100) determines the values of K_o for which the Davydov soliton exists. The stationary soliton ($s = 0, v = 0$) exists for $K_o \leq 2.93$ N/m, and the soliton moving with velocity $v = 928.1$ m/s ($s = 0.25$) exists for $K_o \leq 5.70$ N/m. The profile of the soliton with dimensionless velocity $s = 0.25$ is shown in Fig. 7.10 (right) when $K_o = 5.6$ N/m and, for comparison, when $K_o = 0$. The soliton motion in the free chain with free ends ($K_o = 0$) leads to displacement of the whole chain, in contrast to the motion in a chain with substrate ($K_o > 0$) (a local deformation with constant profile moves along the chain).

The soliton dynamics in an infinite chain is conveniently modeled in a finite cyclic chain consisting of $N = 100$ molecules. This dynamics is described by the dimensionless equations of motion

$$i\phi_n' = -D(\phi_{n-1} + \phi_{n+1}) + \frac{1}{2}g(u_{n+1} - u_{n-1})\phi_n ,$$

$$u_n'' = u_{n+1} - 2u_n + u_{n-1} - \eta u_n + \frac{1}{2}(|\phi_{n+1}|^2 - |\phi_{n-1}|^2) - \gamma u_n' .$$

$$(7.101)$$

where $n = 1, 2, .., N$ ($n + 1 = 1$ for $n = N$ and $n - 1 = N$ for $n = 1$), γ is the coefficient of viscous friction ($\tau_0 = 1/\gamma$ is the dimensionless time of velocity relaxation in the phonon subsystem). The following initial conditions of the system correspond to the soliton (7.101):

$$\phi_n(0) = \varphi_n^0 \exp(ink) , \quad u_n(0) = \frac{v_n^0}{1 - s^2} , \quad u_n'(0) = -s\left[u_{n+1}(0) - u_n(0)\right] ,$$

where $n = 1, \ldots, N$, s is the dimensionless soliton velocity, $\{\varphi_n^0, v_n^0\}_{n=1}^N$ is the soliton solution of the constrained minimum problem (7.99), and $k = \arcsin(s/s_{\text{ex}})$ is the wave number.

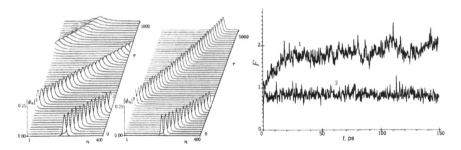

Fig. 7.11 *Left*: Soliton dynamics in the chain with substrate and friction for $K_o = 5$ N/m and $K_o = 2.5$ N/m ($\gamma = 0.2, s = 0.25$). *Right*: Thermalization of the cyclic chain consisting of $N = 50$ PGs at $T = 300$ K. Dependence of the thermalization coefficient $F(t)$ on time t at $t_c = 0.001$ ps (*line 1*, $\overline{F} = 1.97$) and $t_c = 0.0216$ ps (*line 2*, $\overline{F} = 0.843$)

Numerical integration of (7.101) with $N = 100$ and $\gamma = 0$ has shown stability of the Davydov soliton in the chain with the substrate. The soliton moves along the chain with constant velocity, keeping its shape. It has the velocity spectrum $(s_m(K_o), s_{ex})$, where the minimum velocity s_m is found from the equation

$$(3.080865G + 0.16606)^2 \zeta = 1$$

so $s_m = 0$ for $K_o \leq 2.93$ N/m and $s_m > 0$ for $K_o > 2.93$ N/m. With increasing K_o, s_m tends monotonically to s_{ex}. If viscous friction is introduced into the chain model ($\gamma > 0$), the soliton should have a finite lifetime for $s_m > 0$. In this case, friction causes the soliton to slow down to a minimum velocity s_m, and then it should collapse. If $s_m = 0$, the soliton slowdown should not cause its collapse.

To test this, we consider the soliton dynamics in a chain with friction ($\gamma = 0.2$) for $K_o = 5$ N/m ($s_m = 0.23$) and $K_o = 2.5$ N/m ($s_m = 0$). Let the soliton have the velocity $s = 0.25$ at the initial time. Then, as shown in Fig. 7.11 (left), when $K_o = 5$ N/m, friction causes the soliton to slow down to the critical velocity $s_m = 0.23$ and then causes it to collapse. When $K_o = 2.5$ N/m, friction results only in a smooth slowdown of the soliton without breaking it (the soliton eventually stops). This behavior of the soliton in the chain with friction confirms the result concerning the necessity and sufficiency of the condition (7.100) for the existence of a Davydov soliton in a chain with substrate.

7.5 Autolocalized State Dynamics of a Thermalized Chain

Here we consider the dynamics of an autolocalized state of intramolecular excitation (Davydov soliton) in a thermalized chain when its interaction with the acoustic and optical modes of the molecular chain is taken into account [141–146]. We will describe the chain interaction with a heat bath using the Langevin equation and introduce friction and colored noise into the acoustic subsystem which characterizes this interaction. We assume that the higher-amplitude optical subsystem is thermalized through its interaction with the acoustic subsystem.

7.5.1   Acoustic and Optical Vibrations

We consider a one-dimensional chain of hydrogen-bonded peptide groups. The peptide groups, allowing internal motion of the proton, are located along the x-axis at the chain sites $x = na$, where $n = 0, \pm 1, \pm 2, \dots$, is the site number and a is the chain spacing. If y_n is the displacement of the nth PG from its equilibrium position and x_n is the displacement of the hydrogen of this PG, the Hamiltonian takes the form

$$H_{\text{lat}} = \sum_n \left[\frac{m}{2} \dot{x}_n^2 + \frac{M}{2} \dot{y}_n^2 + V(y_{n+1} - y_n) + \frac{K}{2}(y_n - x_n)^2 + \frac{K_\circ}{2} y_n^2 \right] ,$$

where the dot denotes differentiation with respect to time t, m and M are the reduced hydrogen and PG masses, respectively, and K is the elasticity constant of the intrapeptide displacement of hydrogen. The Morse potential $V(\rho) = \epsilon_\circ (e^{-b\rho} - 1)^2$ describes the hydrogen bond between PGs, $\epsilon_\circ$ is the bonding energy, and $b = \sqrt{\kappa / 2\epsilon_\circ}$ is the phenomenological parameter. For small deformations of the bond ρ, we have $V(\rho) \approx \kappa \rho^2 / 2$, where $\kappa = V''(0)$ is the stiffness of the bond. The interaction of the nth PG with the chain substrate is described by the potential $V_\circ(y_n) = K_\circ y_n^2 / 2$, where $K_\circ$ is the stiffness of its interaction. The chain is shown schematically in Fig. 7.1.

In the calculation we use the parameter values

$$\epsilon_\circ = 0.17 \, \text{eV} = 2.72357 \times 10^{-20} \, \text{J} \, [147], \quad m = m_\text{p} = 1.67343 \times 10^{-27} \, \text{kg} ,$$

$$M = 113.2 m_\text{p} = 1.89432 \times 10^{-25} \, \text{kg} , \quad \kappa = 13 \, \text{N/m} , \quad a = 4.5 \, \text{Å} \, [10] .$$

The transverse intrapeptide vibration of hydrogen (the amide-V vibration) is characterized by the frequency $\Omega_V = 640\text{–}800 \, \text{cm}^{-1}$ [148]. The elastic constant here is $K = \Omega_V^2 \mu = 30.5116 \, \text{N/m}$, which corresponds to the average value of Ω_V, viz., $720 \, \text{cm}^{-1} = 1.35625 \times 10^{14} \text{s}^{-1}$, where $\mu = Mm/(M + m)$ is the reduced mass of the vibration. The value of the coefficient $K_\circ$ depends on the structure of the protein molecule characterized by a high packing density. The values of $K_\circ$ are maximum when the PG is located within a protein globule and minimum when the PG is on its surface. In the latter case, structured water surrounding the macromolecule forms a substrate. Let us take the value $K_\circ = 4 \, \text{N/m}$, which corresponds to the stiffness of the interaction with the substrate for structured (ice-like) water.

The small-amplitude vibrations of the chain are described by the equations of motion

$$m\ddot{x}_n = K(y_n - x_n) \,,$$

$$M\ddot{y}_n = \kappa(y_{n+1} - 2y_n + y_{n-1}) - K(y_n - x_n) - K_\circ y_n \,,$$

where $n = 0, \pm 1, \pm 2, \dots$. The dispersion law for the small-amplitude vibrations

$$x_n(t) = A_x \exp\left[i\Omega(q)t + 2qni\right] \,, \qquad y_n(t) = A_y \exp\left[i\Omega(q)t + 2q(n + 1)i\right] \,,$$

has the form

$$\Omega_a(q) = \sqrt{A(q) - \sqrt{A^2(q) - B(q)}} \,,$$

for the acoustic branch, and

$$\Omega_0(q) = \sqrt{A(q) + \sqrt{A^2(q) - B(q)}} \,,$$

for the optical one. Here

$$A(q) = \frac{MK + m(K + K_\circ + 4\kappa \sin^2 q)}{2mM} \,, \qquad B(q) = K\frac{K_\circ + 4\kappa \sin^2 q}{mM} \,,$$

and $q \in [0, \pi/2]$ is the dimensionless wave number. The frequencies of the acoustic vibrations are $\Omega_a(0) \leq \Omega_a(q) \leq \Omega_a(\pi/2)$, while those of the optical vibrations are $\Omega_0(\pi/2) \leq \Omega_0(q) \leq \Omega_0(0)$, with

$$\Omega_a(0) = 24.3 \, \text{cm}^{-1} \,, \quad \Omega_a(\pi/2) = 90.9 \, \text{cm}^{-1} \,,$$

$$\Omega_0(\pi/2) = 719.86 \, \text{cm}^{-1} \,, \quad \Omega_0(0) = 719.91 \, \text{cm}^{-1} \,.$$

The Einstein temperature is $T_E \leq \hbar\Omega_a(\pi/2)/k_B = 130.7$ K for the acoustic branch and $T_E = \hbar\Omega_V/k_B = 1{,}035.8$ K for the optical branch, with $\hbar$ and k_B the Planck and Boltzmann constants, respectively. It is clear that, at $T = 300$ K, the optical vibrations are almost not thermalized. To describe thermalization of the acoustic vibrations, the methods of classical mechanics can be used.

7.5.2 Langevin Equation

We describe thermalization of the vibrational mode with frequency Ω using the Langevin equation

$$\ddot{u} + \Omega^2 u + \Gamma\dot{u} = \xi(t)/\mu \,,$$

where u is the vibration coordinate, $\Gamma = 1/t_r$, with t_r is the relaxation time of the vibration, and $\xi(t)$ is the normally distributed random force describing the interaction of the mode with a heat bath at temperature T. The autocorrelation function of the random force is $\langle \xi(t)\xi(t') \rangle = 2\mu\Gamma k_B T \varphi(t - t')$, where the dimensionless autocorrelation function $\varphi(t)$ is normalized according to the condition $\int_{-\infty}^{+\infty} \varphi(t)dt = 1$.

In thermal equilibrium, the average energy of thermal vibrations of the mode is determined by the relation

$$E = \lim_{t \to \infty} \frac{1}{\tau} \int_0^\tau \frac{\mu}{2}(\dot{u}^2 + \Omega^2 u^2)dt = \int_0^{+\infty} \mu(\omega^2 + \Omega^2)|H(\omega)|^2 F(\omega)d\omega ,$$

where $H(\omega) = [M(\Omega^2 - \omega^2 + i\omega\Gamma)]^{-1}$ is the response function for the Langevin equation and the Fourier transform of the autocorrelation function is

$$F(\omega) = \frac{1}{2\pi} \int_{-\infty}^{+\infty} \langle \xi(t)\xi(0) \rangle \exp(-i\omega t)dt = \frac{\mu\Gamma k_B T}{\pi} \int_{-\infty}^{+\infty} \varphi(t)\exp(-i\omega t)dt .$$

Therefore, $E = K + P$, where the average values of the kinetic K and potential P energies of the vibrations are

$$K = 2k_B T\Gamma \int_0^{+\infty} \frac{\omega^2 \mathscr{F}(\omega)d\omega}{(\Omega^2 - \omega^2)^2 + \omega^2\Gamma^2}, \quad P = 2k_B T\Gamma \int_0^{+\infty} \frac{\Omega^2 \mathscr{F}(\omega)d\omega}{(\Omega^2 - \omega^2)^2 + \omega^2\Gamma^2},$$

$$(7.102)$$

and the Fourier transform of the dimensionless autocorrelation function $\varphi(t)$ is

$$\mathscr{F}(\omega) = \frac{1}{2\pi} \int_{-\infty}^{+\infty} \varphi(t)\exp(-i\omega t)dt .$$

For a delta-correlated random force (the case of the Langevin equation with white noise), we have $\varphi(t) = \delta(t)$ and $\mathscr{F}(\omega) = 1/2\pi$. The integrals (7.102) are easily found by a contour integration method. Finally, we obtain the vibration energy $E = k_B T$, $P = K = k_B T/2$ ($\Omega \neq 0$), and the amplitude $A = \sqrt{k_B T/\mu\Omega^2}$.

The Langevin equation with white noise describes the thermal vibrations of the mode (of the harmonic oscillator) in the classic approximation, where the average energy is $E = k_B T$. In the case of quantum oscillator, the average energy is

$$E = \hbar\Omega \left[\frac{1}{2} + \frac{1}{\exp(\hbar\Omega/k_B T) - 1} \right] .$$

and the heat capacity of the oscillator is $c(\Omega, T) = dE(\Omega, T)/dT = k_B F_E(\Omega, T)$, where the Einstein function is

$$F_E(\Omega, T) = \left(\frac{\hbar\Omega}{k_B T}\right)^2 \frac{\exp(\hbar\Omega/k_B T)}{\left[\exp(\hbar\Omega/k_B T) - 1\right]^2} .$$

As $T \to 0$, the heat capacity $c(\Omega, T) \to 0$, and $c(\Omega, T) = k_B$ for high temperatures. Therefore, at low temperatures $T < T_E$, the vibrations of the quantum oscillator are frozen out. The Einstein temperature $T_E = \hbar\Omega/k_B$ is determined from the equation $F_E(\Omega, T_E) = e/(e - 1)^2 = 0.9206735$. The classical approximation for the description of thermal vibrations can be used only for temperatures $T > T_E$.

Without taking into account zero-point vibrations, the degree of thermalization of the quantum oscillator can be characterized by the coefficient

$$G(\Omega, T) = \frac{\hbar\Omega/k_B T}{\exp(\hbar\Omega/k_B T) - 1} .$$

The coefficient $G(\Omega, T) \to 0$ as $T \to 0$, $G(\Omega, T_E) = 1/(e - 1) = 0.5819767$ at $T = T_E$, and $G(\Omega, T) \to 1$ as $T \to \infty$. Therefore, at temperature T, the vibrations of the frequencies $\Omega > \Omega_E(T) = k_B T/\hbar$ are frozen out. The Langevin equation with white noise cannot be used to describe them.

Partial thermalization of the high-frequency vibrations and total thermalization of the low-frequency ones can be obtained using the Langevin equation with colored noise which contains only low-frequency components. By analogy with light, this colored noise could be called red noise. The autocorrelation function of the noise should depend on the temperature. Let us find this dependence in the case of an exponentially correlated random force, when

$$\varphi(t) = \frac{\lambda}{2} \exp(-|\lambda t|) , \qquad \mathscr{F}(\omega) = \frac{\lambda^2}{2\pi(\omega^2 + \lambda^2)} ,$$

where $\lambda = 1/t_c$, and t_c is the correlation time of the random force.

For the exponentially correlated random force, the mean values of the kinetic and potential energies are $K = k_B T f_K(\Omega, \Gamma, \lambda)/2$ and $P = k_B T f_P(\Omega, \Gamma, \lambda)/2$, respectively, where

$$f_K(\Omega, \Gamma, \lambda) = \frac{\lambda^2}{\lambda^2 + \lambda\Gamma + \Omega^2} , \qquad f_P(\Omega, \Gamma, \lambda) = \frac{\lambda^2(\Omega^2 + \lambda^2 - \Gamma^2) + \lambda\Gamma\Omega^2}{(\Omega^2 + \lambda^2)^2 - \Gamma^2\lambda^2} ,$$

are the dimensionless functions. The dependence of the correlation coefficient λ on the temperature can be found from the equation

$$g(k_B T/\hbar, \Gamma, \lambda) = G(k_B T/\hbar, T) = \frac{1}{e - 1} ,$$

where

$$g(\Omega, \Gamma, \lambda) = \frac{1}{2}\left[f_K(\Omega, \Gamma, \lambda) + f_P(\Omega, \Gamma, \lambda)\right].$$

For $\Gamma \ll k_B T/\hbar$, this equation can be rewritten in the simpler form:

$$\frac{\lambda^2}{\lambda^2 + (k_B T/\hbar)^2} = \frac{1}{e - 1}.$$

From this equation we obtain the linear dependence of the correlation coefficient on the temperature:

$$\lambda = 1/t_c = k_B T/\hbar \sqrt{e - 2}. \qquad (7.103)$$

7.5.3   Chain Thermalization

The chain of PGs in a protein macromolecule interacts with a heat bath via amino acid residues. Therefore, in order to take into account the interaction with the heat bath, it suffices to introduce friction and random external forces acting only on the PGs (intrapeptide vibrations are thermalized through the thermal motion of the peptide groups).

Let us first consider how thermalization of intrapeptide vibrations proceeds in an isolated PG. The PG dynamics is given by the equations of motion

$$m\ddot{x} = K(y - x),$$

$$M\ddot{y} = K(x - y) - \Gamma M\dot{y} + \xi,$$

where the coordinate x describes the displacement of the hydrogen atom in the PG and y is the displacement of the PG as a whole ($x = y$ for a non-deformed bond). Substituting $u = x - y$ and $v = (mx + My)/M_t$, this system takes the form

$$\mu\ddot{u} = -Ku - \frac{\Gamma\mu^2}{M}\dot{u} - \frac{\mu}{M}\xi - \left(\frac{\mu}{M}\right)^2 K_o u + K_o \frac{m}{M_t}v + \Gamma\mu\dot{v},$$

$$M_t\ddot{v} = -K_o v - \Gamma M\dot{v} + \xi + K_o\frac{\mu}{M}u + \Gamma\mu\ddot{u},$$

where $M_t = M + m$ and $\mu = mM/M_t$ are the total and reduced masses of the PG, respectively. Since $m \ll M$ and the displacement of the centre of gravity occurs much more slowly than the changes in the intrapeptide position of the hydrogen atom, the last three terms in the first equation and the last two terms in the second equation of this system can be neglected. The system then splits into the two separate

Langevin equations

$$\mu\ddot{u} = -Ku - \mu\Gamma'\dot{u} + \xi',$$
$$M_t\ddot{v} = -K_o v - M_t\Gamma''\dot{v} + \xi'',$$

where $\Gamma' = 1/t_r' = \mu\Gamma/M = \mu/Mt_r$ and $\Gamma'' = 1/t_r'' = M\Gamma/M_t = M/M_t t_r$ are the friction coefficients, and the random forces $\xi'(t) = \mu\xi(t)/M$ and $\xi''(t) = \xi(t)$ have the autocorrelation functions

$$\langle\xi'(t)\xi'(t')\rangle = 2\mu\Gamma' k_B k_B T \varphi(t - t'), \qquad \langle\xi''(t)\xi''(t')\rangle = 2M_t\Gamma'' k_B T \varphi(t - t'),$$

$$\langle\xi'(t)\xi''(t')\rangle = 0.$$

Therefore, the relaxation time $t_r' = (1 + M/m)t_r$ of intrapeptide vibrations is more than two orders of magnitude greater than the relaxation time $t_r'' = (1 + m/M)t_r$ of acoustic waves.

Next we consider a cyclic chain consisting of N molecules. The dynamics of the PG chain is given by the equations of motion

$$m\ddot{x}_n = K(y_n - x_n), \tag{7.104}$$

$$M\ddot{y}_n = F(y_{n+1} - y_n) - F(y_n - y_{n-1}) - K_o y_n - \Gamma M \dot{y}_n + \xi_n(t), \tag{7.105}$$

where $n = 1, 2, \ldots, N$ ($n+1 = 1$ for $n = N$ and $n-1 = N$ for $n = 1$), $\xi_n(t)$ is the normally distributed random force with autocorrelation functions $\langle\xi_n(t)\xi_m(t')\rangle = 2M\Gamma k_B T \varphi(t - t')\delta_{nm}$, and $F(\rho) = dV(\rho)/d\rho$. To model thermalization of the chain by exponential colored noise $\varphi(t) = \exp(-|t|/t_c)/2t_c$, (7.104) and (7.105) are conveniently completed by the equations

$$\dot{\xi}_n = t_c^{-1}[\eta_n(t) - \xi_n(t)], \tag{7.106}$$

where $\eta_n(t)$ is the independent, normally distributed, delta-correlated random force: $\langle\eta_n(t)\eta_m(t')\rangle = 2M\Gamma k_B T \delta(t - t')\delta_{mn}$.

Let us suppose that the chain have zero temperature at the initial time, i.e., the chain is in the ground state $\{x_n(0) = 0, \dot{x}_n(0) = 0, y_n(0) = 0, \dot{y}_n(0) = 0\}_{n=1}^N$. Then the energy of the acoustic thermal vibrations grows proportionally to the function $\mathscr{E}''(t) = 1 - \exp(-t/t_r'')$, and the energy of the optical thermal vibrations grows proportionally to the function $\mathscr{E}'(t) = 1 - \exp(-t/t_r')$. Therefore, in order to model thermalization of this chain, it suffices to integrate (7.104)–(7.106) over the time interval $0 \leq t \leq 5t_r' = 5(1 + M/m)t_r$.

The friction coefficient is determined by $\Gamma = 6\pi b\eta/M$, where $b = 1–10$ Å is the characteristic linear size of the mobile amino acid residue, and η is the viscosity of the medium. For water, $\eta = 10^{-3}$ kg/m s, so $\Gamma = (1–10) \times 10^{13}$ s^{-1}. We take the minimum value $\Gamma = 10^{13}$ s^{-1}, which corresponds to the relaxation time $t_r =$

0.1 ps. The correlation time of the random force t_c can be found from (7.103). At the physiological temperature $T = 300\,\text{K}$, $t_c = 0.0216\,\text{ps}$.

Note that the correlation time of the random force should roughly match the correlation time of molecular displacements in the heat bath. For protein molecules, the heat bath is formed by the surrounding structured water, whose dynamics can be described by the equation of restricted diffusion: $\varphi(t) = k_B T \exp(-K_o t / M_o \Gamma)/K_o$ [149, 150], where $M_o = 18 m_p$ is the mass of the water molecule. Assuming that, at the physiological temperature, the friction coefficient of the water molecules is $\Gamma = 10^{13}\,\text{s}^{-1}$, the correlation time of the displacements is $t_c' = M_o \Gamma / K_o = 0.075\,\text{ps}$. On the other hand, in their work [151] devoted to a study of the motion of the aromatic ring Tyr–21 in the protein molecule, a trypsin inhibitor, Karplus and McCammon have shown that the average time between significant changes in the random force is 0.07 ps. This approximate estimate does not contradict the value $t_c = 0.0216\,\text{ps}$ we obtained at $T = 300\,\text{K}$. This fact indirectly confirms (7.103).

The character of chain thermalization at time t is conveniently described by the degree of thermalization of optical and acoustic vibrations

$$F_o(t) = \sum_{k=1}^{N} \mu(\dot{x}_n - \dot{y}_n)^2 / N k_B T \,, \qquad F_a(t) = \sum_{k=1}^{N} (m\dot{x}_n - M\dot{y}_n)^2 / M_t N k_B T \,,$$

their amplitudes

$$A_o(t) = \left[\sum_{k=1}^{N} (x_n - y_n)^2 / N \right]^{1/2} \,, \qquad A_a(t) = \left[\sum_{n=1}^{N} (y_{n+1} - y_n)^2 / N \right]^{1/2} \,,$$

and the function $F(t) = H_{\text{lat}} / N k_B T$.

Equations (7.104)–(7.106) with $N = 50$ were integrated numerically by the standard fourth-order Runge–Kutta method with constant grid spacing $\Delta t = 0.001\,\text{ps}$ [123], using the lagged Fibonacci random number generator [152]. The process of chain thermalization is shown in Fig. 7.11 (right). As can be seen from this figure, complete thermal equilibrium is achieved over the time $5 t_{t'} = 57.1\,\text{ps}$. After approaching thermal equilibrium, the mean values of $\bar{F}_o$, $\bar{F}_a$, $\bar{A}_o$, $\bar{A}_a$, and $\bar{F}$ can be found. On the other hand, these values can be found analytically using the relations for the integrals (7.102):

$$\tilde{F}_o = \sum_{n=1}^{N} f_K(\Omega_o(q_n), \Gamma', \lambda) / N \,,$$

$$\tilde{F}_a = \sum_{n=1}^{N} f_K(\Omega_a(q_n), \Gamma'', \lambda) / N \,,$$

Table 7.10 Dependencies of the mean values of $\tilde{F}_o$, $\overline{F}_o$, $\tilde{F}_a$, $\overline{F}_a$, $\tilde{F}$, $\overline{F}$, $\tilde{A}_o$, $\overline{A}_o$, $\tilde{A}_a$, and $\overline{A}_a$ on the temperature T (amplitudes given in angstroms)

T (K)	$\tilde{F}_o$	$\overline{F}_o$	$\tilde{F}_a$	$\overline{F}_a$	$\tilde{F}$	$\overline{F}$	$\tilde{A}_o$	$\overline{A}_o$	$\tilde{A}_a$	$\overline{A}_a$
50	0.003	0.003	0.299	0.230	0.380	0.288	0.003	0.003	0.057	0.042
100	0.013	0.013	0.449	0.444	0.605	0.493	0.008	0.008	0.089	0.073
150	0.028	0.027	0.587	0.580	0.742	0.620	0.014	0.014	0.112	0.098
200	0.049	0.043	0.676	0.681	0.834	0.717	0.021	0.020	0.132	0.116
250	0.078	0.078	0.741	0.741	0.913	0.796	0.030	0.030	0.149	0.133
300	0.104	0.105	0.778	0.767	0.965	0.843	0.038	0.038	0.164	0.146

$$\tilde{A}_o^2 = \sum_{n=1}^{N} k_B T f_P(\Omega_o(q_n), \Gamma', \lambda)/\Omega_o^2(q_n)\mu N \ ,$$

$$\tilde{A}_a^2 = \sum_{n=1}^{N} k_B T f_P(\Omega_a(q_n), \Gamma'', \lambda)/\Omega_a^2(q_n) M_t N \ ,$$

$$\tilde{F} = \sum_{n=1}^{N} [g(\Omega_o(q_n), \Gamma', \lambda) + g(\Omega_a(q_n), \Gamma'', \lambda)]/N \ ,$$

where $q_n = -\pi/2 + \pi n/N$. The dependencies of these mean values on the temperature of the heat bath T are given in Table 7.10. As can be seen from Table 7.10, optical vibrations are weakly thermalized. Chain thermalization occurs due to the energy being pumped into acoustic vibrations.

When taking into account thermal vibrations, the thermalization coefficient of the chain is

$$\hat{F}(T) = \sum_{n=1}^{N} \left[G\big(\Omega_o(q_n), T\big) + G\big(\Omega_a(q_n), T\big) \right]/N \ ,$$

and the dimensionless heat capacity of the chain is

$$C(T) = \sum_{n=1}^{N} \left[F_E\big(\Omega_o(q_n), T\big) + F_E\big(\Omega_a(q_n), T\big) \right]/N \ .$$

The temperature dependence of the functions $\tilde{F}(T)$, $\hat{F}(T)$, and $\overline{F}(T)$, characterizing the degree of chain thermalization, and the functions $d[T\tilde{F}(T)]/dT$ and $C(T)$ characterizing the heat capacity of the chain, is shown in Fig. 7.12. As can be seen from this figure, the functions $\tilde{F}(T)$ and $d[T\tilde{F}(T)]/dT$ are good approximations to the functions $\hat{F}(T)$ and $C(T)$, respectively. One can thus use the Langevin equation with exponentially correlated colored noise, whose correlation coefficient depends on temperature according to (7.103), to carry out molecular dynamics simulation

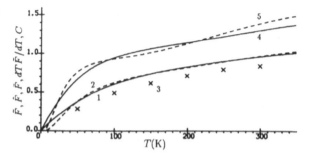

Fig. 7.12 Dependence of the thermalization coefficients $\tilde{F}(T)$, $\hat{F}(T)$, $\overline{F}(T)$ (*lines 1–3*), the heat capacity $d[T\tilde{F}(T)]/dT$, and $C(T)$ (*lines 4 and 5*)

of the pure quantum effect constituted by the decrease in the heat capacity of a molecular system when it is cooled.

7.5.4 Soliton States of Intrapeptide Amide-I Excitation

The dynamics of one quantum of the intrapeptide amide-I excitation is described by the equations of motion (7.7)–(7.9). Let us consider the excitation dynamics in a homogeneous chain. In this case, we can put $J_n = J - \chi_D(y_n - y_{n-1})$ in (7.7) and $M_n \equiv M$ in (7.8). Since for the chain of hydrogen-bonded PGs $\chi_L \ll \chi_R$ and $\chi_D = 3J/a \ll \chi_R$ [122], we assume $\chi_L = 0$ and $\chi_D = 0$. Then the Hamiltonian function takes the form

$$\mathcal{H} = \mathcal{H}_{lat} + \mathcal{H}_{ex}$$

$$= \mathcal{H}_{lat} + \sum_n \left\{ [\varepsilon_0 + \chi_0(y_n - x_n) + \chi_R(y_{n+1} - y_n)]\phi_n\phi_n^* \right. \qquad (7.107)$$

$$\left. - J(\phi_n\phi_{n+1}^* + \phi_n^*\phi_{n+1}) \right\} .$$

The Hamiltonian function of the chain yields the equations of motion

$$i\hbar\dot{\phi}_n = [\varepsilon_0 + \chi_0(y_n - x_n) + \chi_R(y_{n+1} - y_n)]\phi_n - J(\phi_{n+1} + \phi_{n-1}) , \qquad (7.108)$$

$$m\ddot{x}_n = K(y_n - x_n) + \chi_0|\phi_n|^2 , \qquad (7.109)$$

$$M\ddot{y}_n = F(y_{n+1} - y_n) - F(y_n - y_{n-1}) - K(y_n - x_n)$$

$$- K_\circ y_n - \chi_0|\phi_n|^2 + \chi_R(|\phi_n|^2 - |\phi_{n-1}|^2) , \qquad (7.110)$$

where $n = 0, \pm 1, \pm 2, \ldots$.

To find the autolocalized state of the excitation (soliton), we use the numerical method from the previous section. In the long-wave approximation

$$\phi_n(t) = \varphi(na - vt)\exp\left\{i[kn - (\varepsilon_0 + \lambda)t/\hbar]\right\} , \qquad (7.111)$$

where $\varphi(z)$ is a smooth real function and v is the soliton velocity. After substituting in (7.111), the normalization condition (7.5) takes the form

$$\sum_n \varphi_n^2 = 1 , \tag{7.112}$$

and (7.108) is converted into the system of two equations

$$-\hbar\dot{\varphi}_n = J(\varphi_{n+1} - \varphi_{n-1})\sin k , \tag{7.113}$$

$$\left[\chi_0(y_n - x_n) + \chi_R(y_{n+1} - y_n)\right]\varphi_n - J(\varphi_{n+1} + \varphi_{n-1})\cos k = \lambda\varphi_n , \tag{7.114}$$

where $\varphi_n = \varphi(na - vt)$.

Equation (7.113) defines the soliton velocity spectrum. To third order in a, we have $\dot{\varphi}_n = -v\varphi_z(na) = -v(\varphi_{n+1} - \varphi_{n-1})/2a$, so $v = v_{ex}\sin k$, where $v_{ex} = 2aJ/\hbar$ is the maximum exciton velocity in the chain and $k \in [-\pi/2, \pi/2]$ is the dimensionless wave number. The parameter λ in (7.114) is found from the normalization condition (7.112) to be

$$\lambda = \sum_n \left[\chi_0(y_n - x_n) + \chi_R(y_{n+1} - y_n)\right]\varphi_n^2 - 2J\varphi_n\varphi_{n+1}\cos k .$$

Let $x_n(t) = x(na - vt)$ and $y_n(t) = y(na - vt)$. Then to fourth order in a, we have

$$\ddot{x}_n = v^2 x_{zz} = v^2(x_{n+1} - 2x_n + x_{n-1})/a^2 , \quad \ddot{y}_n = v^2 y_{zz} = v^2(y_{n+1} - 2y_n + y_{n-1})/a^2 .$$

Therefore, (7.109) and (7.110) can be written in the form

$$\frac{mv^2(x_{n+1} - 2x_n + x_{n-1})}{a^2} = K(y_n - x_n) + \chi_0\varphi_n^2 , \tag{7.115}$$

$$\frac{Mv^2(y_{n+1} - 2y_n + y_{n-1})}{a^2} = F(y_{n+1} - y_n) - F(y_n - y_{n-1}) - K(y_n - x_n)$$
$$-K_0 y_n - \chi_0\varphi_n^2 + \chi_R(\varphi_n^2 - \varphi_{n-1}^2) . \tag{7.116}$$

In a cyclic chain consisting of N PGs, the solution of (7.112) and (7.114)–(7.116) is equivalent to the solution of the constrained minimum problem

$$\mathscr{L} \longrightarrow \min : \sum_{n=1}^N \phi_n^2 = 1 , \tag{7.117}$$

where the functional is

$$
\mathcal{L} = \sum_{n=1}^{N} \Big\{ - 2J\varphi_n\varphi_{n+1}\cos k + \big[\chi_0(y_n - x_n) + \chi_R(y_{n+1} - y_n)\big]\varphi_n^2
$$
$$
- \frac{mv^2}{2a^2}(x_{n+1} - x_n)^2 - \frac{Mv^2}{2a^2}(y_{n+1} - y_n)^2
$$
$$
+ V(y_{n+1} - y_n) + \frac{K}{2}(y_n - x_n)^2 + \frac{K_\circ}{2}y_n^2 \Big\} .
$$

The problem (7.117) was solved numerically by the method of steepest descent [123]. We considered the chain consisting of $N = 50$ PGs with resonant interaction energy $J = 1.55 \times 10^{-22}$ J, which corresponds to the resonant interaction energy of neighboring PGs in an α-helix protein molecule [120]. The values of the exciton–phonon interaction parameter χ_0, χ_R were varied. The descent algorithm starts at the point

$$
\varphi_{N/2} = 1, \quad \varphi_n = 0, \quad n \neq N/2, \quad x_n \equiv 0, \quad y_n \equiv 0,
$$

which corresponds to the localization of the excitation at one PG in the non-deformed chain.

The autolocalized state $\{\varphi_n^0, x_n^0, y_n^0\}_{n=1}^{N}$ found numerically is characterized by the following quantities: the energy

$$
E = 2J + \sum_{n=1}^{N} \Big\{ - 2J\varphi_n\varphi_{n+1}\cos k + \big[\chi_0(y_n - x_n) + \chi_R(y_{n+1} - y_n)\big]\varphi_n^2
$$
$$
+ \frac{mv^2}{2a^2}(x_{n+1} - x_n)^2 + \frac{Mv^2}{2a^2}(y_{n+1} - y_n)^2
$$
$$
+ V(y_{n+1} - y_n) + \frac{K}{2}(y_n - x_n)^2 + \frac{K_\circ}{2}y_n^2 \Big\} ,
$$

which is defined relative to the energy of the stationary exciton $E_\circ = \varepsilon_0 - 2J$ (the zero level of energy), the width $L = (\sum_{n+1}^{N} \varphi_n^4)^{-1}$ ($L = 1$ for the excitation localized at a single PG, $\varphi_n = 1/\sqrt{N}$ and $L = N$ for the exciton), the amplitude of the optical sublattice deformation $A_\circ = -\max(y_n - x_n)$, and the amplitude of the acoustic sublattice deformation $A_a = -\max(y_{n+1} - y_n)$.

The profile of the autolocalized state (soliton) is shown in Fig. 7.13. The probability distribution for finding the excitation $\{\varphi_n^2\}_{n=1}^{N}$ has a bell-shaped profile. The compression of hydrogen bonds ($y_{n+1} - y_n < 0$) and the N–H bonds ($y_n - x_n < 0$), occurs in the localization region of the excitation, which stabilizes the autolocalized state.

Fig. 7.13 Profile of the soliton ($s = 0.25$, $\chi_0 = 8 \times 10^{-11}$ N, $\chi_R = 4 \times 10^{-11}$ N) at the initial time $t = 0$ (lines 1, 2, 3) and at time $t = 175$ ps, after the passage of 124 chain links in the cyclic chain consisting of 100 PGs. The dynamics is described by (7.108)–(7.110) (*thin lines 4–6*) or by (7.118)–(7.120) (*thick lines 7–9*)

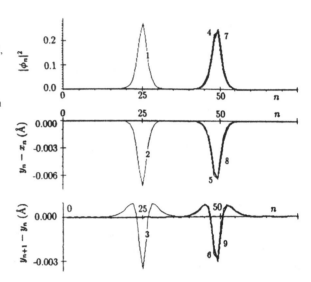

As shown in the previous section, when $\chi_0 = 0$ (the case of the Davydov soliton) and $K_0 > 0$, the soliton exists only for large values of the exciton–phonon interaction parameter χ_R. For low values of χ_R, only a stable delocalized excitation state exists. Note that, when $K_0 = 0$, the Davydov soliton is stable, while the exciton is not stable for any value of the parameter χ_R.

In the chain with $\chi_0 = 0$, the stationary soliton exists only if $\chi_R \geq 0.8 \times 10^{-10}$ N, but when $\chi_0 > 0$, it already exists for all values of the parameter χ_R. With increasing values of χ_R, the soliton energy decreases, the soliton becomes narrower, and the amplitude of the acoustic sublattice deformation increases. At a fixed value of χ_R, the energy and width of the soliton decrease monotonically with increasing values of χ_0, and the deformation amplitudes of both sublattices increase steadily.

The soliton velocity $v = v_{ex} \sin k$ is conveniently represented in the dimensionless form $s = v/v_{ex}$, where $v_{ex} = 2aJ/\hbar = 1{,}322.9$ m/s is the maximum soliton velocity. With increasing velocity s, the soliton energy and the amplitudes A_0 and A_a increase steadily, while the width L decreases monotonically.

The soliton dynamics in the chain is given by (7.108)–(7.110), where, without loss of generality, we may set $\varepsilon_0 = 0$. Let us consider the soliton dynamics in a non-thermalized cyclic chain consisting of $N = 100$ PGs. For this purpose, one must solve (7.108)–(7.110) with the initial conditions

$$\phi_n(0) = \varphi_n^0 \exp(ikn) , \quad x_n(0) = x_n^0 , \quad y_n(0) = y_n^0 ,$$
$$\dot{x}_n(0) = -v(x_{n+1}^0 - x_n^0)/a , \quad \dot{y}_n(0) = -v(y_{n+1}^0 - y_n^0)/a ,$$

where $n = 1, 2, \ldots, N$, $v = sv_{ex}$ is the soliton velocity, $k = \arcsin s$, and $\{\varphi_n^0, x_n^0, y_n^0\}_{n=1}^N$ is the soliton solution of the problem (7.117).

Numerical integration of the equations of motion has shown that the soliton moves along the chain with constant velocity and keeping its shape. For $s = 0.25$, $\chi_O = 0.8 \times 10^{-10}$ N, and $\chi_R = 0.4 \times 10^{-10}$ N, and the soliton passes over 124 chain sites during the time $t = 175$ ps, which corresponds to the dimensionless velocity $s = 0.241$ (see Fig. 7.13).

7.5.5 Vibrational Lattice Model of the Amide-I Excitation Dynamics

A classical vibrational model of the intrapeptide excitation, alternative to the quantum Davydov model considered previously, was suggested by Taneko [100–102]. In this model, the energy of the intrapeptide amide-I vibration in a chain of PGs has the form

$$H_{os} = \sum_n \left\{ \frac{\mu}{2}\dot{q}_n^2 + \frac{\beta}{2}q_n^2 - Q q_n q_{n+1} + \frac{1}{2}[R_O(y_n - x_n) + R_L(y_{n+1} - y_n)]q_n^2 \right\},$$

where q_n is the reduced coordinate of the intrapeptide vibration of the nth PG, μ is the reduced mass of the vibration, β is the stiffness of the C=O bond, Q is the parameter of the resonant interaction of neighboring PGs, and R_O and R_L are the parameters of the vibration interaction with the optical and acoustic subsystems of the chain. For the chain of PGs in an α-helix protein molecule, the frequency of the amide-I vibration $\Omega_1 = \sqrt{\beta/\mu} = 1,650$ cm^{-1}, the energy $\varepsilon_0 = 0.205$ eV, $\beta = 1,540$ N/m, and $\mu = 9.52634 m_p = 1.59416 \times 10^{-26}$ kg.

The Hamiltonian of the chain $H = H_{os} + H_{lat}$ gives the equations of motion

$$\mu \ddot{q}_n = -\beta q_n + Q(q_{n+1} + q_{n-1}) - [R_O(y_n - x_n) + R_L(y_{n+1} - y_n)]q_n, \tag{7.118}$$

$$m \ddot{x}_n = K(y_n - x_n) + \frac{1}{2}R_O q_n^2, \tag{7.119}$$

$$M \ddot{y}_n = F(y_{n+1} - y_n) - F(y_n - y_{n-1}) - K(y_n - x_n) - K_O y_n$$
$$- \frac{1}{2}R_O q_n^2 + \frac{1}{2}R_L(q_n^2 - q_{n-1}^2). \tag{7.120}$$

Making the replacement $\phi_n = q_n \sqrt{\beta/2\varepsilon_0} + i\dot{q}_n \sqrt{\mu/2\varepsilon_0}$ and putting $Q = 2J\beta/\varepsilon_0$, $R_O = 2\chi_O \beta/\varepsilon_0$, and $R_L = 2\chi_R \beta/\varepsilon_0$, we obtain

$$H_{os} = H_{ex} - \sum_n J(\phi_n \phi_{n+1} + \phi_n^* \phi_{n+1}^*)$$

$$+ \sum_n \frac{1}{2}[\chi_O(y_n - x_n) + \chi_R(y_{n+1} - y_n)](\phi_n^2 + \phi_n^{*2}). \tag{7.121}$$

The frequency of the amide-I vibration significantly exceeds the frequencies of the optical and acoustic vibrations. Therefore, the optical and acoustic sublattices of the chain are sensitive to the averaged amplitude of the amide-I vibration. If the function $\phi_n(t)$ in the Hamiltonian (7.121) is replaced by its average over the period of the intrapeptide vibration, the Hamiltonian H_{os} coincides with the Davydov Hamiltonian H_{ex}. Indeed, the averages are

$$\langle \phi_n \phi_{n+1} + \phi_n^* \phi_{n+1}^* \rangle = \frac{\langle \beta q_n q_{n+1} - \mu \dot{q}_n \dot{q}_{n+1} \rangle}{\varepsilon_0} = \frac{1}{\varepsilon_0}\Big[\beta \langle q_n q_{n+1}\rangle - \mu \langle \dot{q}_n \dot{q}_{n+1}\rangle\Big] = 0 \,,$$

$$\langle \phi_n^2 + \phi_n^{*2} \rangle = \frac{\langle \beta q_n^2 - \mu \dot{q}_n^2 \rangle}{\varepsilon_0} = \frac{1}{\varepsilon_0}\Big[\beta \langle q_n^2 \rangle - \mu \langle \dot{q}_n^2 \rangle\Big] = 0 \,.$$

Therefore, for a long period of time, both systems of equations of motion (7.108)–(7.110) and (7.118)–(7.120) should describe the same dynamics. To test this numerically, we consider the dynamics of the soliton autolocalized excitation state in the framework of the vibrational model, i.e., we consider the dynamics of a vibron–soliton.

We integrate (7.118)–(7.120) numerically with periodic boundary conditions ($n = 1, 2, \ldots, N$, $N = 100$) and the initial conditions

$$q_n(0) = \varphi_n^0 \sqrt{\beta/2\varepsilon_0}\cos(kn)\,, \quad \dot{q}_n(0) = -\varphi_n^0 \sqrt{\mu/2\varepsilon_0}\sin(kn)\,,$$

$$x_n(0) = x_n^0\,, \quad y_n(0) = y_n^0\,,$$

$$\dot{x}_n(0) = -v(x_{n+1}^0 - x_n^0)/a\,, \quad \dot{y}_n(0) = -v(y_{n+1}^0 - y_n)/a\,,$$

where $\{\varphi_n^0, x_n^0, y_n^0\}_{n=1}^N$ is the solution of the problem (7.117).

Numerical integration has demonstrated that the vibron–soliton in the vibrational model is characterized by exactly the same dynamics as the Davydov soliton in the quantum model (Fig. 7.13). The stability analysis of the numerical integration method has shown that a much smaller integration step must be chosen for the numerical simulation of soliton motion in the vibrational model than in the quantum model. When using the standard fourth-order Runge–Kutta method [123], the integration step should not exceed the value $\Delta t = 10^{-15}$ s when modeling the Davydov soliton motion and $\Delta t = 0.25\times10^{-15}$ s when modeling the vibron–soliton motion. Therefore, for numerical modeling of the interpeptide amide-I excitation dynamics, it suffices to restrict consideration to the quantum model discussed above.

7.5.6 Autolocalized State Dynamics of the Intrapeptide Amide-I Vibrations in a Thermalized Chain

In a thermalized chain, the autolocalized state has a finite lifetime. Let us first estimate how its lifetime depends on the frequency spectrum of molecular thermal vibrations in the chain. For this purpose, we will analyse a simplified model for which the Hamiltonian function of the excitation has the form

$$H = \sum_n \left[(\varepsilon_0 + \chi u_n)\phi_n\phi_n^* - J(\phi_n\phi_{n+1}^* + \phi_n^*\phi_{n+1}) + \frac{\mu}{2}\dot{u}_n^2 + \frac{\kappa}{2}u_n^2 \right],$$

where χ is the EPI parameter, u_n is the intrapeptide displacement coordinate, and μ is the reduced mass of the intrapeptide vibration. Let us choose $J = 1.55 \times 10^{-22}$ J, $\chi = 10^{-10}$ N, and stiffness $\kappa = 30.5116$ N/m, and vary the mass μ, thereby changing the frequency $\Omega = \sqrt{\kappa/\mu}$ of the intrapeptide vibration. With these parameters, the stationary autolocalized state has width $L = (\sum_{n+1}^{N} \varphi_n^4)^{-1} = 5.22$. The stabilization energy $E = 0.93595 \times 10^{-4}$ eV is more than two orders of magnitude less than the energy of thermal vibrations at $T = 300$ K.

The excitation dynamics in the thermalized chain is given by the Langevin equations

$$i\hbar\dot{\phi}_n = \chi u_n\phi_n - J(\phi_{n+1} + \phi_{n-1}),$$

$$\ddot{u}_n = -\Omega^2 u_n - \chi|\phi_n|^2/\mu - \Gamma\dot{u}_n + \xi_n/\mu,$$

$$\dot{\xi}_n = (\eta_n - \xi_n)/t_c, \quad n = 1, 2, \ldots, N.$$

Let the excitation at the initial time be in the stationary autolocalized state centered at the $N/2$th site of the thermalized chain consisting of $N = 50$ molecules. The destruction of the autolocalized state can be monitored using the autocorrelation function $D(t) = \sum_{n=1}^{N} |\phi_n(0)\phi_{n+m}(t)|$, where the index m is found from the condition $\max_n |\phi_n(t)| = |\phi_{N/2+m}(t)|$. The state lifetime t_e can be obtained from the equation $t_e^{-1} \int_0^{t_e} D(\tau)d\tau = 0.9$. The profiles of the excitation distribution over the chain for $t = 0$ and $t = t_e$ are shown in Fig. 7.14 (left).

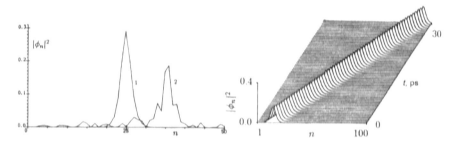

Fig. 7.14 *Left*: Profiles of the excitation distribution in the thermalized chain for $t = 0$ (*line 1*) and $t = t_e$ (*line 2*). Frequency $\Omega = 800\,\text{cm}^{-1}$, $T = 300$ K, $t_c = 0.001$ ps, and $t_e = 355$ ps. *Right*: Soliton dynamics in the thermalized chain for $\chi_0 = 0.5 \times 10^{-10}$ N, $\chi_R = 0$, $K_o = 4$ N/m, $s = 0.9$, and $T = 300$ K

Table 7.11 Dependence of t_e (ps) on the frequency Ω for $t_c = 0.001$ and 0.0216 ps at $T = 300$ K

Ω (cm^{-1})	25	50	100	200	400	800
$t_c = 0.001$	3.9 ± 2.7	5.0 ± 4.9	3.9 ± 2.5	13.4 ± 7.8	71.1 ± 16.3	390.5 ± 69.2
$t_c = 0.0216$	7.4 ± 5.5	4.5 ± 4.2	3.7 ± 2.3	19.5 ± 4.5	107.2 ± 16.8	415.0 ± 41.6

The lifetime t_e depends on the initial realization of the chain thermalized state. In this sense, $t_e = \bar{t}_e \pm \sigma_e$ is a random variable that can be characterized by its mean value $\bar{t}_e$ and the standard deviation σ_e. For each value of the frequency Ω, the lifetime t_e was obtained from ten different realizations of the thermalized state of the chain. Using these values $\{t_{e_i}\}_{i=1}^{10}$, the following estimates can be made:

$$\bar{t}_e \simeq \frac{1}{10}(t_{e_1} + \cdots + t_{e_{10}}) , \qquad \sigma_e^2 \simeq \frac{1}{10}\left[(t_{e_1} - \bar{t}_e)^2 + \cdots + (t_{e_{10}} - \bar{t}_e)^2\right] .$$

The dependence of t_e on the frequency Ω for $T = 300$ K, $t_c = 0.001$ ps (for white noise), and $t_c = 0.0216$ ps (for colored noise) is given in Table 7.11.

The amplitude of thermal vibrations exceeds the amplitude of the deformation caused by the intrapeptide excitation by more than an order of magnitude, so thermal vibrations should lead to the rapid destruction of the soliton state. Indeed, when $\Omega \leq 100$ cm^{-1}, the state lifetime is only a few picoseconds. However, for large values, an increase in the frequency Ω is accompanied by exponential growth of the lifetime. This growth is observed with both colored or white noise. Therefore, the growth in t_e cannot be accounted for by the freezing out of high-frequency vibrations (freezing out results in only a small increase in the lifetime). This growth relates to the fact that, at the frequencies $\Omega > \Omega_e$, where $\Omega_e = 4\pi J/\hbar = 98$ cm^{-1} is the maximum frequency of excitation hopping, the intrapeptide excitation is sensitive to the average amplitude of the vibrations in the phonon subsystem, responding only to a change in the positions of vibration centres. The more the frequency Ω exceeds Ω_e, the stronger the effect. when $\Omega \leq \Omega_e$, the excitation responds to the random, high-amplitude chain deformations caused by thermal molecular vibrations, leading to fast degradation of the autolocalized state.

In the two-component model we considered, the motion of the intrapeptide excitation in a thermalized chain is given by the equations

$$i\hbar\dot{\phi}_n = \left[\chi_0(y_n - x_n) + \chi_R(y_{n+1} - y_n)\right]\phi_n - J(\phi_{n+1} + \phi_{n-1}) , \quad (7.122)$$

$$m\ddot{x}_n = K(y_n - x_n) + \chi_0|\phi_n|^2 , \qquad (7.123)$$

$$M\ddot{y}_n = F(y_{n+1} - y_n) - F(y_n - y_{n-1}) - K(y_n - x_n) - K_0 y_n$$
$$- \chi_0|\phi_n|^2 + \chi_R(|\phi_n|^2 - |\phi_{n-1}|^2) - \Gamma M \dot{y}_n + \xi_n , \qquad (7.124)$$

$$\dot{\xi}_n = (\eta_n - \xi_n)/t_c , \quad n = 1, 2, \ldots, N ,$$

where all the variables and parameters were defined previously. Based on the energy difference between the exciton and soliton states of the amide-I excitation in crystalline acetanilide, viz., $\Delta E = 17 \text{ cm}^{-1}$ [91], the EPI parameter can be estimated as $\chi_O = 0.5 \times 10^{-10} \text{ N}$. The parameter χ_R is varied.

We choose the following initial conditions

$$\phi_n(0) = \varphi_n^0 \exp(ik\,n)\,, \qquad x_n(0) = \tilde{x}_n + x_n^0\,,$$
$$y_n(0) = \tilde{y}_n + y_n^0\,, \qquad \dot{x}_n(0) = \dot{\tilde{x}}_n - v(x_{n+1}^0 - x_n^0)/a\,,$$
$$\dot{y}_n(0) = \dot{\tilde{y}}_n - v(y_{n+1}^0 - y_n^0)/a\,, \qquad \xi_n(0) = \tilde{\xi}_n\,,$$

where v is the soliton velocity, $k = \arcsin(v/v_{ex})$, $\{\varphi_n^0, x_n^0, y_n^0\}_{n=1}^N$ is the solution of the problem (7.117), and $\{\tilde{x}_n, \tilde{y}_n, \dot{\tilde{x}}_n, \dot{\tilde{y}}_n, \tilde{\xi}_n\}_{n=1}^N$ are the solutions of (7.104)–(7.106) at time $t = 50\,\text{ps}$, when the chain is already in thermal equilibrium with the heat bath. Let us consider the excitation dynamics in a chain consisting of $N = 100$ PG.

The interaction of the excitation with thermal acoustic vibrations should lead to the fast destruction of the soliton state because the spectrum of acoustic vibrations lies below the frequency $\Omega_e = 98\,\text{cm}^{-1}$. When $\chi_R = 0.3 \times 10^{-10}\,\text{N}$, the soliton state is completely destroyed in a few picoseconds. On the other hand, the optical vibration spectrum $719.86 \le \Omega_0 \le 719.91\,\text{cm}^{-1}$ lies much higher than the critical frequency Ω_e. Therefore, in the absence of the excitation interaction with acoustic vibrations ($\chi_R = 0$), the soliton state will have a significant lifetime at physiological temperatures ($\sim$1,000 ps). Figure 7.14 (right) shows the soliton dynamics in the thermalized chain when $T = 300\,\text{K}$ and $\chi_R = 0$. As can be seen from this figure, when $s = 0.9$, the soliton moves along the chain with constant velocity. The excitation interaction with thermal optical vibrations does not lead to noticeable changes in the soliton shape.

Quantum-mechanical calculations carried out in [122] gives the following estimate for the EPI parameter: $\chi_R = (3\text{–}5) \times 10^{-11}\,\text{N}$. At this value of χ_R, the soliton is unstable to thermal vibrations. As globular protein macromolecules have a high packing density, the stiffness of the chain substrate within the globule can significantly exceed the value used, viz., $K_0 = 4\,\text{N/m}$. In this case, the frequency of acoustic vibrations will be higher and their amplitude will be an order of magnitude less ($\sim$0.01 Å) than those obtained in the calculation. An increase in the frequency of acoustic vibrations will lead to an increase in the stability of the soliton. The soliton dynamics in the thermalized chain with $K_0 = 400\,\text{N/m}$ and $\chi_R = 3 \times 10^{-11}\,\text{N}$ is shown in Fig. 7.15. For these values, the excitation interaction with thermal acoustic vibrations leads only to minor changes in the soliton shape.

The numerical investigation of the intrapeptide amide-I excitation dynamics performed for the PG chain shows that the thermally stable autolocalized states of the intramolecular excitation can arise within globular protein macromolecules, and that they are stabilized by the coupling of the excitation with intrapeptide displacements. The excitation interaction with the deformation of hydrogen bonds was shown to cause the destabilizing effect. Therefore, on the surface of a globular protein, where the thermal vibrations of hydrogen bonds are characterized by a large amplitude, the autolocalized states turn out to be unstable [153].

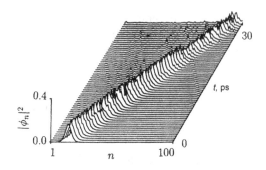

Fig. 7.15 Soliton dynamics in the thermalized chain with $\chi_0 = 0.5 \times 10^{-10}$ N, $\chi_R = 0.3 \times 10^{-10}$, $K_0 = 4$ N/m, $s = 0.9$, and $T = 300$ K

7.6 Electrosoliton Dynamics in a Thermalized Chain

The possibility of electrosoliton formation in α-helical proteins (a localized state of an extra electron bound with the deformation region of the helix arising due to the electron interaction with chain PGs) was predicted by Davydov [26, 83]. Electrosolitons can be formed at the chain end if the value of the electron–phonon interaction (EPI) parameter exceeds a certain threshold [6, 26, 154]. One must therefore estimate the EPI parameters in order to model the electrosoliton dynamics numerically.

7.6.1 Estimation of Electron–Phonon Interaction Parameters

Calculations of the bound electron states in the field of the PG and the conduction band characteristics of an extra electron in the models of polypeptide structures were carried out in a series of studies [155–157]. These calculations only took into account the electron interaction with the electrostatic field of the PG, which was treated as a field of point charges. According to the results of the quantum-mechanical calculations [158], the charges on the PG groups H–N–C=O were assumed to be equal to 0.3 (H), -0.3 (N), 0.4 (C), and -0.4 (O) (in units of the electron charge e). The dipole moment of the PG was reported to be $d \approx 4$ D [159]. It follows from the results in [155–157] that the electron can be in a bound state with this dipole, characterized by an affinity 0.6–1.28 eV. The electron in this state is located at a distance $r_- = 3.79$ a.u. from the negative charge of the dipole and $r_+ = 2.82$ a.u. from the positive charge.

Let us assume that the bound state of an extra electron with the nth PG, and therefore with the chain as a whole, is characterized by the same configuration as in the case of an isolated PG. Consider the electron interaction with the three nearest PG dipoles (see Fig. 7.16). Each dipole is represented by two charges $-q$ and $+q$ ($q = d/R$), located at a distance $R < a$ from each other, where $a = 4.5$ Å is the

Fig. 7.16 Position of an
extra electron relative to the
three nearest PGs

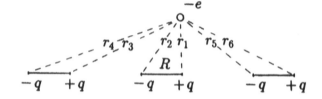

chain spacing. The interaction energy between the electron and the PG is

$$W = -|qe| \left(\frac{1}{r_1} - \frac{1}{r_2} + \frac{1}{r_3} - \frac{1}{r_4} - \frac{1}{r_5} + \frac{1}{r_6} \right) ,$$

where $r_1, \ldots, r_6$ are the distances between the electron and dipole ends (Fig. 7.16),
$r_1 = r_+$, and $r_2 = r_-$.

When the distance between the first and the second PG changes by u, the change
in the interaction energy is

$$W_- = -|qe| \left[\frac{1}{r_3(u)} - \frac{1}{r_4(u)} - \frac{1}{r_3(0)} + \frac{1}{r_4(0)} \right] = -\chi_- u + O(u^2) ,$$

and when the distance between the second and the third PG changes by u, the energy
change is

$$W_+ = -|qe| \left[\frac{1}{r_6(u)} - \frac{1}{r_5(u)} - \frac{1}{r_6(0)} + \frac{1}{r_5(0)} \right] = -\chi_+ u + O(u^2) ,$$

where $\chi_- = -dW_-/du|_{u=0}$ and $\chi_+ = -dW_+/du|_{u=0}$ are the 'left' and 'right' EPI
parameters, respectively.

Changing the length of the dipole R from $r_2 - r_1$ to $r_1 + r_2$, the interval of
permissible values of the EPI parameters $\chi_\pm$ can be obtained. The dependence of $\chi_\pm$
on R is shown in Fig. 7.17. The 'left' EPI parameter χ_- is always negative and the
'right' parameter χ_+ is always positive. The absolute value of χ_+ is approximately
twice the absolute value of χ_-. We can choose the following estimates of the 'left'
and 'right' parameters: $\chi_- = -(1\text{--}7) \times 10^{-10}$ N and $\chi_+ = (3\text{--}15) \times 10^{-10}$ N,
respectively.

7.6.2   Model Hamiltonian

The Hamiltonian function of a cyclic chain of PGs with an extra electron can be
represented as the sum of three terms:

$$H = H_{el} + H_{ph} + H_{int} , \tag{7.125}$$

Fig. 7.17 Dependence of the
EPI parameters χ_+ and χ_-
on the dipole length R

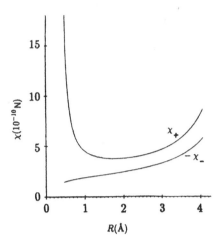

where the first term corresponds to the electron energy

$$H_{\text{el}} = \sum_{n=1}^{N} \left(\varepsilon_0 - \frac{\hbar^2}{ma^2} \right) |\phi_n|^2 - \frac{\hbar^2}{2ma^2} \phi_n^* (\phi_{n+1} + \phi_{n-1}) \, ,$$

the second represents the mechanical energy of the chain

$$H_{\text{ph}} = \sum_{n=1}^{N} \frac{M}{2} \dot{u}_n^2 + V(\rho_n) + \frac{K}{2} u_n^2 \, ,$$

and the third corresponds to the interaction energy between the electron and the chain deformation

$$H_{\text{int}} = - \sum_{n=1}^{N} (\chi_+ \rho_n + \chi_- \rho_{n-1}) |\phi_n|^2 \, .$$

Here, N is the number of GPs in the chain, $\{\phi_n\}_{n=1}^{N}$ is the wave function of the electron in the lattice representation, normalized by the condition

$$\sum_{n=1}^{N} |\phi_n|^2 = 1 \, , \qquad (7.126)$$

u_n is the displacement of the nth PG from its equilibrium position, $\rho_n = u_{n+1} - u_n$ is the lengthening of the nth hydrogen bond, $\varepsilon_0 = 9 \times 10^{-20}$ J is the electron affinity with the PG, $m = 2\hbar^2/\Delta E a^2$ the effective electron mass, $a = 4.5 \times 10^{-10}$ m is the chain spacing, $\Delta E = 1.2$ eV is the width of the conduction band [157], and $M = 114.2 m_{\text{p}}$ is the reduced mass of the PG, with $m_{\text{p}} = 1.67265 \times 10^{-27}$ kg the proton mass. The hydrogen bond energy is described by the Morse potential

$$V(\rho) = \varepsilon_1 \left[\exp(-b\rho) - 1 \right]^2 ,$$

where $\varepsilon_1 = 0.22\,\text{eV}$ is the energy of the hydrogen bond and the phenomenological parameter $b = \sqrt{\kappa/2\varepsilon_1}$ is defined by the elasticity of the bond $\kappa = 13\,\text{N/m}$. The coefficient K corresponds to the stiffness of the PG interaction with the chain substrate.

Note that the difference in signs of the 'right' and 'left' EPI parameters distinguishes the model Hamiltonian H considered here from those considered in the previous sections.

7.6.3   Stationary Autolocalized Electron States

Let us consider the stationary electron states. These can be obtained by solving the constrained minimum problem

$$E = \varepsilon_0 - 2J + \sum_{n=1}^{N} \left\{ -J(\varphi_{n+1} + \varphi_{n-1})\varphi_n - \varphi_n^2 \left[\chi_+(u_{n+1} - u_n) + \chi_-(u_n - u_{n-1}) \right] \right.$$

$$\left. + V(\rho_n) + \frac{K}{2} u_n^2 \right\} \longrightarrow \min : \sum_{n=1}^{N} \varphi_n^2 = 1 , \qquad (7.127)$$

where $\{\varphi_n\}_{n=1}^{N}$ are the amplitudes of the real electron wave function given in the lattice representation for the cyclic chain consisting of N molecules ($n + 1 = 1$ for $n = N$ and $n-1 = N$ for $n = 1$) and $J = \Delta E/4$. Further, let $\chi_+ = 15.0 \times 10^{-10}\,\text{N}$ and $\chi_- = -7.5 \times 10^{-10}\,\text{N}$.

Numerical solution of the problem (7.127) with $N = 100$ shows that, in addition to the stationary electron state in the conduction band $\{\varphi_n = 1/\sqrt{N}, u_n = 0\}$, there exist two stationary autolocalized (soliton) electron states characterized by different degrees of localization: strong (the localization region is approximately 1–2 chain links) and weak (the localization region is approximately 10–20 chain links). An increase in the stiffness K of the chain interaction with the substrate leads to a decrease in the soliton width. When $K > 0$, the band state can be stable.

The strongly localized soliton arises due to the opposite signs of the EPI parameters. In this state, the electron is mainly localized at the right end of the PG. The soliton is stable when $0 \le K \le 10.9\,\text{N/m}$. Figure 7.18 (left) shows the dependence of the profile of the strongly localized soliton on the parameter K.

The weakly localized soliton has a large enough width to respond on average to the FPI as the symmetric interaction with parameter $\chi = -(\chi_- + \chi_+)/2$. This soliton corresponds to the Davydov soliton. It is stable only if $0 \le K \le 0.31\,\text{N/m}$.

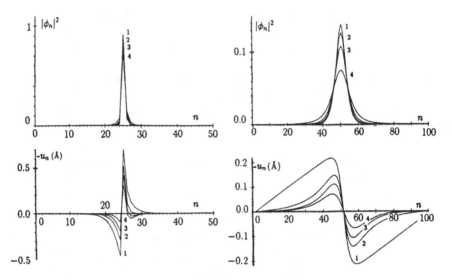

Fig. 7.18 *Left*: Profiles of the strongly localized soliton for $K = 4, 6, 8,$ and $10.8\,\text{N/m}$ (*lines 1–4*). *Right*: Profiles of the weakly localized soliton for $K = 0, 0.1, 0.2,$ and $0.3\,\text{N/m}$ (*lines 1–4*)

An increase in K leads to an increase in the soliton width and a decrease in chain deformation (see Fig. 7.18 right).

In a cyclic chain consisting of $N = 100$ PGs, the electron band state is stable if $K \geq 0.19\,\text{N/m}$ and unstable if $K < 0.19\,\text{N/m}$. Therefore, the following states can exist in the chain:

- If $K < 0.19\,\text{N/m}$, two stable soliton modes,
- If $0.19 \leq K \leq 0.31\,\text{N/m}$, two stable soliton modes and a stable band electron state,
- If $0.31 \leq K \leq 10.9\,\text{N/m}$, stable strongly localized soliton mode and stable band state, and
- If $K \geq 10.9\,\text{N/m}$, only a stable band state.

7.6.4 Dynamics of Autolocalized Electron State in a Thermalized Chain

The equations of motion corresponding to the Hamiltonian (7.125) have the form

$$i\hbar\dot{\phi}_n = -J(\phi_{n+1} + \phi_{n-1}) - |\phi_n|^2(\chi_+\rho_n + \chi_-\rho_{n-1}) ,$$
$$M\ddot{u}_n = V'(\rho_n) - V'(\rho_{n-1}) - Ku_n - \chi_+(|\phi_n|^2 - |\phi_{n-1}|^2) \qquad (7.128)$$
$$-\chi_-(|\phi_{n+1}|^2 - |\phi_n|^2) - \Gamma M\dot{u}_n + \xi_n(t) ,$$

for $n = 1, 2, \ldots, N$. Here, the Langevin approach was used to describe the thermal vibrations of the PGs in the chain. In the second equation of the system (7.128), we introduced viscous friction and the normally distributed external random force $\xi_n(t)$, which satisfies the conditions $\langle \xi_n(t) \rangle = 0$ and $\langle \xi_n(t)\xi_l(s) \rangle = 2M\Gamma k_B T \delta_{nl} \Delta(t-s)$, where Γ is the friction coefficient, $k_B = 1.380622 \times 10^{-23}$ J is the Boltzmann constant, T is the thermostat temperature, $\Delta(t - s) = 1/t_2$ for $|t - s| \leq t_2/2$ and $\Delta(t - s) = 0$ for $|t - s| > t_2/2$, and t_2 is the relaxation time of the random force. In order to apply the Langevin equation, the relaxation time of the velocity must be equal to $t_1 = 1/\Gamma \gg t_2$.

The maximum electron velocity $v_{el} = 2aJ/\hbar = 410,175.5$ m/s in the chain is more than two orders of magnitude greater than the velocity $v_{ph} = a\sqrt{\kappa/M} = 3,712.3$ m/s of the long-wave phonons. Hence, for numerical simulation of the dynamics, it is convenient to introduce the dimensionless time $\tau = 2Jt/\hbar$, whose unit corresponds to the minimum possible time of the electron passage of one chain link, namely, 1.09702×10^{-15} s. Let the relaxation time of the PG velocity t_1 be equal to the time taken by sound to pass one chain link, i.e., $t_1 = \sqrt{M/\kappa} = 1.21216 \times 10^{-13}$ s, while the correlation time of the random force is equal to $t_2 = \hbar/2J$.

The equations of motion (7.128) were integrated using the standard fourth-order Runge–Kutta method with constant integration step $\Delta\tau = 0.1$ (corresponding to $\Delta t = 1.09702 \times 10^{-16}$ s) [123]. The accuracy of the numerical integration was estimated by conservation of the integral (7.126). The initial conditions corresponded to the stationary electron state in a cyclic chain with temperature $T = 0$ K. The thermostat temperature is $T = 300$ K. Electron localization in the chain and chain thermalization are conveniently characterized by the dimensionless width of the electron wave function $L(\tau) = 1/\sum_{n=1}^{N} |\phi_n(\tau)|^4$ and the instantaneous temperature $T(\tau) = \sum_{n=1}^{N} M\dot{u}_n^2/Nk_B$.

The dependence of the width $L(\tau)$ on time for $N = 50$ and $K = 4$ N/m is shown in Fig. 7.19 (left) for the strongly localized soliton. The numerical integration shows the stability of the strongly localized soliton electron state with respect to the thermal vibrations of the PG chain, which lead only to insignificant vibrations of the soliton width. The band electron state is unstable with respect to thermal vibrations of the PG chain, which causes its decay and the appearance of several strongly

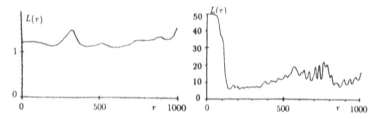

Fig. 7.19 *Left*: Dependence of the width L of the strongly localized soliton on time τ in the thermalized chain at $K = 4$ N/m. *Right*: Dependence of the width L of the band electron state on time τ in the thermalized chain with $K = 4$ N/m

localized soliton states (Fig. 7.19 right). For $K = 0.2$ N/m, the weakly localized soliton is also unstable with respect to thermal vibrations. Thermal fluctuations cause its collapse and the appearance of several strongly localized soliton states.

The numerical simulation carried out shows that electrosolitons can exist in α-helical proteins at the physiological temperature $T \sim 300$ K. The results discussed in this section were published in [160].

7.7 Resonant Effects of Microwaves Due to Their Interaction with Solitons in α-Helical Proteins

7.7.1 Biological Effects of EMR

The dynamic properties of solitons manifest themselves through their interaction with external fields and other degrees of freedom which determine soliton relaxation processes. In the previous sections, we investigated the dynamics of localized states in thermalized soft molecular chains. It was shown that chaotic (thermal) molecular vibrations do not destroy these states. However, it has been reported in a number of studies [54, 161, 162] that soliton decay can be induced by sound waves [162–164] or electromagnetic radiation (EMR). Electromagnetic radiation absorption at frequencies determined by the binding energy may indicate the presence of solitons in molecular chains. This can provide a basis for experiments to test the presence of solitons in biological systems [112, 162, 165, 166]. One experiment to observe electrosolitons [163] could involve registering the occurrence of (or change in) electric current in a sample.

The hypothesis and mechanism of Davydov soliton dissociation in the electromagnetic radiation field, as suggested in [161, 167], can provide a qualitative explanation for the resonant effects of millimeter range EMR on biological systems, as observed in a number of experiments [168, 169]. From a scientific point of view, a study of the mechanisms underlying external field effects (including resonant effects) on soliton processes may lead to an understanding of these phenomena, which still lack a strict explanation, and it may then become possible to use them to solve applied problems [170]. In particular, the biological effects on living organisms of different organization levels observed under exposure to low-intensity electromagnetic fields can be explained as the result of EMR interaction with collective excitations in α-helical proteins – Davydov solitons – which, according to the Davydov theory, are responsible for energy and charge transfer on the molecular level. In this section, we will discuss the properties of Davydov solitons over the temperature range from 0 to 350 K and define the frequency spectrum of EMR effectively interacting with biological systems [171–178].

One of the most important characteristics of the autolocalized state is the coupling energy of a quasi-particle (exciton or electron) with molecular chain deformation, which also determines the soliton stability. Davydov solitons are stable because the probability of their energy dissipating into thermal energy is very small, and this ensures a high efficiency for soliton transport of energy, charge, and conformation changes in biosystems at a physiological temperature of 310 K. However, under electromagnetic radiation, the probability of soliton decay (photodissociation) increases [64, 179–181]. This may explain some of the resonant effects of low-intensity electromagnetic radiation on biological systems, as observed in numerous experiments [182–187]. In this connection, the dependence of resonant frequencies and hence binding energy on the temperature of a molecular chain are of interest.

In addition to the transport of energy and electrical charges by Davydov solitons, excess charges may be transferred along α-helical proteins by being captured by the moving acoustic solitons [188–191]. In this section, we also discuss photodissociation of these acoustic solitons, transferring an excess charge along a molecular chain, and Davydov electrosoliton photodissociation under the influence of electromagnetic radiation at specific frequencies.

7.7.2 Stationary Autolocalized States in Thermalized Molecular Chains

In the previous sections, we investigated the conditions for the formation of various autolocalized states, including Davydov solitons, for the temperature range from 0 to 350 K, and studied the dependence of the binding energy on temperature in this range. Recall that the energy operator of the intramolecular excitation of a discrete molecular chain can be represented as the sum of three terms:

$$H = H_e + H_{ph} + H_{int} , \tag{7.129}$$

where the first term describes the energy of the intramolecular amide-I excitation, viz.,

$$H_e = \sum_{n=1}^{N} \varepsilon_0 B_n^+ B_n^- - J \left(\sum_{n=1}^{N-1} B_n^+ B_{n+1}^- + \sum_{n=1}^{N-1} B_n^+ B_{n-1}^- \right) , \tag{7.130}$$

the second represents the deformation energy of the molecular chain, viz.,

$$H_{ph} = \sum_{n=1}^{N} \frac{1}{2} M \dot{u}_n^2 + \sum_{n=1}^{N} \frac{1}{2} \kappa (u_n - u_{n-1})^2 , \tag{7.131}$$

and the third is the energy of the interaction between the intramolecular excitation and chain deformation, viz.,

$$H_{int} = \chi \sum_{n=1}^{N} B_n^+ B_n^- (u_{n+1} - u_n) . \tag{7.132}$$

Here, B_n^+ and B_n^- are the creation and annihilation operators of the amide-I vibration at the nth molecule, respectively, u_n is the operator of longitudinal displacements from the equilibrium position of the nth molecule, κ is the chain elasticity coefficient, M is the molecular mass, ε_0 is the excitation energy including the non-resonant interaction with neighboring molecules, J is the energy of the resonant dipole–dipole interaction of neighboring molecules, and χ is the exciton–phonon interaction parameter. The wave function of the molecular chain can be represented in the following form (the Davydov ansatz):

$$|\psi\rangle = \sum_{n=1}^{N} \phi_n S_n B_n^+ |0\rangle \otimes |v\rangle , \qquad (7.133)$$

where

$$S_n = \exp \left[\sum_q (\beta_{q,n}^* b_q^- - \beta_{q,n} b_q^+) \right]$$

is the unitary operator of the molecular displacement from its equilibrium position,

$$|v\rangle = |\{v_q\}\rangle = \prod_q |v_q\rangle = (v_q!)^{-1/2} (b_q^+)^v |0\rangle$$

is the phonon wave function, and b_q^+ and b_q^- are the creation and annihilation operators, respectively, of a phonon with wave number q.

In order to take into account the interaction of the chain with a heat bath, we take $v_q = \tilde{v}_q = \left[\exp(\hbar \Omega_q / k_B T) - 1 \right]^{-1}$, where Ω_q and q are the frequency and wave number of a phonon, respectively, and k_B is the Boltzmann constant. To obtain a stationary soliton state, one must solve the constrained minimum problem

$$\mathscr{H} \left(\ldots, \phi_n, \phi_n^*, \beta_{q,n} \beta_{q,n}^*, \ldots \right) \longrightarrow \min : \sum_{n=1}^{N} |\phi_n|^2 = 1 , \qquad (7.134)$$

where $\mathscr{H} = \langle \psi | H | \psi \rangle$ is the Hamiltonian of the molecular chain and H is the energy operator.

This problem was solved numerically by the steepest descent method with varying step [123]. The analysis carried out has shown that the degree of localization of the intramolecular excitation in soliton states increases with increasing temperature T and EPI parameter χ (see Fig. 7.20).

With an increase in T and χ, there is also an increase in the width of the energy gap ΔE which separates the soliton state energy from the exciton zone (see Fig. 7.21). Therefore, the conditions for Davydov soliton photodissociation under EMR depend on both the temperature T and the EPI parameter χ. The Davydov soliton is a bound state of the intramolecular excitation with chain deformation (phonons).

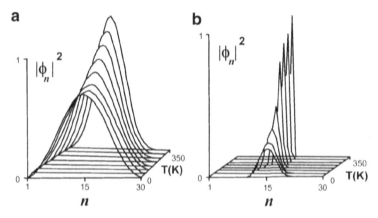

Fig. 7.20 Localization of the intramolecular excitations in soliton states. Dependence of the autolocalized state profile on temperature T and the parameter χ in the chain consisting of 30 molecules, for $\chi = 3 \times 10^{-11}$ N (**a**) and $\chi = 12 \times 10^{-11}$ N (**b**)

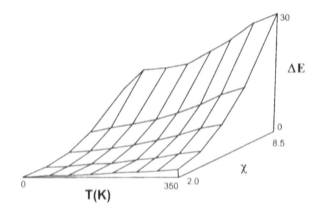

Fig. 7.21 Dependence of the width of the energy gap ΔE (10^{-4} eV) which separates the soliton state from the exciton zone (or conduction band) on the temperature T and the EPI parameter χ (10^{-11} N)

For strong coupling, that is, when the dimensionless nonlinearity parameter $g = \chi^2/J\kappa \gg 1$, the energy required for its breakage is given by

$$\hbar\omega_0 = \frac{hc}{\lambda} = \Delta E + E_{\mathrm{ph}} , \qquad (7.135)$$

where E_{ph} is the deformation energy. For a weak bond $g \ll 1$, this energy is

$$hc/\lambda = \Delta E . \qquad (7.136)$$

The dependencies of the 'resonant' wavelength λ of EMR on T and χ are given in Tables 7.12 and 7.13.

Table 7.12 Dependencies of the resonant wavelength λ (mm) on the temperature T and the EPI parameter χ (N) for a chain with dimensionless nonlinearity parameter $g \ll 1$

T (K)	$\chi = 3 \times 10^{-11}$	5×10^{-11}	7×10^{-11}	9×10^{-11}	12×10^{-11}
250	9.8	3.35	1.65	0.97	0.52
300	8.3	2.88	1.43	0.85	0.47
350	7.5	2.64	1.32	0.79	0.44

Table 7.13 Dependencies of the resonant wavelength λ (mm) on the temperature T and the EPI parameter χ (N) for a chain with dimensionless nonlinearity parameter $g \gg 1$

T (K)	$\chi = 3 \times 10^{-11}$	5×10^{-11}	7×10^{-11}	9×10^{-11}	12×10^{-11}
250	5.8	1.9	0.93	0.53	0.27
300	4.6	1.6	0.80	0.46	0.24
350	4.0	1.4	0.70	0.42	0.23

If energy transport in biosystems is realized by Davydov solitons on the molecular level, soliton decay under EMR can cause resonant biological effects in living organisms in the millimeter wavelength range. This follows from the analysis of the values of λ obtained for physiological temperatures $T \approx 300$ K and values of the EPI parameter in the range $3 \times 10^{-11} \le \chi \le 5 \times 10^{-11}$ N, which correspond to the 'soliton window'. The most significant biological effects caused by Davydov soliton photodissociation can be expected at EMR wavelengths $\lambda = 1.6$–4.6 mm corresponding to the frequency range 65–187.5 GHz and at wavelengths 2.9–8.3 mm corresponding to the frequency range 36–103 GHz. When the temperature increases, the 'resonant' wavelength of EMR also increases.

7.7.3 Stationary States of Excess Electrons in a Chain of PGs

Transport of extra electrons along α-helix proteins is determined by the specific properties of these molecules. Due to the asymmetric distribution of electric charge in PGs, these proteins have a large constant dipole moment. As a result, the chain of PGs is capable of capturing the outer electron from a donor molecule. As has been shown previously by both analytical investigation and numerical simulation, the outer electron can propagate along the thermalized chain of PGs in the form of an electrosoliton (ES) if the value of electron–phonon interaction parameter exceeds a certain threshold.

In this section we investigate the possibility of electrosoliton decay under the action of EMR. The Hamiltonian for a chain of PGs with an excess electron is written, as before, as the sum (7.129), where now

$$H_e = \sum_{n=1}^{N} \left(\varepsilon_0 - \frac{\hbar^2}{ma^2} \right) |\phi_n|^2 - \frac{\hbar^2}{2ma^2} \phi_n^*(\phi_{n+1} + \phi_{n-1}) , \qquad (7.137)$$

$$H_{\text{ph}} = \sum_{n=1}^{N} \frac{M}{2} \dot{u}_n^2 + V(\rho_n) + \frac{K}{2} u_n^2 , \tag{7.138}$$

$$H_{\text{int}} = \sum_{n=1}^{N} (\chi_+ \rho_n + \chi_- \rho_{n-1}) |\phi_n|^2 . \tag{7.139}$$

Here, N is the number of PGs in the chain and ϕ_n is the wave function of the electron in the PG chain, satisfying the normalization condition $\sum_{n=1}^{N} |\phi_n|^2 = 1$, u_n is the displacement of the nth PG from its equilibrium position, $\rho_n = u_{n+1} - u_n$ is the deformation of the nth hydrogen bond, $\varepsilon_0 = 9 \times 10^{-20}$ J is the electron affinity with the PG, $m = 2\hbar^2/\Delta E a^2$ is the effective electron mass, $a = 4.5 \times 10^{-10}$ m is the chain step, $\Delta E = 1.2$ eV is the width of the conduction band, and $M = 114.2 m_p$ is the PG mass ($m_p = 1.67265 \times 10^{-27}$ kg is the proton mass).

The hydrogen bond interaction is described by the Morse potential

$$V(\rho) = \varepsilon_1 \left[\exp(-b\rho) - 1 \right]^2 ,$$

where $\varepsilon_1 = 0.22$ eV is the hydrogen bond energy, the phenomenological parameter $b = \sqrt{\kappa/2\varepsilon_1}$ characterizes the bond elasticity, and $\kappa = 13$ N/m. The coefficient K describes the elastic interaction of the PG with the chain substrate.

As shown previously, the 'right' χ_+ and 'left' χ_- EPI parameters have different signs. Consideration of both this asymmetry and the presence of optical phonons in the PG chain ($K > 0$) distinguishes the Hamiltonian $\mathcal{H}$ from the Davydov Hamiltonian (7.129–7.132), where $\chi_+ = \chi_- > 0$ and $K = 0$.

To find the stationary states of an excess electron in the PG chain, one must solve the constrained minimum problem

$$E = \varepsilon_0 - 2J + \sum_{n=1}^{N} \left\{ - J(\varphi_{n+1} + \varphi_{n-1})\varphi_n + \varphi_n^2 [\chi_+(u_{n+1} - u_n) + \chi_-(u_n - u_{n-1})] \right.$$

$$\left. + V(\rho_n) + \frac{1}{2} K u_n^2 \right\} \longrightarrow \min : \sum_{n=1}^{N} \varphi_n^2 = 1 , \tag{7.140}$$

where $\{\varphi_n\}_{n=1}^{N}$ are the real wave function amplitudes of the electron in the chain consisting of N PGs and $J = \Delta E/4$.

Let us assume further that $\chi_+ = 15.0 \times 10^{-10}$ N and $\chi_- = 7.5 \times 10^{-10}$ N. Numerical solution of the problem (7.137) with $N = 100$ has shown that, in addition to the stationary electron state in the conduction band $\{\varphi_n = 1/\sqrt{N}, u_n = 0\}$, there are two stationary autolocalized (soliton) states of an excess electron (electrosoliton), which are characterized by the different degrees of localization: strong (with localization region spanning $\sim$1–2 chain links) and weak (with localization region spanning $\sim$10–20 links). An increase in the elastic interaction of the chain with the substrate atoms leads to a decrease in the electrosoliton width.

Table 7.14 Dependence of the binding energy ΔE of the strongly localized electrosoliton state on the parameter K (N/m)

K	4	6	8	18.8
ΔE	0.08	0.11	0.15	0.37

Table 7.15 Dependence of the binding energy ΔE of the weakly localized electrosoliton state on the parameter K (N/m)

K	0	0.1	0.2	0.3
ΔE	0.0081	0.017	0.0265	0.0341

The electrosoliton is stable if $0 \leq K \leq 10.9\,\text{N/m}$. Figure 7.18 (right) shows the dependence of the profile of the strongly localized soliton on the parameter K, while Tables 7.14 and 7.15 give the dependencies of the energy gaps of strongly and weakly localized solitons on the parameter K.

The weakly localized soliton is broad enough to allow us to treat the electron–phonon interaction as symmetrical, with the EPI parameter $\chi = (\chi_- + \chi_+)/2$. This soliton corresponds to the Davydov electrosoliton (DES), which is stable only if $0 \leq K \leq 0.31\,\text{N/m}$. With increasing K, the soliton width increases and the chain deformation decreases (see Fig. 7.18 right).

In a cyclic chain consisting of 100 PGs, the bound state of an excess electron remains stable if $K \geq 0.19\,\text{N/m}$ and becomes unstable if $K < 0.19\,\text{N/m}$. Therefore, two stable soliton modes exist in the chain for $K < 0.19\,\text{N/m}$. If $0.19 \leq K \leq 0.31\,\text{N/m}$, there can exist two stable soliton modes together with the stable band state of an excess electron in the chain. For $K \geq 10.9\,\text{N/m}$, only one stable band state exists in the chain. It follows from the data given in Tables 7.14 and 7.15 that there can be frequency-dependent bioeffects of microwave EMR on the electrosoliton states in α-helix proteins in the following frequency ranges: 19.3–89.5 and 1.96–8.25 THz, depending on the substrate properties.

7.7.4 Photodissociation of the Electrosoliton Under EMR

As shown in the previous section, transport of electric charges along a protein molecule can proceed as a result of excess electron capture by the acoustic soliton moving along the chain. The presence of the soliton leads to an additional interaction energy (the deformation potential) for this extra electron in a periodic chain of PGs interacting with the electron. As a result, the electron can be captured by the acoustic soliton and can then move together with the soliton in the form of an electrosoliton.

In this section, we consider electrosoliton decay under EMR. To obtain analytical results, the continuum approximation is usually used. The investigation of a finite discrete chain can be carried out only by numerical simulation. The state of an extra electron interacting with PGs in the chain is defined by the Hamiltonian (7.129),

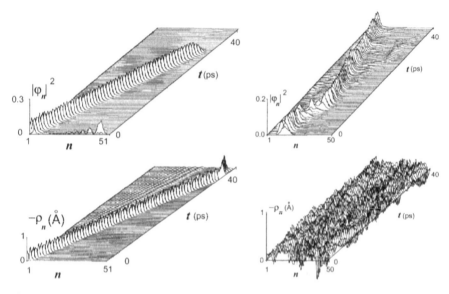

Fig. 7.22 *Left*: Excess electron capture by the acoustic compression soliton with energy 0.49 eV. (At the initial time the electron was located near the first chain molecule.) *Right*: Electrosoliton stability to thermal vibrations of the chain. The soliton dynamics in the thermalized chain is shown for $T = 300$ K

with terms H_e, H_{ph}, and H_{int} determined by (7.137)–(7.139). The supersonic soliton ($v_{sol} \geq v_{sound}$ and $E_{sol} = 0.49$ eV) and electrosoliton were obtained by numerical integration of the equations of motion (see Fig. 7.22 left). Analysis of the solutions shows that the electrosoliton is stable with respect to thermal vibrations of the PG chain at $T = 300$ K. The thermal vibrations cause only insignificant changes in the soliton width (Fig. 7.22 right).

The electron state is characterized by the wave functions $\{\phi_n\}$ (see Fig. 7.23 left) and the spectrum of negative energies $\{E_n\}$ (see Fig. 7.23 right).

The absorption of microwave radiation interacting with electrosolitons can result both from the electron transition between its energy levels and, directly, from the electrosoliton dissociation into a free electron and acoustic soliton under the EMR. If the Frank–Condon condition holds (chain atoms do not change their positions during electron interaction with EMR), the absorption spectrum is defined by the usual relationship $h\nu = E_m - E_n$ (in the case of electrosoliton dissociation, $E_m - E_n \geq 0$).

Numerical investigation of the electrosoliton dynamics has revealed two specific frequencies: $\nu_1 \approx 1.8$ THz and $\nu_2 \approx 2.7$ THz, which are determined by the discrete nature of the PG chain. To take into consideration internal electrosoliton oscillations, a semiclassical approach was used.

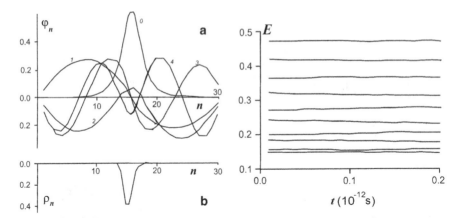

Fig. 7.23 *Left*: Wave functions $\{\phi_n\}$ of the first five electron states in the potential well formed by the acoustic soliton (**a**) and the chain displacement ρ_n (Å) in the excitation region (**b**). The soliton energy is 0.49 eV. *Right*: Energy levels of the electron $-E_n$ (eV) captured by the acoustic soliton. Small changes in the electrosoliton energy with time are due to the chain discreteness

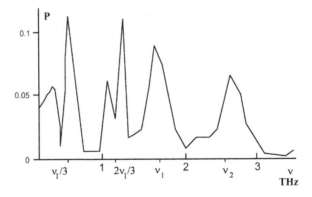

Fig. 7.24 Dependence of the probability P of electrosoliton photodissociation on the frequency of the external field ν

The equations of motion for the wave function $\{\phi_n\}$ corresponding to the Hamiltonian of the electrosoliton in the presence of EMR have the form

$$\hbar\dot{\phi}_n = eE_0\cos(2\pi\nu t)na\phi_n - J(\phi_{n+1}+\phi_{n-1})+|\phi_n|^2(\chi_+\rho_n+\chi_-\rho_{n-1}) , \quad (7.141)$$

where E_0 is the electric field strength and ν is its frequency.

Figure 7.24 shows the dependence of the probability of electrosoliton photodissociation on the frequency ν, viz.,

$$P = 1 - \sum_{n=1}^{N+1} |\varphi_n(t)|^2 . \quad (7.142)$$

By numerical integration of the equations of motion (7.141), the four resonant frequencies in the absorption spectrum of microwave EMR were found to be $\nu_1/3$, $2\nu_1/3$, ν_1, and ν_2.

Therefore, the theoretical analysis of cooperative processes has shown that the frequency-dependent, resonance-type biological effects of microwaves on α-helix macromolecules, interacting with the environment (thermostat), are defined by the behavior of the (bio)system as a whole, and hence can depend on its biological activity. This approach provides a qualitative explanation for the resonant bioeffects caused by low-intensity EMR on living organisms if the energy transfer of ATP hydrolysis and/or electron transport from donor to acceptor molecules in real protein molecules are realized by different types of solitons. Then, as a result of the soliton state decay, the efficiency of the energy and electron transport should decrease. Assuming this mechanism, the resonance frequencies were obtained at which the absorption spectrum of EMR and corresponding bioeffects should be observed.

References

1. Green, D.E.: Mechanism of energy transduction in biological systems. Science **181**, 583 (1973)
2. Davydov, A.S., Kislukha, N.I.: Solitary excitations in one-dimensional molecular chains. Phys. Stat. Sol. (b) **59**, 465 (1973)
3. Davydov, A.S.: Theory of Molecular Excitons. Pergamon Press, New York (1970)
4. Agranovich, V.M.: Theory of Excitons. Nauka, Moscow (1968)
5. Davydov, A.S.: Solid State Theory. Nauka, Moscow (1976)
6. Hyman, J.M., McLaughlin, D.W., Scott, A.C.: On Davydov's alpha-helix solitons. Physica D **3**, 23 (1981)
7. Scott, A.C.: Dynamics of Davydov solitons. Phys. Rev. A **26**, 578 (1982)
8. Scott, A.C.: The vibrational structure of Davydov solitons. Phys. Scr. **25**, 651 (1982)
9. Lomdahl, P.S., Layne, S.P., Bigio, I.J.: Solitons in biology. Los Alamos Sci. **10**, 2 (1984)
10. Scott, A.C.: Davydov solitons in polypeptides. Philos. Trans. R. Soc. Lond. A **315**, 423 (1985)
11. Landau, L.D.: Uber die Bewegund der Electronen in Kristallgotter. Phys. Z. Sowjetunion **3**, 664 (1933)
12. Pekar, S.I.: Autolocalization of the electron in the dielectric inertially polarized media. J. Exp. Theor. Phys. **16**, 335 (1946)
13. Pekar, S.I.: A Study on the Electron Theory of Crystals. Gostekhizdat, Moskva (1951)
14. Firsov, Y.A. (ed.): Polarons. Nauka, Moscow (1975)
15. Toda, M.: Theory of Nonlinear Lattices, 2nd edn. Springer, Berlin (1989)
16. Zabusky, N.J., Kruskal M.D.: Interaction of solitons in a collisionless plasma and the recurrence of initial states. Phys. Rev. Lett. **15**(6), 240 (1965)
17. Gardner, C.S., Greene, J.M., Kruskal, M.D., Miura, R.M.: Method for solving the Korteweg–de Vries equation. Phys. Rev. Lett. **19**, 1095 (1967)
18. Shabat, A., Zakharov, V.: Exact theory of two-dimensional self-focusing and one-dimensional self-modulation of waves in nonlinear media. Sov. Phys. JETP **34**, 62 (1972)
19. Zakharov, V.E., Faddeev, L.D.: Korteweg–de Vries equation: a completely integrable Hamiltonian system. Funct. Anal. Appl. **5**, 280 (1971)
20. Zakharov, V.E., Shabat, A.B.: A scheme for integrating the nonlinear equations of mathematical physics by the method of the inverse scattering problem. I. Funct. Anal. Appl. **8**, 226 (1974)

21. Zakharov, V.E., Manakov, S.V., Novikov, S.P.: Theory of Solitons: The Method of the Inverse Scattering Problem. Nauka, Moscow (1980)
22. Makhankov, V.G.: Dynamics of classical solitons (in non-integrable systems). Phys. Rep. C **35**, 1 (1978)
23. Bullaf, R., Codri, F.: Solitons. Mir, Moscow (1983)
24. Tahtadjian, L.A., Faddeev, L.D.: Hamiltonian Approach in Soliton Theory. Nauka, Moscow (1986)
25. Ablowitz, M., Sigur, H.: Solitary Waves and the Inverse Problem Method. Mir, Moscow (1982)
26. Davydov, A.S.: Solitons in Molecular Systems. Naukova Dumka, Kiev (1988)
27. Davydov, A.S.: Solitons in Bioenergetics. Naukova Dumka, Kiev (1986)
28. Zakharov, V.E.: Collapse of Langmuir waves. Sov. Phys. JETP **35**, 908 (1972)
29. Kampan, V.I.: Nonlinear Waves in Dispersive Medium. Nauka, Moscow (1973)
30. Kadomtsev, B.B., Karpman, V.I.: Nonlinear waves. Phys.-Usp. **14**, 40 (1971)
31. Scott, A.C., Chu, F.Y.F., McLaughlin, D.W.: The soliton: a new concept in applied science. Proc. IEEE **61**, 1443 (1973)
32. Wizem, J.: Linear and Non-Linear Waves. Mir, Moscow (1977)
33. Bishop, A.R., Krumhansl, J.A., Trullinger, S.E.: Solitons in condensed matter: a paradigm. Physica D **1**, 1 (1980)
34. Longren, K., Scott, A.C. (eds.): Solitons in Action. Academic, New York (1978)
35. Kosevich, A.M., Ivanov, B.A., Kovalev, A.S.: Nonlinear Magnetization Waves: Dynamical and Topological Solitons. Naukova Dumka, Kiev (1983)
36. Kosevich, A.M.: Dislocations in Elasticity Theory. Naukova Dumka, Kiev (1978)
37. Kosevich, A.M.: Physical Mechanics of Real Crystals. Naukova Dumka, Kiev (1981)
38. Davydov, A.S.: Solitons in quasi-one-dimensional molecular structures. Sov. Phys. Usp. **25**, 898 (1982)
39. Makhankov, V.G., Fedyanin, V.K.: Nonlinear effects in quasi-one-dimensional models of condensed matter theory. Phys. Rep. C **104**, 1 (1984)
40. Ovchinnikov, A.A., Ukrainskii, I.I.: Electronic processes in 1-d systems. Sov. Sci. Rev. Sect. B **9**, 123 (1987)
41. Kosevich, A.M.: Nonlinear mechanics of crystals (one-dimensional problem). In: Proceedings of Ural Scientific Centre of RA USSR, Institute of Metal Physics, Sverdlovsk (1975)
42. Petrina, D.Ya., Enol'skii, V.Z.: On oscillations of one-dimensional systems. Dokl. Akad. Nauk Ukr. SSR Ser. A **8**, 759 (1976)
43. Davydov, A.S., Kislukha, N.I.: Solitons in one-dimensional molecular chains. Sov. Phys. JETP **44**, 571 (1976)
44. Rashba, E.I.: The theory of strong interactions of electron excitations with the lattice vibrations in molecular solids. Opt. i Spektrosk. **2**(1), 75 (1957)
45. Rashba, E.I.: Exciton interaction with molecular lattice. Izv. AS USSR Phys. **21**(1), 37 (1957)
46. Toyozawa, Y.: Self-trapping of an electron by the acoustic mode of lattice vibration. I. Prog. Theor. Phys. **26**, 29 (1961)
47. Davydov, A.S., Nitsovich, B.M.: Exciton–phonon interaction in one-dimensional molecular chains. Fiz. Tverd. Tela. **9**(1), 2230 (1967)
48. Fedyanin, V.K., Yakushevich, L.V.: On exciton–phonon interaction in one-dimensional molecular chains. I. Linear approximation. Theor. Math. Phys. **30**(1), 133 (1977)
49. Fedyanin, V.K., Yakushevich, L.V.: Elementary excitations in one-dimensional systems with resonance interaction. Theor. Math. Phys. **37**, 1081 (1978)
50. Iordansky, S.V., Rashba, E.I.: Continuum model of tunnel autolocalization. ZhETF74 **74**(5), 1872 (1978)
51. Ioselevich, A.S., Rashba E.I.: Theory of autolocalization rate. Zh. Eksp. Teor. Fiz. **88**(5), 1873 (1985)
52. Degtiarev, L.M., Nakhankov, V.G., Rudakov L.I.: Dynamics of the formation and interaction of Langmuir solitons and strong turbulence. Sov. Phys. JETP **40**, 264 (1975)

53. Schulman, E.I.: On the integrability of long and short wave resonant interaction equations. Dokl. Acad. Nauk USSR **259**, 579 (1981)
54. Malomed, B.A.: Soliton radiation stimulated by a sound wave or an external field. Sov. Phys. Phys. Plasma **13**(6), 662 (1987)
55. Davydov, A.S., Zolotaryuk, A.V.: Autolocalized collective excitations in molecular chains with cubic anharmonicity. Phys. Stat. Sol. (b) **115**, 115 (1983)
56. Davydov, A.S., Zolotaryuk, A.V.: Solitons in molecular systems with nonlinear nearest-neighbor interactions. Phys. Lett. A **94**, 49 (1983)
57. Davydov, A.S., Zolotaryuk, A.V.: Electrons and excitons in nonlinear molecular chains. Phys. Scr. **28**, 249 (1983)
58. Karabovsky, A.V., Kislukha, N.I.: A variational approach to the problem of solitons in molecular chains. Ukr. Fiz. Jurn. **27**(4), 496 (1982)
59. Skrinjiar, M.J., Kapor, D.V., Stojanovic, S.D.: Solitons in molecular chains with cubic anharmonicity. Phys. Lett. A **117**, 199 (1986)
60. Primatarova, M.T.: Solitary excitations in molecular chains due to the anharmonicity of the intermolecular vibrations. Phys. Stat. Sol. (b) **138**, 101 (1986)
61. Ablowitz, M.J., Ladik, J.F.: A nonlinear difference scheme and inverse scattering. Stud. Appl. Math. **55**(3), 213 (1976)
62. Ablowitz, M.J., Ladik, J.F.: Nonlinear differential-difference equations and Fourier analysis. J. Math. Phys. **17**(6), 1011 (1976)
63. Kuprievich, V.A.: Variational study of stationary localized states of molecular chains in a discrete model. Theor. Math. Phys. **64**, 269 (1985)
64. Kuprievich, V.A.: On autolocalization of the stationary states in a finite molecular chain. Phys. D: Nonlinear Phenom. **14**, 395 (1985)
65. Kislukha, N.I.: Strong autolocalization of exciton in monomolecular film. Ukr. Fiz. Zh. **23**(2), 209 (1983)
66. Kislukha, N.I.: Effect of discreteness in the theory of Davydov solitons. Ukr. Fiz. Zh. **31**, 1323 (1986)
67. Vakhnenko, A.A., Gaididei, Yu.B.: On the motion of solitons in discrete molecular chains. Theor. Math. Phys. **68**, 873 (1986)
68. Kyslukha, N.I., Karbovsky, A.V.: Davydov soliton under consideration of acoustic wave dispersion. Ukr. Fix. Zh. **27**(3), 328 (1982)
69. Eremko, A.A., Sergeenko, A.I.: On soliton theory in molecular chains. Ukr. Fiz. Zh. **24**(12), 3720 (1982)
70. Eremko, A.A., Sergeenko, A.I.: Excitons and soliton excitations in molecular chains. Ukr. Fiz. Zh. **28**, 338 (1983)
71. Sergeenko, A.I.: Solitons in closed molecular chains. Nonlinear Turbul. Process. Phys. **1**, 1051 (1984)
72. Kapitanchuk, O.L., Kudritskaia, Z.G., Shramko, O.V.: Numerical study of the soliton dynamics in discrete cyclic chains. Ukr. Fiz. Zh. **32**(4), 498 (1987)
73. Tran, P.X., Reichl, L.E.: The effect of finiteness on the selftrapping of an electron in a one-dimensional lattice. Phys. Lett. A **121**, 135 (1987)
74. Tran, P.X.: Bistability transition in the self-trapping of an electron in a one-dimensional molecular chain. Phys. Lett. A **123**, 231 (1987)
75. van Velzen, G.A., Tjon, J.A.: Numerical studies on the stability of Davydov solitons. Phys. Lett. A **116**, 167 (1986)
76. Zolotaruk, A.V., Savin, A.V.: O dinamicheskoi ustoichivosti nelineinyh kollektivnyh vozbuzhdenij v molekulyarnyh tsepochkah. Preprint ITF-87-129R, Kiev (1987)
77. Davydov, A.S., Eremko, A.A., Sergeenko, A.I.: Solitons in α-helix protein molecules. Ukr. Fis. Zh. **23**, 983 (1978)
78. Kuprievich, V.A., Klimenko, V.E., Shramko, O.V.: Stationary autolocalized states in α-helix polypeptide. Ukr. Fiz. Zh. **30**(8), 1158 (1985)
79. Zabusky, N.J.: Solitons and energy transport in nonlinear lattices. Comput. Phys. Commun. **5**, 1 (1973)

80. Makhankov, V.G.: On stationary solutions of Schrödinger equation with a self-consistent potential satisfying Boussinesq's equation. Phys. Lett. A **50**(1), 42 (1974)
81. Ktitorov, S.A., Siparov, S.V.: Condensons in thin filaments. Fiz. Tverd. Tela. **19**, 3562 (1977)
82. Zmnidzinas, J.S.: Electron trapping and transport by supersonic solitons in one-dimensional systems. Phys. Rev. B **17**, 3919 (1978)
83. Davydov, A.S.: Influence of the electron–phonon interaction on electron motion in a one-dimensional molecular chain. Theor. Math. Phys. **40**, 408 (1979)
84. Kislukha N.I., Karbovskii, A.V.: Influence of anharmonicity and dispersion properties on Davydov soliton. Ukr. Fiz. Zh. **28**, 515 (1983)
85. Davydov, A.S., Zolotaryuk, A.V.: Subsonic and supersonic solitons in nonlinear molecular chains. Phys. Scr. **30**, 426 (1984)
86. Mel'nikov, V.K.: Integration method of the Korteveg–de Vries equation with a self-consistent source. Phys. Lett. A **133**, 493 (1988)
87. Zolotaryuk, A.V.: Multi-particle Davydov solitons. In: Physics of Multi-Particle Systems, vol. 13, p. 40. Naukova Dumka, Kiev (1988)
88. Bolterauer, H., Henkel, R.D.: Solitons in the alpha-helix. Phys. Scr. **13**, 314 (1986)
89. Perez, P., Theodorakopoulos, N.: Competing mechanisms for the transport of energy in the α-helix. Phys. Lett. A **124**, 267 (1987)
90. Zolotaryuk, A.V., Pnevmatikos, St., Savin A.V.: Self–trapping in a molecular chain with substrate potential. In: Christiansen, P.L., Scott, A.C. (eds.) Davydov's Soliton Revisited, p. 191. Plenum Press, London (1990)
91. Careri, G., Buontempo, U., Carta, F., Gratton, E., Scott, A.C.: Infrared absorption in acetanilide by solitons. Phys. Rev. Lett. **51**, 304 (1983)
92. Careri, G., Buontempo, U., Galluzzi, F., Scott, A.C., Gratton, E., Shyamsunder, E.: Spectroscopic evidence for Davydov-like solitons in acetanilide. Phys. Rev. B **30**, 4689 (1984)
93. Eilbeck, J.C., Lomdahl, P.S., Scott, A.: Soliton structure in crystalline acetanilide. Phys. Rev. B **30**, 4703 (1984)
94. Holstein, T.: Studies of polaron motion. Part I. The molecular crystal model. Ann. Phys. **8**, 325 (1959)
95. Holstein, T.: Dynamics of large polarons in quasi-1-d solids. Mol. Cryst. Liq. Cryst. **77**, 235 (1981)
96. Appel, J.: Polarons. Solid State Phys. **21**, 193 (1968)
97. Davydov, A.S., Enol'skii, V.Z.: Motion of an excess electron in a molecular lattice under consideration of interacting with optical phonons. Zh. Eksp. Teor. Fiz. **79**, 1888 (1980)
98. Zolotaruk, A.V., Savin, A.V.: Solitony v molekulyarnyh tsepochkah s opticheskimi kolebaniyami i angarmonizmom. Preprint ITF-87-68R, Kiev (1987)
99. Zolotaryuk, A.V., Savin, A.V.: Solitons in molecular chains with intramolecular nonlinear interactions. Physica D **46**, 295 (1990)
100. Takeno, S.: Vibron solitons in one-dimensional molecular crystals. Prog. Theor. Phys. **71**, 395 (1984)
101. Takeno, S.: Vibron solitons and coherent polarization in an exactly tractable oscillator-lattice system. Prog. Theor. Phys. **73**, 853 (1985)
102. Takeno, S.: Vibron solitons and soliton-induced infrared spectra of crystalline acetanilide. Prog. Theor. Phys. **75**, 1 (1986)
103. Kosevich, A.M., Kovalev, A.S.: Selflocalization of vibrations in a one-dimensional anharmonic chain. Zh. Eksp. Teor. Fiz. **67**, 1793 (1974)
104. Oraevskii, A.N., Sudakov, M.Yu.: Anharmonicity and solitons in molecular chains. Zh. Eksp. Teor. Fiz. **92**(4), 1366 (1987)
105. Davydov, A.S.: The motion of a soliton in a one-dimensional molecular lattice taking into account thermal vibrations. Zh. Eksp. Teor. Fiz. **78**, 789 (1980)
106. Davydov, A.S.: Quantum theory of quasi-particle motion in a molecular chain taking into account thermal vibrations. I. Non-localized states. Ukr. Fiz. Zh. **32**, 170 (1987)
107. Davydov, A.S.: Quantum theory of quasi-particle motion in a molecular chain taking into account thermal vibrations. I. Localized states. Ukr. Fiz. Zh. **32**, 352 (1987)

108. Kadantsev, V.N., Lupichev, L.N., Savin, A.V.: Intramolecular excitation dynamics in a thermalized chain: I. Formation of autolocalized states in a cyclic chain. Phys. Stat. Sol. (b) **143**, 569 (1987)

109. Kadantsev, V.N., Lupichev, L.N., Savin, A.V.: Intramolecular excitation dynamics in a thermalized chain: II. Formation of autolocalized states in a chain with free ends. Phys. Stat. Sol. (b) **147**, 155 (1988)

110. Kadantsev, V.N., Lupichev, L.N., Savin, A.V.: Formation of soliton states in a molecular chain taking into account quantum thermal vibrations. Ukr. Fiz. Zh. **33**, 1135 (1988)

111. Kadantsev, V.N., Lupichev, L.N., Savin, A.V.: Intramolecular dynamics of excitation in a thermalized molecular chain. Formation of localized states in a cyclic chain. In: Physics of Many-Particle Systems, vol. 15, p. 40. Naukova dumka, Kiev (1989)

112. Kadantsev, V.N., Lupichev, L.N., Savin, A.V.: Intramolecular dynamics of excitation in a thermalized molecular chain. In: Proceedings of the Methods of Analytical and Numerical Modeling of Many-Particle Systems, Moscow, vol. 102 (1989)

113. Kadantsev, V.N., Savin, A.V.: Numerical investigation of Davydov soliton dynamics depending on the parameters of α-helix proteins. In: Proceedings of the Conference on Implementation of Mathematical Methods Using Computers in Clinical and Experimental Medicine, Moscow (1986)

114. Kadantsev, V.N., Savin, A.V.: Intramolecular excitation dynamics in a thermalized chain II. Formation of autolocalized states in a chain with free ends. Phys. Stat. Sol. (b) **47**, 155–161 (1988)

115. Kadantsev, V.N.: Dynamic properties of the one-dimensional Hamiltonian of a thermalized molecular chain. In: Lupichev, L.N. (ed.) Proceedings of Institute of Physical and Technological Problems. A Study of Dynamic Properties of Distributed Media, Moscow, vol. 23 (1991)

116. Cruseiro, L., Halding, J., Christiansen, P.L., Skovguard, O., Scott, A.C.: Temperature effects on the Davydov soliton. Phys. Rev. A **37**, 880 (1988)

117. Lomdahl, P.S., Kerr, W.C.: Do Davydov solitons exist at 300 K? Phys. Rev. Lett. **55**, 1235 (1985)

118. Lomdahl, P.S., Kerr, W.C.: Finite temperature effects on models of hydrogen-bonded polypeptides. In: Physics of Many-Particle Systems, vol. 12, p. 20. Naukova Dumka, Kiev (1987)

119. Lawrence, A.F., McDaniel, J.C., Chang, D.B., Pierce, B.M., Birge, R.R.: Dynamics of the Davydov model in α-helical proteins: effects of the coupling parameter and temperature. Phys. Rev. A **33**, 1188 (1986)

120. Nevskaya, N.A., Chirgadze, Yu.N.: Infrared spectra and resonance interactions of amide-I and II vibrations of α-helix. Biopolymers **15**, 637 (1976)

121. Chirgadze, Yu.N.: Resonant interaction of amide oscillations in polypeptide structures. In: Modern Problems of Solid State Physics and Biophysics, vol. 246. Naukova dumka, Kiev (1982)

122. Kuprievich, V.A., Kudritskaya, Z.G.: Davydov solitons and evaluation of the exciton–phonon interaction parameters. In: Modern Problems of Solid State Physics and Biophysics, vol. 96. Naukova dumka, Kiev (1982)

123. Press, W.H., Teukolsky, S.A., Vetterling, W.T., Flannery, B.P.: Numerical Recipes in Fortran 77: The Art of Scientific Computing. Press Syndicate of the University of Cambridge, Cambridge/New York (1992)

124. Rolfe, T.J., Rice, S.A., Dancz, J.: A numerical study of large amplitude motion on a chain of coupled nonlinear oscillators. J. Chem. Phys. **70**, 26 (1979)

125. Rolfe, T.J., Rice, S.A.: Simulation studies of the scattering of a solitary wave by a mass impurity in a chain of nonlinear oscillators. Physica D **1**, 375 (1980)

126. Halding, J., Lamdahl, P.S.: Coherent excitations of a 1-dimensional molecular lattice with mass variation. Phys. Lett. A **124**, 37 (1987)

127. Yomosa, S.: Solitary excitation in muscle proteins. Phys. Lett. A **32**, 1752 (1985)

128. Perez, P., Theodorakopoulos, N.: Solitary excitation in the α-helix: viscous and thermal effects. Phys. Lett. A **117**(8), 405 (1986)

129. Layne, S.P.: The Modification of Davydov Solitons by the Extrinsic N–H–C=O Group. Nonlinear Electrodynamics in Biological Systems. Plenum Press, New York (1986)
130. Layne, S.P.: A possible mechanism for general anesthesia. Los Alamos Sci. **10**, 23 (1989)
131. Savin, A.V.: Davydov soliton dynamics in heterogeneous molecular chain. Ukr. Fiz. Zh. **34**(9), 1300–1305 (1989)
132. Förner, W.: Quantum and disorder effects in Davydov soliton theory. Phys. Rev. A **44**, 2694 (1976)
133. Chirgadze, Yu.N., Nevskaya, N.A.: Infrared spectra and resonance interaction of amide-I vibration of the antiparallel-chain pleated sheet. Biopolymers **15**, 607 (1976)
134. Chirgadze, Yu.N., Nevskaya, N.A.: Infrared spectra and resonance interaction of amide-I vibration of the parallel-chain pleated sheet. Biopolymers **15**, 627 (1976)
135. Kadantsev, V.N., Lupichev, L.N., Savin, A.V.: Synergetics of molecular systems. I. Dynamics of an one-dimensional nonlinear lattice. In: Lupichev L.N. (ed.) Proceedings of Nonlinear Phenomena in Open Systems, Moscow, vol. 3 (2002)
136. Kadantsev, V.N., Lupichev, L.N., Savin, A.V.: Synergetics of molecular systems. II. Bistability induced by extension of a homogeneous chain. In: Proceedings of Nonlinear Phenomena in Open Systems, Moscow, vol. 3 (2004)
137. Kadantsev, V.N., Lupichev, L.N., Savin, A.V.: Synergetics of molecular systems. III. Localisation of nonlinear vibrations in one-dimensional lattice. In: Proceedings of Nonlinear Phenomena in Open Systems, Moscow, vol. 3 (2006)
138. Fletcher, R., Reeves, C.M.: Function minimization by conjugate gradients. Comput. J. **7**, 149 (1964)
139. Zolotaryuk, A.V., Spatschek, K.H., Savin, A.V.: Bifurcation scenario of the Davydov–Scott self-trapping mode. Europhys. Lett. **31**, 531–536 (1995)
140. Zolotaryuk, A.V., Spatschek, K.H., Savin A.V.: Supersonic mechanisms for charge and energy transfers in anharmonic molecular chains. Phys. Rev. B **54**, 266 (1996)
141. Kadantsev, V.N.: Features of auto-localised steady states in finite thermalized molecular chains. In: Lupichev, L.N. (ed.) Proceedings of a Study of Dynamic Properties of Distributed Media, Moscow, vol. 73 (1990)
142. Kadantsev, V.N., Savin, A.V.: The capture of an extra electron by supersonic acoustic solitons in a molecular chain in the presence of thermal vibrations. Phys. Stat. Sol. (b) **161**, 769 (1990)
143. Kadantsev, V.N., Lupichev, L.N.: Extra electron capture by supersonic acoustic soliton in a molecular chain in the presence of thermal vibrations. In: Lupichev, L.N. (ed.) Proceedings of Applied Aspects of Distributed System Analysis, Moscow, vol. 3 (1990)
144. Kadantsev, V.N., Lupichev, L.N., Savin, A.V.: Dynamics of electro-soliton in a thermalized chain. In: Lupichev, L.N. (ed.) Proceedings of the Dynamical Processes in Complex Systems, Moscow, vol. 3 (1991)
145. Kadantsev, V.N., Lupichev, L.N., Savin, A.V.: Dynamics of a localised state of intramolecular excitation in a molecular chain with thermalized acoustic and optic vibrations. In: Lupichev, L.N. (ed.) Proceedings of the Processes and Structures in Open Systems, Moscow, vol. 3 (1992)
146. Kadantsev, V.N., Lupichev, L.N., Savin, A.V.: Auto-localisation of a quantum quasi-particle in a biomolecular chain with substrate. In: Lupichev, L.N. (ed.) Proceedings of the Nonlinear Phenomena in Distributed Systems, Moscow, vol. 3 (1994)
147. Popov, E.M.: Structural Organization of Proteins. Nauka, Moscow (1989)
148. Volkenstein, M.V.: Biophysics. Nauka, Moscow (1988)
149. Shaitan, K.V., Rubin, A.B.: Conformational mobility and Mossbauer effect in biological systems. Mol. Biol. (USSR) **14**, 1323 (1980)
150. Rubin, A.B.: Biophysics, vol. 1. Vyshaia shkola, Moscow (1987)
151. Karplus, M., McCammon, J.A.: The internal dynamics of globular proteins. CRC Crit. Rev. Biochem. **9**, 293 (1981)
152. Petersen, W.P.: Lagged Fibonacci random number generators for the NEC SX-3. Int. J. High Speed Comput. **6**, 387 (1993)

153. Savin, A.V., Zolotaryuk, A.V.: Dynamics of the amide-I excitation in a molecular chain with thermalized acoustic and optical modes. Physica D **68**, 59 (1993)
154. Brizhik, L.S., Davydov, A.S.: Soliton excitation in one-dimensional molecular systems. Phys. Stat. Sol. (b) **115**, 615 (1983)
155. Turner, I.E., Anderson, V.E., Fox, K.: Ground-state energy eigenvalues and eigenfunctions for a electron in an electric-dipole field. Phys. Rev. **174**, 81 (1968)
156. Ukrainskii, I.I., Mironov, S.L.: The origination of the conducting band in polypeptide chains. Teor. Eksper. Chem. **15**, 144 (1979)
157. Mironov, S.L.: Calculation of the conduction band for excess electron in proteins. Teor. Eksper. Chem. **18**, 155 (1982)
158. Hol, W.G.J., Van Duijnen, P.T., Berendsen, H.J.C.: The alpha-helix dipole and the properties of proteins. Nature **273**, 443 (1978)
159. Bates, W.W., Hobbs, M.E.: The dipole moment of some amino acids and the structure of amino group. J. Am. Chem. Soc. **73**, 2151 (1951)
160. Kadantsev, V.N., Lupichev, L.N., Savin, A.V.: Electrosoliton dynamics in a thermalized chain. Phys. Stat. Sol. (b) **183**, 193 (1994)
161. Eremko, A.A.: Dissociation of Davydov's solitons in the field of electromagnetic wave. Dokl. Ac. Nauk UkrSSR **3**, 52 (1984)
162. Eremko, A.A., Gaididei, Yu., Vakhnenko, A.A.: Dissociation-accompanied Raman scattering by Davydov solitons. Phys. Stat. Sol. (b) **127**, 703 (1985)
163. Eremko, A.A., Vakhnenko, A.A.: The movement of Davydov soliton in a periodic potential. Preprint MTF, AS USSR. 88-I3P, Kiev (1988)
164. Kadantsev, V.N., Savin, A.V.: Interaction of Davydov solitons with periodic perturbations of a molecular chain. In: Proceedings of the Study of Dynamic Properties of Distributed Systems, Moscow, vol. 23 (1989)
165. Sitko, S.P., Sugakov, V.I.: Rol spinovyh sostoyanii belkovyh molekul. Dokl. Acad. Nauk USSR B **6**, 63 (1984)
166. Webb, S.J.: Laser Raman spectroscopy of living cells. Phys. Rep. **60**, 201 (1980)
167. Kadantsev, V.N., Mogilevskiy, V.D.: Dynamics of the protein α structure. Scientific Report. Institute of Control Problems, Moscow (1983)
168. Deviatkov, N.D.: Influence of electromagnetic millimeter waves on biological objects. Usp. Fiz. Nauk. **110**, 453 (1973)
169. Webb, S.J., Stoneham, A.M.: Resonances between 10 and 1012 Hz in active bacterial cells as seen by laser Raman spectroscopy. Phys. Lett. A **60**, 267 (1977)
170. Del Giudice, E., Doglia, S., Milani, M., Vitiello, G.: Electromagnetic interaction and cooperative effects in biological systems. In: Guttman, F., Reyzer, H. (eds.) Modern Bio-electrochemistry. Plenum, New York (1986), and Nucl. Phys. B **275** (1986)
171. Kadantsev, V.N.: Soliton solutions in biosystems. In: Proceedings of Conference Physics and Application of Microwave, Krasnovidovo, May 1991, pp. 22–27
172. Kadantsev, V.N., Kononenko, K.M., Fisun, G.O.: Bioeffects of EMF are caused by interaction with collective excitations in alpha-helix proteins. In: Proceedings of III International Congress of the European Bio-Electromagnetics Association (EBEA), Nancy, 29 Feb–2 Mar 1996
173. Kadantsev, V.N., Savin, A.V.: Resonance effects of microwaves are caused by their interactions with solitons in alpha-helical proteins. J. Biol. Chem. **16**, 95 (1997)
174. Kadantsev, V.N., Goltsov, A.N.: The mechanism of elastic wave excitation by electromagnetic radiation. In: Proceedings of VI Workshop of Liquid Crystal State in Biosystems and Their Models, Puschino (1988)
175. Kadantsev, V.N.: Collective dynamics of quasi-one-dimensional structures in lipid membranes. In: Lupichev, L.N. (ed.) Proceedings of Nonlinear Phenomena in Distributed Systems, Moscow, vol. 68 (1994)
176. Kadantsev, V.N., Goltsov, A.N.: A study of the influence of lipid composition of cellular membranes on domain structure of a bilayer and lateral transport. Hum. Physiol. **21**, 113 (1995)

177. Goltsov, A.N., Kadantsev, V.N.: Self-organization of surface and channels of communication within lipid membranes and mechanisms of protein and ion lateral transport. In: Proceedings of International Workshop of Information Processing in Cells and Tissues (IPCAT), University of Liverpool, Liverpool, 6–8 Sept 1995, p. 601

178. Kadantsev, V.N., Kadantsev, V.V., Tverdislov, V.A., Yakovenko, L.V.: Cooperative dynamics of quasi-1D lipid structures and lateral transport in biological membranes. Gen. Physiol. Biophys. **16**, 311 (1998)

179. Kadantsev, V.N., Lupichev, L.N., Savin, A.V.: Stationary autolocalized states in thermalized molecular chains. Phys. Stat. Sol. (b) **162**, K33 (1990)

180. Kadantsev, V.N., Savin, A.V.: Photodissociation of Davydov solitons in thermalized chains. In: Proceedings of Symposium on the Mechanisms of Biological Action of Electromagnetic Radiation, Puschino, p. 11 (1987)

181. Kislukha, N.I.: Discreteness effects in the theory of Davydov solitons. Preprint ITF-85-ZR, Institute of Theoretical Physics, Kiev (1985)

182. Webb, S.J., Booth, A.D.: Microwave absorption by normal and tumor cells. Science **174**, 72 (1971)

183. Devyatkov, N.D.: Influence of electromagnetic radiation of millimeter range on biological objects. Usp. Phys. Nauk. **116**, 453 (1973)

184. Belyaev, I.Ya., Kravchenko, V.C.: Resonance effect of low-intensity millimeter waves on the chromatin conformational state of rat thymocytes. Z. Naturforsch. **49**, 352 (1994)

185. Belyaev, I.Ya., Alipov, Ye.D., Shcheglov, V.S., Polunin, V.A., Aizenberg, O.A.: Cooperative response of Escherischia coli cells to the resonance effect of millimeter waves at super low intensity. Electro-Magnetobiol. **13**, 53 (1994)

186. Motzkin, S.M., Benes, L., Block, N., Israel, I., May, N., Kuriyel, J., Birenbaum, L., Rosenthal, S., Han, Q.: Effects of low-level millimeter waves on cellular and subcellular systems. In: Coherent Excitations in Biological Systems, p. 47. Springer, Berlin (1983)

187. Didenko, N.P., Zelentsov, V.I., Cha, V.A.: Conformational changes in biomolecules induced by electromagnetic radiation. In: Devyalkov, N.D. (ed.) Effects of Nonthermal Millimeter Waves on Biological Objects, p. 63. IRE AN SSSR, Moscow (1983)

188. Kadantsev, V.N.: Interaction of solitons with phonons in DNA molecule. In: Lupichev, L.N. (ed.) Proceedings of the Study of Dynamic Properties of Distributed Media, Moscow, p. 17 (1991)

189. Kadantsev, V.N.: Effect of collective excitations in electrolyte on the electron transport along protein molecule. In: Lupichev, L.N. (ed.) Proceedings of the Study of Dynamic Properties of Distributed Media, Moscow, p. 31 (1993)

190. Kadantsev, V.N., Lupichev, L.N., Savin, A.V.: Cooperative effects in an α-helix protein. In: Lupichev, L.N. (ed.) Proceedings of the Methods of Complex System Analysis, Moscow, p. 3 (1993)

191. Kadantsev, V.N., Lupichev, L.N.: Collective excitations in an α-helix protein molecule interacting with environment. In: Lupichev, L.N. (ed.) Proceedings of Nonlinear Phenomena in Open Systems, Moscow, p. 3 (1995)

Index

© Springer International Publishing Switzerland 2015
L.N. Lupichev et al., *Synergetics of Molecular Systems*, Springer Series
in Synergetics, DOI 10.1007/978-3-319-08195-3

Printed in the United States
By Bookmasters